Trends in Plant Biotechnology

Siddra Ijaz • Imran Ul Haq •
Hayssam Mohamed Ali
Editors

Trends in Plant Biotechnology

Springer

Editors
Siddra Ijaz
CABB
University of Agriculture Faisalabad
Faisalabad, Pakistan

Imran Ul Haq
Department of Plant Pathology
University of Agriculture Faisalabad
Faisalabad, Pakistan

Hayssam Mohamed Ali
Department of Botany and Microbiology
King Saud University
Riyadh, Saudi Arabia

ISBN 978-981-97-0813-0 ISBN 978-981-97-0814-7 (eBook)
https://doi.org/10.1007/978-981-97-0814-7

This Springer imprint is published by the registered company Springer Nature Singapore Pte Ltd.
The registered company address is: 152 Beach Road, #21-01/04 Gateway East, Singapore 189721, Singapore

Preface

Molecular biology has grown faster among the many disciplines in biology. The knowledge of molecular biology boosts the competency of the scientific community in basic and applied biosciences to unlock the complexities of biological processes. This book is aimed at providing basic and advanced concepts of plant biotechnology. It covers the advances, recent trends, and current developments in plant biotechnology and provides up-to-date and comprehensive knowledge of this emerging field for enhancing the repertoires of professional colleagues. This book comprehends the knowledge of biotechnology and appraises the application of biotechnology in all fields of life. The bounteous depiction of scientific principles underlying plant cell, tissue, and organ culture protocols is provided. The applications of plant tissue culture technology for clonal propagation, germplasm conservation, assisting plant breeding, and plant improvement are comprehensively described in this book. We have tried to demonstrate the skills in protoplast-to-plant techniques, including protoplast fusion and regeneration of somatic hybrids. It attempts to provide an understanding of the scientific principles behind plant transformation and various plant gene transfer technology protocols. It describes the methods and principles for validating transgene integration into the plant genome. It provides an overview of the potentials and applications of transgenic plant technology and up-to-date knowledge of the latest innovations in the biotechnology field. This book also covers the proteomic principle and its applications of modern methods and strategies to analyze the proteome. This book articulates how proteomics fits into the broader picture of systems biology, biological research, and medical biotechnology. It would enable the researchers to identify proteins and peptides using mass spectra and genome databases. In this book, we have tried to provide a comprehensive knowledge regarding the tools in bioinformatics and an understanding of the computational challenges (and their solutions) in analyzing large biological datasets. This book describes molecular networks underlying the myriad phenotypes of living cells. In a nutshell, we have tried to present a precise and reasonable picture of this dynamic field by covering competitive areas of biotechnology.

Faisalabad, Pakistan — Siddra Ijaz
Faisalabad, Pakistan — Imran Ul Haq
Riyadh, Saudi Arabia — Hayssam Mohamad Ali

Contents

Editors and Contributors

About the Editors

Siddra Ijaz with a vibrant career in agriculture and biotechnology, obtained PhD from the University of Agriculture, Faisalabad, Pakistan, and postdoc from Plant Reproductive Biology Laboratory, University of California Davis, USA. She is currently serving as Associate Professor in the Centre of Agricultural Biochemistry and Biotechnology (CABB), University of Agriculture, Faisalabad, Pakistan. She is also serving as Managing Editor of an international impact factor journal, Pakistan Journal of Agricultural Sciences. She has supervised more than 60 MPhil and PhD students. She had been the member of National Curriculum Revision Committee during 2013–2014. She has published more than 70 articles, 9 books, 1 laboratory manual, and several book chapters. She is in charge of "Molecular Biology of Plant Disease Resistance Laboratory," CABB, UAF. She is the curator of Fungal Molecular Biology Laboratory Culture Collection (FMB-CC-UAF), an affiliated member of the World Federation for Culture Collections (WFCC). Her research focuses on plant biotechnology, including genomics, genome probing for identifying disease resistance genes, genome engineering for fungal disease resistance in plants, and disease management through nanobiotechnology.

Imran Ul Haq with a bright career in agriculture, plant pathology, and fungal molecular biology, obtained postdoc from the University of California Davis, USA. He is currently serving as Associate Professor in the Department of Plant Pathology, University of Agriculture, Faisalabad, Pakistan. He has supervised more than 40 graduate students and established Fungal Molecular Biology Laboratory Culture Collection (FMB-CC-UAF), an affiliated member of the World Federation for Culture Collections (WFCC). He has published more than 60 articles, 8 books, 4 laboratory manuals, and several book chapters. He has made colossal contributions in fungal taxonomy by reporting novel species of a fungal pathogen in plants. His research interests are fungal molecular taxonomy and nanotechnology integration with other control strategies for sustainable plant disease management.

Hayssam Mohamad Ali works at the Botany and Microbiology Department, College of Science, King Saud University. Dr. Hayssam M. Ali has a "world ranking of top 2% scientists" according to the Stanford Scientists' Ranking released by Stanford University in 2021 and 2022. He has more than 300 published papers and 10 published books, and he has reviewed more than 1000 papers in highly impacted journals. He is interested in studying plant physiology, biotic and abiotic stress, sustainability, molecular biology, environmental stresses, plant biotechnology, crop improvement, plant breeding, forestry, and plant biology.

Contributors

Amjad Abbas Department of Plant Pathology, University of Agriculture, Faisalabad, Pakistan

Elsayed Fathi Abd-Allah Plant Production Department, College of Food and Agricultural Sciences, King Saud University, Riyadh, Saudi Arabia

Hafiz Muhammad Ahmad Department of Bioinformatics and Biotechnology, Government College University, Faisalabad, Pakistan

Hayssam Mohamad Ali Department of Botany and Microbiology, College of Science, King Saud University, Riyadh, Saudi Arabia

Muhammad Amjad Ali Department of Plant Pathology, University of Agriculture, Faisalabad, Pakistan

Sadia Ali Centre of Agricultural Biochemistry and Biotechnology (CABB), University of Agriculture, Faisalabad, Pakistan

Sidra Altaf Department of Pharmacy, University of Agriculture, Faisalabad, Pakistan

Muhammad Anas Department of Plant Pathology, University of Agriculture, Faisalabad, Pakistan

Haseeb Anwar Department of Pharmacy, University of Agriculture, Faisalabad, Pakistan

Mahrukh Babar Institute of Physiology and Pharmacology, University of Agriculture, Faisalabad, Pakistan

Maria Babar Centre of Agricultural Biochemistry and Biotechnology (CABB), University of Agriculture, Faisalabad, Pakistan

Askim Hediye Sekmen Cetinel Department of Biology, Science Faculty, Ege University, Bornova-Izmir, Turkey

Barbaros Çetinel Republic of Türkiye Ministry of Agriculture and Forestry, Directorate of Plant Protection Research Institute, Bornova, Bornova-Izmir, Turkey

Amer Habib Department of Plant Pathology, University of Agriculture, Faisalabad, Pakistan

Zakia Habib Centre of Agricultural Biochemistry and Biotechnology (CABB), University of Agriculture, Faisalabad, Pakistan

Imran Ul Haq Department of Plant Pathology, University of Agriculture, Faisalabad, Pakistan

Abeer Hashem Botany and Microbiology Department, College of Science, King Saud University, Riyadh, Saudi Arabia

Faiza Hassan Institute of Physiology and Pharmacology, University of Agriculture, Faisalabad, Pakistan

Abdalmenem I. M. Hawamda Department of Agricultural Biotechnology, Faculty of Agricultural Science and Technology, Palestine Technical University-Kadoorie (PTUK), Tulkarm, Palestine

Arslan Iftikhar Department of Pharmacy, University of Agriculture, Faisalabad, Pakistan

Misbah Ijaz Department of Clinical Medicine and Surgery, University of Agriculture, Faisalabad, Pakistan

Siddra Ijaz Centre of Agricultural Biochemistry and Biotechnology, University of Agriculture Faisalabad, Faisalabad, Pakistan

Hafiza Maimoona Ikram Department of Pharmacy, University of Agriculture, Faisalabad, Pakistan

Zobia Jabeen Balochistan Agriculture College, Quetta, Pakistan

Muhammad Jabran State Key Laboratory for Biology of Plant Diseases, Institute of Plant Protection, CAAS, Beijing, China

Shahnaz Jamil Department of Plant Pathology, University of Agriculture, Faisalabad, Pakistan

Sukhwinder Kaur University of California Davis, Davis, CA, USA

Nasir Ahmad Khan Department of Plant Pathology, University of Agriculture, Faisalabad, Pakistan

Riffat Malik Centre of Agricultural Biochemistry and Biotechnology, University of Agriculture Faisalabad, Faisalabad, Pakistan

Muhammad Tahir Mohyuddin Department of Clinical Medicine and Surgery, University of Agriculture, Faisalabad, Pakistan

Ghalia Nadeem Centre of Agricultural Biochemistry and Biotechnology, University of Agriculture Faisalabad, Faisalabad, Pakistan

Bukhtawer Nasir Government College University, Lahore, Pakistan

Zeenat Niaz Graduate School of Agricultural Science, Tohoku University, Sendai, Japan

Graciela Dolores Ávila Quezada Facultad de Ciencias Agrotecnológicas, Universidad Autónoma de Chihuahua, Chihuahua, Chihuahua, Mexico

Hafiza Arooj Razzaq Centre of Agricultural Biochemistry and Biotechnology, University of Agriculture, Faisalabad, Faisalabad, Pakistan

Faiza Saleem Department of Pharmacy, University of Agriculture, Faisalabad, Pakistan

Ikhlas Shafique Department of Plant Pathology, University of Agriculture, Faisalabad, Pakistan

Muhammad Huzaifa Tanveer Department of Plant Pathology, University of Agriculture, Faisalabad, Pakistan

Muhammad Usama Department of Plant Pathology, University of Agriculture, Faisalabad, Pakistan

Muhammad Waris Balochistan Agriculture College, Quetta, Pakistan

Adil Zahoor Department of Biotechnology, Chonnam National University, Yeosu, Republic of Korea

Chapter 1
Tissue Culture and Somatic Fusion in Plants

Sukhwinder Kaur, Siddra Ijaz, and Bukhtawer Nasir

Abstract Molecular biology is a great avenue to combat climatic risks and challenges in agriculture. Interventions have improved the seasonal fluctuation in the supply of agricultural produce, and now, the availability of off-season produce is a very common global phenomenon. Plant cell culture is the sterile growth and multiplication of cells in an artificial nutrient media and axenic-controlled environment. It enhances the nutritional value of crop plants, provides disease-free plant material and somatic hybrids, makes germplasm preservation and conservation feasible, and promotes the synthesis of secondary metabolites and compounds of industrial interest. It provides the best quality planting material and compounds to farmers and industry. Somatic cell fusion is the induced fusion of protoplasts acquired from different or the same species. The new hybrid contains traits of fused species; hence, they have different genetics from their parents. This chapter entails comprehensive knowledge, including history and advancement in plant tissue culture and somatic hybridization.

Keywords Micropropagation · Callogenesis · Gametogenesis · Somatic embryogenesis · MS media

---

S. Kaur
University of California Davis, Davis, CA, USA

S. Ijaz (✉)
Centre of Agricultural Biochemistry and Biotechnology, University of Agriculture Faisalabad, Faisalabad, Pakistan

B. Nasir
Government College University Lahore, Lahore, Pakistan

S. Ijaz et al. (eds.), *Trends in Plant Biotechnology*,
https://doi.org/10.1007/978-981-97-0814-7_1

Introduction

Genetically modified plants grown in tissue culture offer new insights into the molecular biology and gene regulation of plants. Regarding the development of somaclonal and gametoclonal variations, plant tissue culture is regarded as a good approach for crop improvement. Higher quality materials could be cultivated using micropropagation. Additionally, it can be utilized to introduce genetic variety and select advantageous high-yield variants with enhanced resistance to biotic or abiotic challenges (Hussain et al., 2012; Brown & Thorpe, 1995). Tissue culture techniques are already available for the majority of crop species although they still need to be optimized for many other species, such as woody plant species (Brown & Thorpe, 1995).

Plant cells retain their totipotency in the differentiated state and can change during development. Under optimum culture conditions, they can dedifferentiate, multiply, and regenerate as complete plants. Haberlandt was the first to propose the idea of in vitro growth of plant cells (Haberlandt, 2003). Recently, this idea has developed into a potential tool used across the plant sciences. Plants can begin cell proliferation and development from many tissues in response to ideal culture conditions. Plant cell and tissue culture research has many basic and applied applications encompassing plant cell division, development, and biochemistry.

The development of genetic engineering has given rise to several genetic tools that can be used to expand the applications of tissue culture research. These tools can investigate molecular and physiological pathways and somaclonal variation during in vitro development. In vitro technology is applied to different types of research to investigate tissue morphogenesis, biochemical states, and somatic cell genetics. The maximum use of this technology is made possible by exploring and categorizing such wide-ranging applications. Plant tissue culture is a broad field that has a wide range of applications.

Plant tissue culture has several advantages as well as disadvantages. Work at the molecular to mass level is possible. Long-term preservation is possible for the regeneration of plant material. Production can be maintained all year and is not dependent on the season. It is the appropriate method for introducing temporal characteristics, examining the impacts of various mutagens, attempting to immobilize cells for biotransformation or biochemical reactions, and analyzing the effects of various mutagens. It could cultivate plant cells on a large scale and offer a consistent source of phytopharmaceuticals. Plant tissue culture can create clones with improved agronomic traits, including drought tolerance, disease-free plants, plants with high nutritional value, and plants with improved secondary metabolite production. The technique delivers consistent biomass that is always available, the chemical synthesis of therapeutic chemicals that are difficult or impossible to synthesize, and independence from changes in climatic circumstances and soil conditions. However, certain drawbacks include requiring various high-priced equipment, methods, and skilled expertise. The development of a callus is not obligatory within a short time.

Calli mass does not appear in optimized culture media or takes months for some explants.

Types of Plant Tissue Culture

Organized Cell Culture

The development of a defined structure is enhanced by organized growth. The differentiation, specialization, and modification of organs that enable them to perform their fundamental jobs is a requirement for the growth of higher plants. Higher plants also have more complex organ systems. A good example of a structured system is an organ culture. Cultures that can maintain an organized form of growth for a long time, generally organ culture, are used. These are isolated plant organ cultures made from root tips, stem tips, leaf primordia, developing sections of flowers and fruits, and cultures made from other isolated plant organs. Determinate and indeterminate organs are the two categories of organized organ cultures.

Determinate Organ Culture

When a plant enters its stage of maturity, it stops growing, which indicates determinate growth. Determinate organ cultures refer to plant organs that exhibit determinate growth and stop growing at maturity. Determinate organ cultures such as leaves, flowers, and fruits typically have specific sizes and shapes. Wheat seedling organ cultures were first developed in 1922 (Kotte, 1922; Robbins, 1922).

Advancements in organ culture include the successful culture of tomato roots in root culture medium in 1963 (White, 1963). LaRue (1942) first cultured *Kalanchoe globulifera* and *Nemesia strumosa* flower buds. Another scientist, Nitsch, successfully cultured the ovaries of *Lycopersicon esculentum* and many other plants (Nitsch, 1951). In 1952, Morel developed virus-free plants from *Dahlia* shoot tips. Organ culture was categorized into vegetative and reproductive organ culture (Morel, 1952). Vegetative organ culture includes the root, tips, and shoots as an explant. However, reproductive organ culture includes complete flowers, isolated ovaries, ovules, and embryos.

Leaf culture: Leaf culture is the growth and development of a removed primordial leaf. A primordium is the smallest group of cells that can initiate an organ's growth. Its developmental stages are crucial for investigating phyllotaxy (Cleland, 2001).

Stamen culture: A significant contribution to the development of stamen culture was made by the Canadian scholar Hicks (1975). However, the ability of petals, stamens, and pistils from five different Sedum species to regenerate in vitro was compared by Wojciechowicz (2007), who concluded that petals, stamens, and pistils cannot regenerate into new plant parts unless they are regenerated through indirect organogenesis.

Flower culture: One of the most crucial stages in a plant's life cycle is flowering. It is suggested that both internal and external stimuli influence this process. The in vitro flowering system is vital for studying particular aspects of flowering, floral initiation, organ development, and senescence. Numerous plant species have successfully flowered in vitro when cytokinins, sucrose concentrations, photoperiods, and subculturing times are well managed (Henrickson, 1954).

The aseptic culture of a flower bud on a chemically defined nutritional medium, which continues to take on its full shape in a culture vessel, is known as flower culture. In vitro flowering is a useful technique in tissue culture for various factors. The potential to shorten plant life cycles is one of the most significant goals, and other goals include research on flower induction and initiation and floral development. It is possible to manipulate the various factors that impact these processes by manipulating the environment and media elements. Anthers, which carry microspores, are found in flower stamens. These microspores form a pollen grain-like exine around it (Casanova et al., 2005). In other words, pollen grains are formed from mature microspores. Explants such as pollen or anthers are typically used for in vitro conditions to produce haploid tissues. These pollen grains are capable of developing both calli and embryos. On solid media, anthers can also be cultivated. Anther cultivation involves aseptically removing developing anthers at a specific stage from the closed flower bud and cultivating them in nutritional media. The microspores in the cultivated anther transform into an embryoid or callus-like tissue. By organogenesis or embryogenesis, this tissue subsequently produces haploid plantlets.

Indeterminate Organ Culture

A plant with indeterminate growth will continue to grow until the environment can support it. Therefore, plant organs with indeterminate growth by adding modules are known as indeterminate organ cultures.

Meristem culture: Meristem culture has been demonstrated to be an effective technique for eradicating viruses from infected plants (Al-Taleb et al., 2011). To develop virus-free plants, mostly the shoot tip of the plant is cultured (meristem culture). When the objective is to develop virus-free plants (or to prevent other harmful microorganism contaminants), shot tips up to 10 mm are typically used. It is important to remember that in these experiments (virus-free plants), the sample must be properly extracted with the least amount of surrounding plant tissue when the apical meristem is used as an explant. Explants can be taken from lateral or apical buds with a tiny stem apex (Grout, 1999).

Embryo culture: This involves isolating and maintaining immature or mature embryos. Zygotic or seed embryos are frequently employed in plant tissue culture as explants. Embryo culture can be used for a variety of purposes, including the development of viable embryos from nonviable seeds produced by interspecific and intergeneric crossings, the breaking of dormancy, the production of haploid plants, and the germination of obligatory parasite seeds without the host (Sugii, 2011).

Seed culture: This is an important approach when explants are obtained from in vitro-generated plants to multiply orchids. Each plant must undergo an appropriate sterilizing technique to prevent contamination and damage to plant regeneration. The ideal approach in this condition is cultivating seeds to produce sterile seedlings. In addition, several plant species have tiny, poorly germinating seeds. One of the most effective instances of this is the orchid. One of the largest families of flowering plants is Orchidaceae, and many of its members have great value in the horticultural industry and herbal medicine. In vitro seed germination has grown numerous uniform seedlings to satisfy commercial objectives and save natural resources. Only a few seeds germinate naturally despite being produced in vast quantities (millions per capsule). The symbiotic interaction between orchids and fungi, which is necessary for germination, is a significant barrier to in vitro production. Asymbiotic germination refers to these plants' in vitro seed culture that depends on the added growth medium. For large-scale, quick multiplication and seedling development, seed culture is frequently used. Ma et al. used the best MS medium enriched with activated carbon to grow *Capparis spinosa* seedlings on a wide scale after sterilizing the seeds with 0.1% $HgCl_2$ for 12 min (Ma et al., 2010).

Root culture: Root tips are cultured in an agitated liquid medium with optimized auxin and other micro- and macronutrient concentrations. Tropane alkaloids are present in larger concentrations in the roots of solanaceous species. The production of alkaloids is dramatically increased when such roots are regenerated from undifferentiated calli.

Unorganized Cell Culture

Callus Culture

Callus comprises an undifferentiated mass of loosely clustered, thin-walled parenchymatous cells developed from the parent tissue's proliferating cells. It is more or less an unorganized tissue with differentiated and undifferentiated cells (beginning material for callus induction, such as leaves, roots, and shoots). Plant cells multiply on suitable media to generate calli or clumps of undifferentiated cells. When differentiated cells from isolated explants are present in the callus, cell division cannot occur until the cells have undergone dedifferentiation. Cell division occurs without dedifferentiation if isolated explants have only meristematic tissue. During the dedifferentiation process, mature adult cells can momentarily switch from the adult to the juvenile stage. The capacity for growth and division is higher in rejuvenated cells. Under specific conditions, these cells can regenerate organs and embryos.

Therefore, during plant in vitro propagation, dedifferentiation is a crucial phase to consider. Suitable callus induction hormones are often supplemented at the time of callus development. Its growth and development require a sterile environment and growth-supporting media. Exogenously administered growth regulators influence callus production from explants in nutritional media (Iwase et al., 2011).

Suspension Culture

Muir et al. (1958) demonstrated the cultivation of plant cells in liquid (suspension) cultures. However, plant cells may exhibit cytogenetic and metabolic differences depending on the growth cycle and the culture parameters. Spatial heterogeneity is a term used to describe this type of variation (Lindsey & Yeoman, 1983). Isolated cells are cultured in liquid media in a process known as suspension culture. Cells can be removed mechanically or through enzymatic digestion from in vivo plant material, and calli can be generated from any explant. Similar to callus culture, it can be started from almost any plant part. It is the most typical method for starting suspension cultures from calli that are already developing in culture. Callus cultures can be broadly categorized into two types: (1) friable calli (cells are loosely grouped) and (2) compact (cells are densely grouped). Suspension culture from friable calli requires less force than compact culture. The liquid media used to generate cell suspension culture typically has a chemical composition similar to solid media.

Batch and continuous cell suspension cultures are the two subtypes of suspension culture. A closed culture technique, batch culture, uses few nutrients. In batch culture, cells are generally cultured in conical flasks on orbital shakers at a speed of 80–120 rpm. Batch cultures have further categories, including stirred, shaken, slowly rotating, and spinning cultures. Continuous cell suspension culture requires a constant supply of the product in the medium or the cell suspension and is further categorized into closed continuous culture and open continuous culture (Panda et al., 1989).

Protoplast Culture

Plant protoplasts are important for research on breeding, genetics, somatic cell fusion, and plant cell culture. In vitro plant regeneration also plays a significant role in developmental biology and has the potential to be used in biotechnology and agriculture. Because of somatic hybridization and the potential for protoplast transformation, especially for recalcitrant species, plant breeding and the development of varieties with desirable characteristics through regeneration technology are becoming more significant (Umate et al., 2005). However, developing a successful protoplast isolation, culturing, and regeneration procedure is a crucial first step in breeding and plant genetic engineering (Duquenne et al., 2007). Developing new varieties with desirable qualities, such as quality and yield, can be accomplished by transferring genes between species and varieties via the protoplast fusion technique

(Prabavathy et al., 2006). To produce parasexual protoplast hybrids, such a technique involves the fusion of two genetically different protoplasts produced from various somatic cells (Nassour & Dorion, 2002). A well-established tissue culture method that supports protoplast isolation and regeneration is required to perform protoplast fusion (Makonkawkeyoon et al., 1995).

Mechanical and enzymatic techniques are the two approaches used to isolate protoplasts. From a plasmolyzed *Stratiotes aloides* cell, Von Klercker (1892) was the first to isolate protoplasts mechanically. These investigations were then expanded to isolate protoplasts from onion bulb tissues. The scales were immersed in 1.0 M sucrose until the protoplasts dissociated from their surrounding wall, and the tissue was then plasmolyzed. The tissue was then cut using a sharp knife at a thickness where only the cell walls were affected and protected from protoplast damage. Protoplasts are released via osmotic pressure when these tissue strips are submerged in a low-concentration sucrose solution. Enzymes are used in the enzymatic approach to breakdown the cell wall and release protoplasts. Initial research was performed on isolating yeast protoplasts by breaking down the cell wall with gastric juice from *Helix pomatia*. To breakdown the cell wall and release protoplasts from tomato roots in 1960, Cocking used a crude cellulase preparation from the fungus *Myrothecium verrucaria*. It has been widely used in recent studies with modified procedures and employing pure enzymes. The enzymatic technique can be applied directly or sequentially. The direct or one-step method involves protoplasts isolated directly from the tissue using two enzymes (cellulase and pectinase). The sequential method (two-step method) involved using pectinase enzyme in which the protoplasts were first isolated from callus or other tissues of plants, and then the cell wall was broken down to release the protoplast by adding cellulose enzyme. Different factors affect protoplast viability and isolation. Standardizing the amount and concentration of osmoticum, such as mannitol, glucose, sucrose, sorbitol, and other sugars, is required to obtain a high yield of viable protoplasts. Enzymes such as cellulase, pectinase, pectin, hemicellulase, and meicellase are the other enzymes that need to be optimized, whether these enzymes are used in combination or separately with osmoticum.

Isolated protoplasts are cultured using different methods, including suspension, hanging drop, agar plating, microculture, and multidrop array techniques. Multiple factors, from the condition of the explant plant to the culture condition, affect the growth and regeneration of protoplasts. With the aid of this technology, fundamental scientific knowledge of cell biology, plant compatibility, membrane functions, cell organelle investigations, and cell wall regeneration has been produced. Furthermore, it transmits cytoplasmic male sterility (Tomar & Dantu, 2010). In addition, somatic hybridization has created new opportunities for crop improvement through in vitro genetic manipulation of plants. Organogenesis and embryogenesis are two processes that enable plant regeneration from newly separated protoplasts or after their union. Takebe et al. provided the first account of plant regeneration using isolated *Nicotiana tabacum* protoplasts (1971). Before cultivation, the density of the purified protoplasts was measured using a hemocytometer and adjusted to the desired value. Several techniques comparable to those used for cell culture have been employed to cultivate newly isolated protoplasts and those following fusion therapy. Protoplasts lose their distinctive spherical form and synthesize a wall in 24 h when growth conditions are ideal. A healthy somatic cell wall is often required for the cell to divide by normal mitosis. Nuclear divisions may occur without a suitable wall, but cytokinesis does not follow them. In protoplast cultures, the time it takes for the first cell to divide depends on the species, genotype, source of the protoplasts, technique used to isolate them, viability, makeup of the culture media, and culture conditions.

Tissue Culture Regime

Depending on the species, a tissue may need different nutrients to develop optimal in vitro conditions. Plant tissues may require different conditions for successful growth (Murashige & Skoog, 1962). Since several plant tissues and organs exist, no single medium can be recommended as completely suitable. It is crucial to determine a medium that will satisfy the different requirements of that tissue before starting with a new system.

Every nutrient needed for a plant's regular growth and development is in the medium used for in vitro plant growth. For the in vitro vegetative propagation of several plant species, Murashige and Skoog media (MS medium) is most frequently used in plant tissue culture. Some of the earliest plant tissue culture media, including root culture medium and Gautheret's (1939) callus culture medium were also in use. The media's pH is a crucial factor that influences both plant growth and the function of plant growth regulators. Both solid and liquid media can be used for cell culture, with the pH value adjusted to a range of 5.4–5.8. The nitrogen source and plant hormones, in particular, as well as the nature of the medium, significantly impact how the explant responds (Murashige & Skoog, 1962).

Components of Plant Tissue Culture Media

Inorganic Nutrients

Macronutrients: One major component promoting plant growth in vitro and in vivo is nitrogen. It comprises proteins, amino acids, certain hormones, and chlorophyll. Both organic and inorganic materials could be used as nitrogen sources in vitro. Nitrogen's influence on the medium's pH indirectly impacts tissue growth (Dougall, 1980; Congard et al., 1986). Phosphorus is essential for plant energy storage, transport, and cell division. It also plays a crucial role in photosynthesis. According to Kozai et al. (1991), PO_4^{3-} is absorbed significantly more efficiently than other minerals in autotrophic strawberry cultures. Plants that receive insufficient phosphorus become aberrant and unhealthy. Plant cell walls contain calcium as calcium pectate, which is essential for maintaining the integrity of the membrane. It has been demonstrated that high-calcium levels encourage the buildup of callose, which prevents cells from proliferating (Eklund & Eliasson, 1990). Stomata were more open in plants if cultivated in high-calcium environments (Atkinson et al., 1991). Calcium controls hormone responses and environmental stimuli such as temperature and light. Under calcium deficiency, the developing points and leaf tips tend to die back. Other macroelements, such as magnesium, potassium, and sulfur, play important roles in plant tissue culture medium.

Micronutrients: Microelements play a crucial role as catalysts in different metabolic processes. The symptoms of microelement deficiency in the medium include decreased rosetting (Zn, Mn), lignification (Cu, Fe), shoot tip necrosis (B), and leaf chlorosis (Fe, Zn, Mn) (Williams, 1995).

Organic Nutrients

1. *Vitamins*: Vitamins play a catalytic role in enzymatic processes. Thiamine (B1) is a vitamin crucial for plant cells. Other vitamins, nicotinic acid (B3) and pyridoxine (B6), are added to cell culture conditions because they may stimulate the cellular response. Myo-inositol is frequently used vitamin in tissue culture. Inositol in the form of inositol phosphate, supports auxins' main function in plants by acting as a second messenger (Verma & Dougall, 1979).

   Vitamin C is helpful to avoid blackening after explant isolation, while vitamin E is used as an antioxidant. Plant tissue cultures are affected by vitamin D's effect on growth regulation. According to Drew and Smith (1986), riboflavin inhibits callus development and enhances shoot growth and quality (Drew & Smith, 1986).

   Thiamine is a crucial co-factor in the metabolism of carbohydrates and is involved in the direct production of several amino acids. The need for thiamine increases with subsequent passages; however, some plant cultures seem to be able to survive without thiamine. Nevertheless, most cultures benefit from small amounts of it. Cytokinins and thiamine may interact synergistically (Digby & Skoog, 1966).
2. Complex nutritional combinations with undefined compositions employed to encourage the growth of various calli and organs. However, it is best to avoid using these natural extracts as much as possible. Straus (1960) suggested replacing tomato juice and yeast extract with only L-asparagine for the maize endosperm callus.
3. In culture, green cells usually do not participate in photosynthetic activity and need a carbon source. In cell culture, sucrose or glucose at 2% and 5% (wt/vol.) are often used as carbon sources. Additionally, starch and fructose are alternative forms of carbohydrates. Lower quantities of carbohydrates may be employed for

protoplast culture, while much higher levels may be used for embryo or anther culture. Several metabolic processes require the presence of sucrose in media (Aloni, 1993). If autoclaved for a long period, sugars caramelize and react with amino compounds (Maillard reaction) (Peer, 1971). Melanoidins are brown, high-molecular-weight chemicals produced during the decomposition of sugars during caramelization. Melanoidins can suppress cell proliferation. An autoclaved medium's yellow to light brown tint indicates too much time in the autoclave.

Plant Growth Regulators

The development pathway of plant cells and tissues in culture media is generally influenced by plant growth regulators (PGRs). The type and amount of hormones employed are mostly determined by the plant species, Explant, and the objective of the experiment. PGRs in the culture medium determine the developmental pathway (Singh et al., 2017). In tissue culture media, abscisic acid, indole-3-acetic acid, naphthaleneacetic acid, 2,4-dichloro phenoxy acetic acid, indole-3-butyric acid, 6-furfurylaminopurine kinetin, 6-benzyl amino purine, N6 (2-isopentenyl)-adenine, trans-6-(4-hydroxyl-3-methylbut-2-enyl) amino purine, zeatin, gibberellic acid, and thidiazuron are important plant growth regulators. The most commonly used PGRs are auxin and cytokinins.

Hormones (auxin) play a role in abscission, rooting, apical dominance, tropism, and other biological processes. Auxins have been employed in tissue cultures for root differentiation and cell division. Auxin (IAA, NAA, 2,4-D, or IBA) is essential for most plant cells to divide and initiate roots. Auxin can inhibit morphogenesis at high concentrations. Auxin 2,4-D is frequently used to induce calluses: For root induction, IAA, IBA, and NAA are applied in the medium (Huang & Murashige, 1976).

Cell division, shoot proliferation, and shoot morphogenesis are all enhanced by cytokinins (Miller & Skoog, 1953; Miller et al., 2007). Cytokinins are mostly added to tissue culture media for cell division and the development of adventitious shoots from calli and organs. By releasing axillary buds from apical dominance, cytokinins are also employed to promote shoot multiplication. Thidiazuron (TDZ; *N*-phenyl-*N′*-1,2,3-thiadiazol-5-ylurea) is a synthetic chemical that economically defoliates cotton. Low concentrations of TDZ are effective in promoting the production of shoots (Santana-Buzzy et al., 2005). The rarely used PGR is gibberellin. Gibberellin (GA3) prevent callogenesis (Jarret et al., 1980) and helpful in morphogenesis research studies. In embryo culture, the plant hormone abscisic acid (ABA) is useful and is involved in the dormancy and abscission of leaves and fruits. Abscisic acid is light sensitive but heat stable.

pH Value

Generally, pH 5.5–6 is adjusted for plant tissue culture media. The agar will not properly solidify if the pH is below 5.5; if it is above 6.0, the media can be too hard. After autoclaving, the media pH typically decreases by 0.6–1.3 units (Sharma et al., 2014). The formation of organic acids or nitrogen nitrification in various plant tissue cultures over time results in a pH decrease. The effects of inorganic salts, carbohydrate sources, gelling agents, activated charcoal, and medium storage techniques on media pH were investigated by Owen et al. (1991). These elements impacted the pH of the tissue culture media and affected callus formation and regeneration.

Solidifying Agents

A gelling agent is most frequently utilized in tissue culture investigations on solid medium. The type of agar used to gel the media may impact our experimental results. When liquid media is employed in static cultures, the tissue becomes submerged and eventually dies because it lacks oxygen. Therefore, a gelling agent is usually used to avoid this issue. Agar, agarose, and Gelrite are the three most common gelling agents found in plant tissue culture media, and they are all biological products (Kacar et al., 2010). Agar is a seaweed-derived high molecular-weight polymer that easily binds with water. It is a complex mixture of interrelated polysaccharides derived from galactose. The gel's strength is provided by the neutral polymer fraction, agarose, and the viscosity of agar is provided by the highly charged anionic polysaccharides, agaropectins (Araki & Arai, 1967). Gelrite is a good substitute for agar because of its numerous benefits and low cost per liter of medium. Gelrite hardens into a transparent gel, allowing for simple monitoring of cultures and any potential contamination. According to Bonga and Aderkas (1992), unlike agar, gelrite's strength is unaffected by a wide range of pH values. Many different plant species have performed just as well on Gelrite as on agar, and in some cases, Gelrite has even performed better. However, certain plants exhibit hyperhydricity on gelrite, possibly due to more easily accessible water. Combining moderate agar with Gelrite could solve this issue (Pasqualetto et al., 1986).

Applications of Plant Tissue Culture

Improving plant growth and production through genetic and breeding approaches and plant tissue techniques can be highly useful. Transgenic plants are still being grown along with considerable propagation for worldwide agriculture to enhance crop production and growth. The vulnerability of plants to harmful organisms is eliminated by tissue culture because it enables the development and growth of genetically homogenous plant material free from disease. Due to its effectiveness, genetic transformation is among the most widely applied, well-developed, and significant tissue culture procedures. Two approaches are used in genetic transformation; one transfers desired genetic features into the recipient cell (Qaim, 2016). This technique may or may not use a biological vector to deliver genetic data. Another method involves using a biologically derived vector to indirectly deliver the component into the cell through natural mechanisms, which enables its integration into the plant's DNA.

Somaclonal variation is a helpful technique that aids in producing subtle variation or differentiation within species in other applications of plant cultivation. Such methods are essential when exposing the explants to specific concentrations to generate the desired mutation. Modifying ploidy levels in plants can affect the expression of various traits and enable the establishment of homozygous haploid lines, genetic transformation strategies, and hybridization (Sahni et al., 2017). After fertilization, embryo rescue can play a part in genotype selection and breeding because otherwise, individuals might undergo an early-stage abortion. Using embryo rescue, species that have lost the ability to reproduce and prevent seed germination can be grown. Conserving genetic material makes it possible to reverse the rapid extinction of species and protect the world's floral heritage (Veloccia et al., 2016).

Genetic or epigenetic changes frequently cause genetic diversity. Identifying somaclonal variants or mutants with specific agronomic or industrial characteristics that will develop at the cell or plant level is essential for identifying genetic diversification in callus tissue and cell culture. Different selective pressures of different stresses are frequently applied to a small piece of callus or a cell suspension, which is made up of thousands or hundreds of thousands of cells, to find resistant cells under controlled conditions. When grown in an optimized culture medium, the recovered resistant cells may regenerate whole resistant plants. By employing this technique, it is possible to make plants resistant to biotic stress, which reduces agricultural production and abiotic stress caused by cold, salinity, and drought (Loyola-Vargas & Ochoa-Alejo, 2018).

The ability to manipulate media compositions to obtain desired results is essential to use tissue culture (Gamborg, 2002) successfully. Tissue can be manipulated to grow shoots, roots, calluses, or somatic embryos or engineered to inhibit growth for long-term storage by changing the components of the media. Micropropagation is the most typical use of tissue culture and often entails growing plants in an agar-solidified nutrient medium. Plant species can be produced and multiplied rapidly with the help of micropropagation. Micropropagation is generally perceived as the development of more established plant propagation techniques. By using clonal multiplication, it aims to quickly increase the number of superior pest- and disease-free plant genotypes. Buds from a selected plant are cultured on an appropriate medium under optimized growth conditions to support the development of axillary shoots. Until many plants are produced with the same genetic characteristics, a subculture of the buds and shoots is repeatedly carried out (Hussey, 1983). Additionally, germplasm can be stored through micropropagation to maintain disease-free plant material (Wilkins & Dodds, 1983).

Explants are grown extensively in liquid culture to increase the production of biopharmaceuticals and secondary metabolites obtained from plants through recombinant proteins. This is considerably more effective when it is necessary to regenerate novel hybrids or fuse protoplasts to develop related species (Tonosaki et al., 2016).

Meristem culture is the most popular tissue technique for isolating pathogen-free plants from systemically diseased plants. Morel and Martin used meristem culture to create virus-free Dahlia plants from infected individuals for the first time in 1952. The method for growing the grape (*Vitis vinifera* L.) cultivar utilizing single node culture from virus-free mother plants was developed by San Pedro et al. (2017). Meristem culture is also used to regenerate secondary metabolite-rich aromatic and medicinal plants. A successful procedure for the micropropagation of Aloe vera by meristem culture was established in 2014 by Molsaghi et al. (2014).

The in vitro establishment of cellular totipotency, callus differentiation, and vegetative propagation have broadened the scope of plant sciences. Organogenesis (differentiation of calluses) and somatic embryogenesis are further micropropagation processes. It is also possible to extensively reproduce rare and endangered plant species using seedlings produced from mature seeds. The apical meristems of virus-infected plants are commonly used to propagate virus-free plants. A typical method for creating homozygous plants in a single generation is through the diploidization of haploid cells, such as pollen grains. Protoplast technology has enabled the emergence of somatic hybrids and cybrids of distantly related plant species. Similar to gene transfer into bacteria, protoplasts are a suitable medium for plant gene splicing. Plant tissue culture may also improve the production and selection of cell variations to develop novel, economically relevant plant varieties (Yancheva & Kondakova, 2018).

Haploid Breeding

Organisms with gametic chromosome numbers (half the numbers of somatic cells) are haploids. Haploid plant cells can double their chromosome numbers spontaneously and are induced by colchicine treatment. Haploid would normally be derived from the pollen or ovule of the original heterozygote (F1 hybrid) and, within one generation, produce homogeneous progeny. If haploid cells are treated with colchicine to chromosome doubling artificially, then it is double haploid. The resulting double haploid is homozygous for all genes and, therefore, true-breeding, making the selection process of desirable traits more efficient.

Immature pollen or microspores at a specific stage of their development are transferred aseptically to a nutrient medium to initiate embryo and, subsequently, plant development. Ovules also provide haploid megaspores, which can be used as a source of haploid tissue and are treated similarly to the microspore. Haploids carry only one allele of each gene. Hence, plants with lethal genes are eliminated. Haploid breeding curtails the time frame for inbreeding to superior genotype production. Haploid breeding may also produce male or female plants from dioecious species.

However, the core loophole in haploid breeding is its genotype dependency. Likewise, unwanted traits of albinism may be fixed with the double haploid method. Raising double haploids for plant breeding is expensive; however, the benefits outweigh the cost of production.

An alternative approach is hybridization between widely separated parents by pollination, where it is known that the parental chromosomes of one parent will abort in the first new division of the zygote. The embryo remains viable with the maternal chromosome intact and is efficiently haploid. This uniparental genome elimination (GE) involves the post-zygotic loss of one parental chromosome set. Hybridization between distantly related parents results in uniparental genome elimination.

Methods of Haploid Production

Anther culture: Pollen-producing anthers are collected from the flower and cultured on a sterile nutrient medium. Mature pollens develop from haploid microspores. Microspore possesses a single nucleus; however, mature pollens contain three nuclei. Individual microspore forms callus when anther that contains microspores (middle to the late uninucleate stage) are cultured on an artificial nutrient regime. Plants with half a set of chromosomes can be regenerated from the callus by embryogenesis or organogenesis. Alternatively, microspores may give rise to embryoids capable of differentiation into embryos.

Microspore culture: It implicates the isolation of microspores from anthers. Anthers are homogenized subsequently to a suitable pre-treatment. Microspores are isolated and purified from homogenate by filtration and centrifugation steps. Purified microspores are then cultured in a medium. Small embryos that develop from the microspores (microspore embryogenesis) shift to the regeneration medium for plantlet development. In micropore culture, embryos are developed directly from the microspores rather than the callus phase resulting in a reduced risk of genetic anomalies.

Ovule culture: Ovules are removed from the ovary under aseptic conditions and are cultured on a tissue culture regime aseptically. An ovule is a megasporangium possessing a megaspore (egg cell).

Wide crosses: An alternate method for haploid development involves intergeneric crossing followed by embryo rescue. Wide crosses are mostly incompatible with no viable seed development. The resulting haploid embryo can be rescued by excision and tissue culture.

Somatic Cell Fusion (Somatic Hybridization)

It involves the fusion of two distantly to closely related protoplasts at interfamily, intergeneric, interspecific, and intraspecific levels with subsequent regeneration of hybrid cells into hybrid plants. Protoplast fusion must be done prior to the regeneration of new cell wall that may start to begin within few hours after protoplast isolation. Wall formation is visualized either by using UV-microscopy or by calcoflour white florescent stain. In isolated protoplasts, mitosis occur after 2–7 days of isolation.

The chromosome numbers of the somatic hybrids generally exceed the total number of parental protoplasts. If the chromosome number is the sum of the chromosome of the two parental protoplasts, then they are *symmetric hybrids*. If they have wide variation in the chromosome number than sum of parental protoplasts' chromosomes, then it is *asymmetric hybrids*.

Aspects of Somatic Hybridization

Protoplast Isolation

Plant protoplast is a cell from which the cell wall is removed and consists of original cellular contents bounded by the plasma membrane.

Methods of Plant Protoplast Isolation

There are broadly two methods of protoplast isolation: (1) mechanical isolation and (2) enzymatic digestion.

Mechanical Isolation of Protoplast

In 1892, Klercker first isolated plant protoplast by plasmolysing and slicing the tissue. This method results in limited yield and excessive damage. The major limitation of this approach is tissue specificity. In the mechanical method, cells are kept in a suitable plasmolyticum (in the plasmolysed cell, the protoplast shrinks away from the cell wall). The plasmolysed tissues are subjected to sharp cuts followed by deplasmolysis to release the protoplasts from the cut ends. This method is used only when cell wall degrading enzymes have an unavoidable deleterious effect on the protoplasts or the cells are large, and a small quantity is required.

Enzymatic Digestion

Large-scale isolation of plant protoplast using cell wall degrading enzymes gives impulse to research in protoplast. There are two basic approaches for enzymatic isolation of protoplasts.

One-step method: This method was developed by Power and Cocking. This method treats plant tissue with a mixture of pectinases and cellulases to simultaneously macerate or separate cells and degrade their walls.

Two-step method: It is a sequential treatment given to cells from liberating protoplast. This method was developed by Nagata and Takebe, by which tissues are macerated or separated by breaking the connection of cells in tissue using macerozyme (pectinase) and then treated with cellulase to isolate protoplast. Cellulase breaks the cell wall, and the protoplast is delivered.

Protoplast yield is higher in the one-step method, but protoplast isolated by the sequential method is better for culture studies. The enzyme mixture's pH and incubation temperature influence the release of the protoplast.

Cell wall degrading digest the pectins, cellulase, and hemicellulose (xylans) constituting the plant cell wall. Macerozyme degrades the galacturonic acid residue of pectins, which confers cell-to-cell adhesion and deliquesce the tissue to single cells. However, cellulases degrade the cellulose imparting the spherical shape of the protoplast. Hemicellulases contribute to the breakdown of the xylans.

Procedural Steps of Protoplast Isolation

Following procedural steps are involved in isolating protoplasts from cells or tissues:

Source material selection: Protoplast is isolated from diverse tissues, including roots, leaves, calli, in vitro regenerated shoots, cell suspension, and pollens. Cell suspension culture of young and juvenile tissues is an ideal target for *en masse* isolation of plant protoplasts though leaf tissue is the most commonly used organ.

Tissue sterilization: Sterilization is done by tissue dipping in 70% ethyl alcohol for 1–2 min, followed by sodium hypochlorite (2%) treatment for half an hour. After this, tissue washing is done thrice with nuclease-free water. The sterilized tissue's lower epidermis is gently peeled off, and stripped leaves are excised into pieces. All operation is done under axenic condition.

Enzyme preparation and tissue incubation: Enzymes are prepared in cell protoplast washing (CPW) solution. Tissue is transferred into CPW solution containing mannitol (13% w/v) and incubated for an hour. It is done because the protoplast does not have a cell wall, so the osmotic pressure of the solution is adjusted by adding non-metabolized sugars and sugar alcohol such as mannitol and sorbitol. As mannitol is an osmoprotectant, it is used to prevent the cell devoid of cell wall from loss and gain of water. This preliminary step is pre-plasmolysis incubation. Subjected tissue is exposed to osmoticum containing lytic enzymes as a sequential or mixed treatment followed by an overnight incubation under a dark regime.

Protoplast harvesting: After this treatment, the protoplasts sink to the bottom of the petri dish. For removing the undigested tissue, the protoplast solution is filtered through the autoclaved mesh, a nylon sieve used to filter the protoplasts, followed by the freezing of harvested protoplasts. Then CPW media is added to the filtered, frozen protoplast solution before centrifugation. Centrifugation is done to remove enzymes and fine debris. After centrifugation, debris is settled down, and the upper phase containing protoplast is transferred into a new tube and resuspended in a culture medium.

Cell wall removal results in pressure on the cell. Therefore, isolated protoplasts maintained hypertonic plasmolyticum provided by a balanced inorganic salt medium or monosaccharides sugar solution. For example, mannitol, which does not readily transport across the plasma lemma, provides a stable osmotic environment to the protoplast; however, an imbalance of osmotic pressure in the medium results in a protoplast burst.

Counting and Viability Determination of Protoplast

For estimating and adjusting suitable plating density (final density), protoplast counting is an integral step. For this purpose, a hemocytometer is used with marking (0.2 mm chamber depth and volume of each small subunit 1/16 mm). Isolated protoplasts are adjusted in a 10 mL CPW solution. The first edges of the hemocytometer are moistened, and the cover glass is placed over it firmly. A drop of protoplast suspension is dispensed onto the counting chamber of the hemocytometer, and several protoplasts are counted in five triple-ruled squares.

Protoplast yield (Y) is calculated using the formula given below:

$$Y = n \times 10^3 \times X$$

X = volume of protoplast suspension; n = number of protoplasts counted in five triple-ruled squares.

After knowing the protoplast yield in a fixed volume, the protoplast suspension is adjusted to a desired culture density by either dilution or concentration of existing suspension.

Fluorescein diastase (FDA), a viability probe, is relatively non-polar. It does not pass freely across the plasma membrane. Exposing the viable protoplasts to the FDA results in hydrolysis due to the catalytic action of esterase. This reaction releases polar fluorescein molecules that cannot cross the plasma membrane and start accumulating in the cell and staining the protoplast greenish white under UV illumination. Therefore, metabolically active protoplasts are visualized. An equal volume of protoplast suspension is mixed with an FDA solution for viability determination. A drop of this mixture is placed onto the chamber of the hemocytometer and visualized under UV illumination within 10 min of staining.

Protoplast Fusion

Spontaneous fusion: During enzymes mediated cell wall degradation for protoplast isolation, some adjacently joined protoplasts get fuse without inducing agents, due to coalescence plasmodesmatal connection. These homokaryons or homokaryocytes cells are plurinucleate with 2–40 nuclei. More chances of homokaryons formation resides when source material is dividing cultured cells; however, exposure to strong plasmolyticum reduces the chances of spontaneous fusion.

Mechanical fusion: In this method, protoplasts are forcefully pushed to fuse without inducing agent. However, this approach, protoplasts get injured.

Induced fusion: Irrespective of source tissue or material, protoplasts are treated with inducing agents to get them fuse and yield somatic hybrids. Induced fusion is achieved though sodium nitrate ($NaNO_3$) treatment, polyethylene glycol (PEG) treatment, high pH/Ca^{2+} treatment and electrofusion, etc.

Selection and Regeneration of Somatic Hybrids

When the protoplasts from two different sources are mixed, 20–25% chances exist for the fusion event. Protoplast suspension subjected to fusion treatment may contain, unfused protoplasts, homokaryons, and heterokaryons. It needs selection steps including manual selection, complementation, dual florescent labeling system, and flow cytometry. Once hybridity confirms, then somatic hybrids are encouraged to regenerate into hybrid plants.

References

Aloni, R. (1993). The role of cytokinin in organised differentiation of vascular tissues. *Functional Plant Biology, 20*, 601–608.

Al-Taleb, M. M., Hassawi, D. S., & Abu-Romman, S. M. (2011). Production of virus free potato plants using meristem culture from cultivars grown under Jordanian environment. *American-Eurasian Journal of Agricultural & Environmental Sciences, 11*, 467–472.

Araki, C., & Arai, K. (1967). Studies on the chemical constitution of agar-agar. XXIV. Isolation of new disaccharide as a reversion product from acidic hydrolysate. *Bulletin Chemical Society of Japan, 40*, 1452–1456.

Atkinson, N. S., Robertson, G. A., & Ganetzky, B. (1991). A component of calcium-activated potassium channels encoded by the Drosophila slo locus. *Science, 253*, 551–555.

Bonga, J. M., & Aderkas, P. V. (1992). Media preparation. In *In vitro culture of trees*. Springer.

Brown, D. C. W., & Thorpe, T. A. (1995). Crop improvement through tissue culture. *World Journal of Microbiology and Biotechnology, 11*, 409–415.

Casanova, E., Trillas, M. I., Moysset, L., & Vainstein, A. (2005). Influence of rol genes in floriculture. *Biotechnology Advances, 23*, 3–39.

Cleland, R. E. (2001). Unlocking the mysteries of leaf primordia formation. *Proceedings of the National Academy of Sciences of the United States of America, 98*, 10981–10982.

Congard, B., Beaujard, F., & Viemont, J. D. (1986). Les bruyères in vitro. VI. Croissance de Calluna vulgaris sur milieu strictement nitrique ou ammoniacal et cinétique du pH en fonction du développement des plantes. *Canadian Journal of Botany, 64*, 959–964.

Digby, J., & Skoog, F. (1966). Cytokinin activation of thiamine biosynthesis in tobacco callus cultures. *Plant Physiology, 41*, 647–652.

Dougall, D. K. (1980). The use of tissue cultures in studies of metabolism. In *Metabolism and respiration* (pp. 627–642). Academic Press.

Drew, R. A., & Smith, N. G. (1986). Growth of apical and lateral buds of papaw (Carica papaya, L.) as affected by nutritional and hormonal factors. *Journal of Horticultural Science, 61*, 535–543.

Duquenne, B., Eeckhaut, T., Werbrouck, S., & Van Huylenbroeck, J. (2007). Effect of enzyme concentrations on protoplast isolation and protoplast culture of Spathiphyllum and Anthurium. *Plant Cell, Tissue and Organ Culture, 91*, 165–173.

Eklund, L., & Eliasson, L. (1990). Effects of calcium ion concentration on cell wall synthesis. *Journal of Experimental Botany., 41*, 863–867.

Gamborg, O. L. (2002). Plant tissue culture. Biotechnology. Milestones. *In Vitro Cellular and Developmental Biology, 38*, 84–92.

Gautheret, R. (1939). Sur la possibilité de réaliser la culture indéfinie des tissues de tubercules de carotte. *Comptes rendus des Société de Biologie (Paris), 208*, 118–120.

Grout, B. W. W. (1999). Meristem-tip culture for propagation and virus elimination. In *Plant cell culture protocols* (Methods in molecular biology) (Vol. 111, pp. 115–125). Humana Press.

Haberlandt, G., 2003. Culturversuche mit isolierten Pflanzenzellen. In Plant tissue culture: 100 years since Gottlieb Haberlandt (pp. 1–24). Vienna: Springer Vienna.

Henrickson, C. E. (1954). The flowering of sunflower explants in aseptic culture. *Plant Physiology, 29*, 536–538.

Hicks, G. S. (1975). Carpelloids on tobacco stamen primordia in vitro. *Canadian Journal of Botany, 53*(1), 77–81.

Huang, L., & Murashige, T. (1976). Plant tissue culture media: Major constituents, their preparation and some applications. *Tissue Culture Manual, 3*, 539–548.

Hussain, A., Qarshi, I. A., Nazir, H., & Ullah, I. (2012). Plant tissue culture: Current status and opportunities. In A. Leva & L. Rinaldi (Eds.), *Recent advances in plant in vitro culture* (pp. 1–28). IntechOpen.

Hussey, G. (1983). In vitro propagation of horticultural and agricultural crops. In *Seminar series—Society for Experimental Biology*.

Iwase, A., Mitsuda, N., Koyama, T., Hiratsu, K., Kojima, M., & Arai, T. (2011). The AP2/ERF transcription factor WIND1 controls cell dedifferentiation in Arabidopsis. *Current Biology, 21*, 508–514.

Jarret, R. L., Hasegawa, P. M., & Erickson, H. T. (1980). Factors affecting shoot initiation from tuber discs of potato (Solanum tuberosum). *Physiologia Plantarum, 49*(2), 177–184.

Kacar, Y. A., Biçen, B., Varol, I., Mendi, Y. Y., Serçe, S., & Çetiner, S. (2010). Gelling agents and culture vessels affect in vitro multiplication of banana plantlets. *Genetics and Molecular Research, 9*, 416–424.

Von Klercker, J., 1892, A method for isolating living protoplasts. Ofvers Vetensk. Akad. Forh. Stockholm 49: 463.

Kotte W (1922) Kulturversuche mit isolierten Wurzelspitzen. Beitr Allg Bot 2:413–43.

Kozai, T., Iwabuchi, K., Watanabe, K., & Watanabe, I. (1991). Photoautotrophic and photomixotrophic growth of strawberry plantlets in vitro and changes in nutrient composition of the medium. *Plant Cell, Tissue and Organ Culture, 25*, 107–115.

LaRue, C. D. (1942). The rooting of flowers in sterile culture. *Bulletin of the Torrey Botanical Club, 1*, 332–341.

Lindsey, K., & Yeoman, M. M. (1983). The relationship between growth rate, differentiation and alkaloid accumulation in cell cultures. *Journal of Experimental Botany, 34*, 1055–1065.

Loyola-Vargas, V. M., & Ochoa-Alejo, N. (2018). An introduction to plant tissue culture: Advances and perspectives. *Methods in Molecular Biology, 1815*, 3–13.

Ma, S. J., Lu, T., Zhang, A. Q., Wang, Y., & Zhou, L. (2010). Tissue culture and rapid propagation of seeds of Uyghur traditional herbal Capparis spinosa. *Zhong Yao Cai, 33*(12), 1833–1836.

Makonkawkeyoon, S., Smitamana, P., Hirunpetcharat, C., & Maneekarn, N. (1995). Production of mouse immunoglobulin G by a hybrid plant derived from tobacco-mouse cell fusions. *Experientia, 51*, 19–25.

Miller, F. J., Jr., Filali, M., Huss, G. J., Stanic, B., Chamseddine, A., Barna, T. J., & Lamb, F. S. (2007). Cytokine activation of nuclear factor κB in vascular smooth muscle cells requires signaling endosomes containing Nox1 and ClC-3. *Circulation Research, 101*, 663–671.

Miller, C., & Skoog, F. (1953). Chemical control of bud formation in tobacco stem segments. *American Journal of Botany, 1*, 768–773.

Molsaghi, M., Moieni, A., & Kahrizi, D. (2014). Efficient protocol for rapid Aloe vera micropropagation. *Pharmaceutical Biology, 52*, 735–739.

Morel, G. T. (1952). Guérison de dahlias atteints d'une maladie à virus. *Comptes rendus des séances de l'Académie des Sciences (Paris), 233*, 1324–1235.

Muir, W. H., Hildebrandt, A. C., & Riker, A. J. (1958). The preparation, isolation, and growth in culture of single cells from higher plants. *American Journal of Botany, 45*, 589–597.

Murashige, T., & Skoog, F. (1962). A revised medium for rapid growth and bio-assays with tobacco tissue cultures. *Physiologia Plantarum, 15*, 473–497.

Nassour, M., & Dorion, N. (2002). Plant regeneration from protoplasts of micropropagated Pelargonium × hortorum 'Alain': Effect of some environmental and medium factors on protoplast system efficiency. *Plant Science, 163*, 169–176.

Nitsch, J. P. (1951). Growth and development in vitro of excised ovaries. *American Journal of Botany, 38*, 566–577.

Owen, H. R., Wengerd, D., & Miller, A. R. (1991). Culture medium pH is influenced by basal medium, carbohydrate source, gelling agent, activated charcoal, and medium storage method. *Plant Cell Reports, 10*, 583–586.

Panda, A. K., Mishra, S., Bisaria, V. S., & Bhojwani, S. S. (1989). Plant cell reactors—A perspective. *Enzyme and Microbial Technology, 11*, 386–397.

Pasqualetto, P. L., Zimmerman, R. H., & Fordham, I. (1986). Gelling agent and growth regulator effects on shoot vitrification of 'Gala' apple in vitro. *Journal of the American Society for Horticultural Science., 111*(6), 976–980.

Peer, H. G. (1971). Degradation of sugars and their reactions with amino acids. In *Effects of sterilization on components in nutrient media* (pp. 105–115). H. Veenman & Zonen NV.

Prabavathy, V. R., Mathivanan, N., Sagadevan, E., Murugesan, K., & Lalithakumari, D. (2006). Intra-strain protoplast fusion enhances carboxymethyl cellulase activity in Trichoderma reesei. *Enzyme and Microbial Technology, 38*, 719–723.

Qaim, M. (2016). *Asian Biotechnology and Developmental Review, 18*, 95–101.

Robbins, W. J. (1922). Cultivation of excised root tips and stem tips under sterile conditions. *Botanical Gazette, 73*, 376–390.

Sahni, S., Prasad, B. D., & Kumar, P. (2017). *Plant biotechnology: Transgenics, stress management, and biosafety issues*. CRC Press.

San Pedro, T., Peiró, R., Villanova, J., Olmos, A., & Gisbert, C. (2017). In vitro propagation of Vitis vinifera L. cv. 'Monastrell'. *Electronic Journal of Biotechnology, 27*, 80–83.

Santana-Buzzy, N., Canto-Flick, A., Barahona-Pérez, F., del Carmen Montalvo-Peniche, M., Zapata-Castillo, P. Y., Solís-Ruiz, A., Zaldívar-Collí, A., Gutiérrez-Alonso, O., & de Lourdes Miranda-Ham, M. (2005). Regeneration of Habanero pepper (Capsicum chinense Jacq.) via organogenesis. *HortScience, 40*(6), 1829–1831.

Sharma, V., Kamal, B., Srivastava, N., Dobriyal, A. K., & Jadon, V. S. (2014). In vitro flower induction from shoots regenerated from cultured axillary buds of endangered medicinal herb Swertia chirayita H. Karst. *Biotechnology Research International, 2014*, 264690.

Singh, K., Sharma, M., & Singh, S. K. (2017). Effect of plant growth regulators on fruit yield and quality of guava (Psidium guajava) cv. Allahabad Safeda. *Journal of Pure and Applied Microbiology, 11*, 1149–1154.

Straus, J. (1960). Maize endosperm tissue grown in vitro III. Development of a synthetic medium. *American Journal of Botany, 47*, 641–647.

Sugii, N. C. (2011). The establishment of axenic seed and embryo cultures of endangered Hawaiian plant species: Special review of disinfestation protocols. *In Vitro Cellular & Developmental Biology, 47*, 157–169.

Tomar, U. K., & Dantu, P. K. (2010). Protoplast culture and somatic hybridization. In *Cellular and biochemical science* (pp. 876–891). IK International House Pvt. Ltd..

Tonosaki, K., Osabe, K., Kawanabe, T., & Fujimoto, R. (2016). The importance of reproductive barriers and the effect of allopolyploidization on crop breeding. *Breeding Science, 66*, 333–349.

Umate, P., Rao, K. V., Kiranmayee, K., Sree, T. J., & Sadanandam, A. (2005). Plant regeneration of mulberry (Morus indica) from mesophyll-derived protoplasts. *Plant Cell, Tissue and Organ Culture, 82*(3), 289–293.

Veloccia, A., Fattorini, L., Della Rovere, F., Sofo, A., D'angeli, S., Betti, C., Falasca, G., & Altamura, M. M. (2016). Ethylene and auxin interaction in the control of adventitious rooting in Arabidopsis thaliana. *Journal of Experimental Botany, 67*(22), 6445–6458.

Verma, D. C., & Dougall, D. K. (1979). Biosynthesis of myo-inositol and its role as a precursor of cell-wall polysaccharides in suspension cultures of wild-carrot cells. *Planta, 146*, 55–62.

White, P. R. (1963). *The cultivation of animal and plant cells* (2nd ed.). Ronald Press.

Wilkins, C. P., & Dodds, J. H. (1983). The application of tissue culture to plant genetic conservation. *Sci Progress (Oxford), 68*, 259–284.

Williams, R. R. (1995). The chemical micro-environment. In J. Aitken-Christie, T. Kozai, & M. A. L. Smith (Eds.), *Automation and environmental control in plant tissue culture* (pp. 405–439). Kluwer Academic.

Wojciechowicz, M. K. (2007). Comparison of regenerative potential of petals, stamens and pistils of five sedum species in vitro. *Biodiversity: Research and Conservation, 5*, 87–94.

Yancheva, S., & Kondakova, V. (2018). *Plant tissue culture technology: present and future development* (Vol. 1, pp. 39–63). Springer.

Chapter 2
Genetic Transformation in Plants: Methods and Applications

Graciela Dolores Ávila Quezada, Siddra Ijaz, and Riffat Malik

Abstract Plant genetic transformation is a technique that involves genome tailoring by adding or deleting a DNA segment. Alteration of a plant's genetic background by adopting non-conventional strategies is done for quantitative and qualitative trait improvement. This approach introduces transgenes (genes from different sources) into the plant's genome for nutritional improvement and makes them resilient against biotic and abiotic stresses. It has also broken species-, genera-, and even kingdom barriers by delivering genes into a plant from an unrelated source. Transgenic technology has expanded the possibilities available to plant breeders for improving breeding efficiency and plant protection. This chapter aims to provide unequivocal and comprehensive knowledge of the philosophy of genetic transformation in plants, the methods involved, and its applications.

Keywords Transgenic plants · Genetic transformation · Biolistic method · Expression vector

Introduction

Approximately 1 million people worldwide suffer from malnutrition, and our agricultural systems are deteriorating due to biodiversity loss, climate change, and other factors. By 2050, the population will increase to approximately 9 billion, and agriculture will confront enormous problems, with the demands for greater yield and quality crops requiring fewer inputs (Tilman et al., 2011). Genetically modified

G. D. Á. Quezada
Facultad de Ciencias Agrotecnológicas, Universidad Autónoma de Chihuahua, Chihuahua, Chihuahua, Mexico

S. Ijaz (✉) · R. Malik
Centre of Agricultural Biochemistry and Biotechnology, University of Agriculture Faisalabad, Faisalabad, Pakistan

S. Ijaz et al. (eds.), *Trends in Plant Biotechnology*,
https://doi.org/10.1007/978-981-97-0814-7_2

crops with the desired characteristics are developed by transferring the genes into elite crop varieties (Prado et al., 2014).

Genetic transformation in plants allows the transfer of genes related to useful agronomic traits into plants and provides an important tool in breeding by producing genetically diverse and novel plant materials. The direct transfer of desirable genes from one organism to another and the incorporation and expression of exogenous genes in that organism is known as "genetic transformation." The gene that is transferred is known as a transgene, and the organism in which the gene has transformed successfully is referred to as a "transgenic" (Babaoglu et al., 2000).

Due to genetic transformation techniques based on recombinant DNA technology, it is now possible to introduce foreign genes that confer resistance against biotic and abiotic stresses into plant genomes (Okpe & Nkaa, 2021). Transgenic technology has expanded the possibilities available to plant breeders for improving breeding efficiency (Danilova, 2007). Developing transgenic plants with various insecticidal genes is one of the many novel opportunities made possible by advances in biotechnology. Other opportunities include access to novel and powerful molecules, the ability to alter gene expression levels and patterns, and various plant transformation techniques.

Genetic modification can introduce exogenous genes that improve plant performance (Hossain et al., 2013). Each transformation procedure requires preparing and manipulating plant cells before permitting genomic modifications (i.e., transformation, identification, and selecting transformants) and effective regeneration of transformants (Siemens & Schieder, 1996). Utilizing the approaches for transformation necessitates a high level of technical proficiency and experience and frequently expensive tools (Yang et al., 2009).

There are two types of expressions. In temporary expression, a gene is transcribed but not integrated into the genome. It occurs 2–4 days after transfer although it is not passed on to progeny. In stable transformation, the gene is passed to the next progeny. DNA is integrated into the host genome within 10–14 days (Tóth et al., 2006). Many advancements have been made in transformation techniques over the past 20 years, which has sped up genetic engineering manipulation. Modern biotechnology methods have overcome incompatibility, sterility, fertilization hurdles, postfertilization barriers, and the risk of transferring gene sequences between gene pools and unrelated creatures, related species, and wild species (De Jeu, 1998).

Agrobacterium has now been widely used for gene transfer to plants, but its limitation to monocots has led to the development of other gene transfer techniques. Different direct methods, such as gene gun or biolistic, electroporation, laser treatments, liposome-mediated, silicon carbide whiskers, ultrasonics, and PEG-mediated transformation, are used for plant gene transformation (Rivera et al., 2012).

Vaccines against disease are present in the market but are costly, and most people cannot afford them in developing countries. Efforts are being made to create inexpensive edible vaccinations that offer numerous advantages over commercially available vaccines in transgenic plants. Transgenic plants can express recombinant proteins such as bacterial and viral antibodies and antigens. Carrots, tomatoes, bananas, rice, and other common food plants have been employed to make vaccines against various illnesses, including HIV, cholera, and hepatitis B (Ahmad et al., 2012).

The basic gene construct has a promoter, terminator, and gene-coding regions. Some special regions, such as a silencer, enhancer, and reporter sequences, are present in some gene constructs. Transformation in plants begins with the construction of a transgene. Transgene constructions have the same components except for selectable markers and target genes. A suitable gene is important for successfully producing transgenic lines (Debnath et al., 2010).

Promoters

Promoters are at the 5′ end upstream of protein-coding regions and are involved in gene expression. Transcription factors recognize the promoter sequence during DNA transcription. Transcription factors attach to the promoter at the consensus region and activate RNA polymerase. DNA transcription starts when the RNA polymerase transcription complex is formed (Hernandez-Garcia & Finer, 2014). There are three categories of promoters: tissue-specific promoters, constitutive promoters, and inducible promoters. At most phases of development, constitutive promoters are active, and they directly contribute to sustaining a constant and moderate level of gene expression. CaMV 35S is a constitutive promoter obtained from *Cauliflower mosaic virus*. If it is used in intact form, it is active in every tissue of numerous species of plants (Odell et al., 1985). CaMV 35S is often employed in dicot transformation rather than monocot transformation because this promoter is more effective in these plants, and this can be due to the availability of certain transcription factors in these plant families (Porto et al., 2014).

Many tissue-specific promoters have been characterized for expression in a particular tissue. The heterologous protein must be absent from the other tissues. The production of exogenous proteins has been linked to several diseases and developmental irregularities, including programmed cell death and dwarfism. When using constitutive promoters, there is a significantly greater risk to the plant's metabolism (Sharma & Sharma, 2009). The specificity of the promoter may differ depending on the type of host organism. On rare occasions, using a promoter thought to be specific can cause nonspecific expression (Park et al., 2010). For instance, wheat transformation using the promoters of glutenin genes resulted in the effective expression of genes in the endosperm. However, in tobacco using the same promoter, the expression was found to be decreased, and heterologous protein was detected in leaves (Potenza et al., 2004). The E8 promoter obtained from tomatoes controls the genes involved in the maturation of fruits. E8 is used to produce fruit antigens and enhance the flavor and aroma of tomatoes (Biłas et al., 2016).

Particular environmental conditions or substances, such as herbicides, metal ion concentrations, steroids, temperature, light, and alcohol, influence inducible promoters. These controlled expression systems are useful because they allow the proliferation of cells and the synthesis of proteins to be optimized. This strategy especially appeals when product production harms cell viability and growth (Vaucheret & Fagard, 2001). A chemically inducible promoter system based on the estrogen receptor (XVE) has recently been developed in plants. The inducible promoter, XVE, is activated by estrogen and permits careful regulation of the transcription of target genes, leading to an increase in GFP expression over the CaMV35S constitutive promoter (Zuo et al., 2000). In plant cells, other chemically inducible promoter systems, such as dexamethasone and ethanol, have been established (Huang & McDonald, 2009).

Enhancers and Repressors

Enhancers (50–150 bp) are among the eukaryotic promoter elements that increase transcriptional efficiency, mostly independent of their orientation and position relative to neighboring genes. Sometimes, enhancers can activate gene transcription present on different chromosomes (Potenza et al., 2004). Short sections of a gene called enhancers can be identified and linked with activator proteins or transcription factors. They attach to enhancers and create a complex of enhancer-bound transcription factors that later bind with the mediator complex (TFIID) to help activate RNA polymerase II (Ong & Corces, 2011).

A repressor inhibits the expression of genes by binding them to related silencers or operators. It prevents the attachment of RNA polymerase to the promoter, thus preventing the transcription of DNA into mRNA. The mRNA is bound by an RNA-binding repressor, which stops the mRNA from being translated into protein. Furthermore, simple species such as yeast are home to most known eukaryotic repressors, which interact with activators directly to exert their effects (Clark et al., 2019).

Reporter Gene

Reporter genes are connected to genes of interest or regulatory sequences to enable the identification of transgenic expression and the location of expressed proteins (Seymour & Sander, 2007). Reporter gene sequences produce proteins or the products of proteins that have been catalyzed for detection (Low et al., 2018).

Green fluorescent protein (GFP), luciferase (Luc), beta-glucuronidase (GUS), chloramphenicol acetyltransferase (CAT), and beta-galactosidase (LacZ) are examples of typical reporter genes used for monitoring transgenic plant expression. These reporter genes allow the identification of transgene location and control of the tagged and expressed proteins. They also enable discrimination between nontransformed and transformed cells (Santi et al., 2003; Koo et al., 2007).

Plasmid Vectors

The Ti plasmid would expand further with the desired genes and selectable markers. Plasmids of large size have low copy numbers and are difficult to handle. However, later, this shortcoming encouraged the creation of cointegration systems in conjunction with binary vector systems, providing a solution for the issue with large plasmids (Van Montagu & Zambryski, 2013). The right and left borders are the repetitive sequences surrounding the area on either end. The right border is a vital component for DNA transfer that causes tumors (Christie, 2004).

Binary and Super Binary Vectors

The binary vector system was developed because T-DNA transmission is still possible even when virulence genes and T-DNA are placed on different replicons. T-DNA is built within plasmids, which can replicate in *Agrobacterium* and *E. coli* (Hoekema et al., 1983). After completing plasmid production in *E. coli*, the vector is delivered to *Agrobacterium* to create a strain suitable for plant transformation. These binary vectors are distributed widely. Employing extra virulence genes, such as virB, virE, and virG, which exhibit specific gene dosage effects, is one of the strategies for increasing the transformation frequency using binary vectors (Komari et al., 2006).

In the super binary vector, a DNA fragment containing virB, virC, and virG from pTiBo542 was inserted into a tiny DNA-carrying plasmid. Strains of *A. tumefaciens* with pTiBo542 have higher transformation efficiency and a wider range of hosts than other Ti plasmids, such as pTiT37 and pTiA6 (Vain et al., 2004). Super binary vectors are extremely effective in transforming various plants, and in the 1990s, the host range was increased to important cereal crops by *A. tumefaciens*-mediated transformation. The last step in super binary vector construction is the integration of intermediate and acceptor vectors in *A. tumefaciens*, and the plasmid is the final product (Komari et al., 2006).

Biological Transformation Methods

Agrobacterium-*Mediated Gene Transfer*

The phytopathogen *Agrobacterium tumefaciens* causes crown gall disease and mostly affects dicotyledonous plants. The disease is characterized by forming tumors in stem-like plant tissues (Mohammed & Abalaka, 2011). In dicotyledonous plants, *Agrobacterium rhizogenes*, a species closely linked to *Agrobacterium tumefaciens*, causes hairy root disease (Ijaz & Haq, 2019).

The virulence (Vir) proteins the bacterium utilizes to manufacture its T-DNA and deliver it to the plant cell are encoded in the vir region, which is present on the *Agrobacterium* Ti plasmid. Then, specific host signals cause the vir region to be expressed. The joint activity of the bacterial proteins VirD1 and VirD2 then results in the production of single-stranded (ss) T-DNA molecules (T-strand). One VirD2 molecule is covalently bonded to the 5′ end of the immature T-complex that exists as T-DNA in bacterial cells (Kunik et al., 2001; Sood et al., 2011). The VirB/D4 type IV secretion system exports this complex to host cells with numerous other viral proteins (Vergunst et al., 2000). *Agrobacterium*-mediated transformation has also been focused on improving trees with commercially valuable features, such as increased timber production and shorter generation times. *Nicotiana tabacum* was the earliest species to be transformed by *Agrobacterium*. Later, it was switched to the simpler-to-transform *Arabidopsis thaliana*. Since then, *Agrobacterium* transformation has successfully improved fruits, legumes, vegetable crops, medicinal plants, forest trees, biodiesel plants, oilseed crops, and orchids (Abu-Arish et al., 2004; Hwang & Gelvin, 2004). *Agrobacterium tumefaciens* with the Ti plasmid is virulent and can cause neoplastic growth in hosts with the aid of vir genes (virulence genes) (Zhang et al., 2006).

Opines are used as food for the growth of bacteria, whereas cytokinins and auxin cause uncontrolled cell proliferation that leads to cancer formation. The vir area comprises vir genes (virulence genes) with approximately 11 operons that can code for 35 vir proteins, 20 of which are implicated in the formation of tumors. By introducing the desired sequence in place of T-DNA, this natural method transferred an important gene to the target host cell (Desfeux et al., 2000).

The vir gene is activated by different mechanisms. The chemotaxis mechanism directs the bacteria to the target site as a sensor for chemicals emitted from the plant's damaged region. The VirG gene is activated by the chromosomal virulence gene system chvG/chvI, which measures acidity. As a defense mechanism against plant phenols, the VirA/VirG system, which is a crucial participant, promotes the coding of the vir protein. When chvE attaches to monosaccharides produced by plants, maximal expression of vir genes occurs. The vir gene is activated by the signaling cascade of acidity, phenols, and sugar (Nester, 2015).

The membrane receptor vir A (histidine kinase) has four domains (periplasmic domain, which senses slight acidity and monosaccharides; linker domain, which recognizes phenolic compounds; kinase, which acts as a catalyst for phosphorylation; and receiver, which sings up virG, which regulates the intercellular response). To start the transcription of vir operons, virA autophosphorylates a particular histidine residue and transfers phosphate to aspartic acid residues of virG (Meyer et al., 2018; Kavipriya et al., 2019). Transgenic bananas have been developed by *Agrobacterium*-mediated transformation (Nester, 2015). Immature embryos co-cultivated with *Agrobacterium tumefaciens* successfully produced transformants of maize (Lacroix & Citovsky, 2018). The *Agrobacterium* transformation system has been used for wheat genetic transformation (Guo et al., 2019).

Virus-Mediated Transformation

DNA or RNA viruses that infect plants are used as vectors for transferring genes into targets. The gene that has to be transferred is integrated into the genome of viruses, which now act as vectors for transferring foreign genes. The virus carrying the foreign gene infects the target plants and causes successful transformation. The main drawbacks of this process are that a high copy number per cell is transferred, only transient transformation can be produced, and stable transformation cannot be achieved because they cannot transfer the gene to the next generation (Okpe & Nkaa, 2021).

While many viral genomes have been modified to function as vectors, the Tobacco Mosaic Virus (TMV) has advanced most in developing a general-purpose viral-based vector. Cloned cDNA representing the entire TMV genome can be transcribed in vitro to create infectious RNA. The TMV genome has been successfully modified as a vector using two approaches. The first technique entails the development of hybrid RNA, in which the coat-protein promoter controls subgenomic RNA expression containing promoter sequences and exogenous genes from related but nonhomologous viruses, producing significant amounts of foreign protein. The second approach involves positioning the sequence necessary for the 130 kDa replicase protein readthrough downstream from the coat-protein-encoding gene and upstream from the exogenous gene, and infected plants produce coat protein and a coat protein fusion with the exogenous peptide. Using that vector has been feasible to obtain plants to manufacture large amounts of the angiotensin I-converting enzyme inhibitor protein (Walden & Wingender, 1995).

Due to the infectious nature of viruses, the maximum amount of exogenous genes can be obtained from the viral genome within a short period, mostly 1 or 2 weeks after inoculation. Numerous plant viruses are easily transmissible and may be employed economically for quick mechanical inoculation of vast acreages of crop plants. Once an appropriate temporary gene vector is found, it is possible to quickly assess the expression of exogenous genes in various plants that serve as the virus's hosts. This characteristic is also helpful for early studies of genes that produce quantifiable phenotypic effects (Scholthof et al., 1996).

Nonbiological Transformation Method

Electroporation

Nucleic acids cannot enter cells by themselves; they need help to cross physical boundaries, such as the plasma membrane, and reach cell sites where they can be expressed or replicated. When cells are exposed to electrical potential, their membranes become reversibly destabilized, and pores are induced through which DNA can enter the cell easily. Electroporation has evolved as a quick, easy, and effective way to transfer DNA into bacteria, plants, yeast, and cultured mammalian cell lines (Kumar et al., 2019). It is a transient in vitro technique to transfer other molecules and DNA into cells (Meijerink et al., 2018).

Electric field pulses at a microsecond to millisecond scale time are applied to induce a transient increase in membrane potential. Usually, 200 mV–1 V creates 108 V/m across a plasma membrane approximately 5 nm thick (Chen et al., 2006).

When an electric field is applied to the lipid bilayer, the TMV (transmembrane voltage) produced by this exposure is composed of two responses: one is the dielectric response that results from the dipole realignment, and the electric field serves to align them, and the other results from the redistribution of charges (ions in the surrounding solutions), which the electric field drives to accumulate them on both sides of the bilayer and thus charge it. Eventually, these clusters merge to form a hydrophobic pore that spans the bilayer. As a result, lipids close to water molecules within the pore begin to reorient with their polar headgroups, enabling additional water along with other ions and polar molecules to enter. However, only hydrophobic pores were found in some bilayers composed of archaeal lipids containing significant sugar moieties. Pore closing takes tens to hundreds of nanoseconds to complete (Kotnik et al., 2019).

Some large hydrophobic pores become hydrophilic as the energy decreases and the TMV increases. Hydrophilic pores have a longer half-life than hydrophilic pores, as hydrophilic pores require less energy. Molecules of DNA up to 150 kb in size can flow through pores, indicating that size is not an obstacle (Mahnič-Kalamiza et al., 2012; Kumar et al., 2019).

As electrophore formation is related to the duration and amplitude of the applied electric field, this heating effect can be reduced by utilizing short-duration pulses with relatively high amplitude pulses (Vorobiev & Lebovka, 2017). Electroporation is reversible at a moderate electric field, electro pores are resealable, and cells maintain viability. In contrast, at a higher electric field strength, electro pores cannot reseal, there is extensive cell membrane disruption and DNA leakage, and cells eventually die (Kotnik & Weaver, 2016). Based on the duration of electric pulses, resealing is a long process requiring a time longer than 1 s (>1 s) compared to the formation of pores (ms or μs) (Kotnik et al., 2012).

An electroporator device is used to carry out the electroporation process. This device normally has three basic components: electrodes, electroporation cuvettes, and pulse power supplies (Reberšek et al., 2007). The minimum apparatus required for electroporation is an electroporator and micropipette puller to prepare micropipettes to inject DNA. All control units are in the power supply (electrical pulse settings). First, the suspension of target cells is transferred to a glass or plastic cuvette. Second, cells are subjected to electrical pulses through electrodes and electrical conductors, which requires a direct connection between the electrodes and cell suspension (Yarmush et al., 2014).

Every cell type possesses a unique field strength depending on the applied pulse characteristics (such as resistance capacitance and voltage), and the optimum field strength induces a transmembrane voltage, resulting in electropermeabilization. Direct contact between a membrane patch and an electrode-containing micropipette allows for measuring changes in the membrane's electrical properties. The membrane patch conductance increases several orders of magnitude during electroporation, and the capacitance changes measurably (Kotnik et al., 2012). Electroporators with three different electrical wave pulses, such as square waves, time constants, and exponential decay, are frequently used in the electrotransformation of nucleic acids (Escoffre et al., 2009). In a time-constant pulse, the cells that will be electrotransformed are given a constant pulse and voltage at a specific time (Siddiqui et al., 2016). The applied voltage, duration and number of the pulses, and the lengths of gaps between the pulses identify square wave pulses. The adjusted voltage delivered from the capacitor in an exponential decay pulse quickly degrades over time (Weaver et al., 2012).

Medium hyperosmolarity is commonly obtained by dissolving 0.5–1.5 M mannitol or mannisorbitol and can also increase the electrotransformation efficiency in bacteria (Kotnik, 2017). The time constant and strength of the field (kV/cm) are crucial factors that define the delivered pulse. Thus, these parameters can be changed to obtain the appropriate transfection efficiency of different cell types (Escoffre et al., 2009).

Electropores on a reversible scale provide a brief transport passage and are then recovered and resealed. On a nonthermal irreversible scale, the cells rupture and discharge their contents without being thermally damaged since the electropores do not progressively close or reseal. On an irreversible scale, applied current results in thermal damage, such as denaturation of proteins over 50 °C and melting of DNA above 70 °C (Markelc et al., 2017).

Electroporation is a stochastic process influenced by cell size, type, electrical conductivity, solutes, mediated medium, and osmolarity; the above-described ranges also partially overlap (Kotnik et al., 2019). Field strengths and pulse lengths typically fall within 1–20 kV/cm and 1–30 ms for optimal transformation efficiency, but the ideal parameters are found empirically. This efficiency decreases as the recipient DNA is covered by more or thicker membranes (Aune & Aachmann, 2010). The maximum efficiency is found in Gram-negative bacteria, yielding 10^7–10^{10} transformants per mg of DNA. However, Gram-positive bacteria and archaea have lower efficiency because of their thick cell walls (10^5–10^7), and yeasts and microalgae have even lower efficiency (10^4–10^7) because of their nuclear membranes (Dzul et al., 2011).

After being electroporated, *Solanum dulcamara* L. protoplasts displayed increased morphogenesis compared to untreated protoplasts. Additionally, compared to shoots from untreated protoplasts, regenerated shoots rooted more easily (Markelc et al., 2017). They created more prolific root systems by electroporating protoplasts derived from cell suspensions with plasmids expressing the NPTII gene under the CaMV 35S promoter; transgenic rice plants were created from somatic embryogenesis. The production of kanamycin-resistant colonies was maximized by the heat shock of protoplasts before electroporation (Kumar et al., 2019). Tobacco protoplast absorption and cucumber mosaic virus (CMV) RNA expression were studied through square wave and exponential wave pulses. Over 60% of the surviving protoplasts could integrate and express RNA when these electropulses were given for a crucial time and at a sufficient field strength (Vorobiev & Lebovka, 2017).

Gene Gun

Sanford et al. developed the biolistic particle method in 1982. In biolistic delivery, genes or DNA are coated and dehydrated on heavy metal particles. Gold and tungsten are frequently used as microcarriers of genes because they can penetrate cells easily due to their high specific gravity of approximately 19.3 and small diameter of approximately 0.5–5 μm. Gold particles are mostly used because they are inert and uniform in size. High-pressure helium pulses accelerate the particles and are launched into plant cells at high speeds. Helium is widely used because it is extremely diffusible, inert, and low density, and particles can achieve high speed (Ijaz & Haq, 2019).

Usually, plant cell epidermal tissue is the target. Depending on the experimental conditions, DNA can enter the nucleus and pass across the plant cell wall plasma membrane. Crucial experimental factors that impact penetration effectiveness, overall levels of gene transfer in plants, and toxicity include helium gas pressure, dosage frequency, and net particle size. Gene delivery and genome editing are made easier by biolistic particle delivery, which transmits many DNA molecules with sizes up to 150 kilobases (kb) (Ueki et al., 2009). Researchers are concerned about DNA integrity after the transfer, low-level and short-term expression of the given genes in plants, and cell damage due to the high pressure that plant tissues encounter despite the affordable and simple approach (Demirer & Landry, 2017).

The unique advantage of this technique is that it is not dependent on the characteristics of cells, such as ligand receptors, electrical potential, and structural components of the cell surface. This technique efficiently transforms different targets and organisms, such as bacteria, fungi, mammals, and intracellular organelles, such as chloroplasts (Carsono, 2007). In the early gene gun procedure, the sample is placed in a sealed chamber in a vacuum and bombarded with DNA-coated or RNA-coated microparticles. This method has limited applications because a small sample was used and cannot be easily damaged by gene sun under a vacuum. Gene gun systems such as PDS1000\He are used in which no vacuum chamber is needed, thus removing the limitations on the size and target of the sample. This system can be used for various applications in gene transfer, including whole animals, cultured cells, and tissues, and is even applicable for both in vivo and in vitro transfections. This technique is most commonly used, easy, and simple. It overcomes barriers such as the epidermis and cell wall of plants (Uchida et al., 2009). The Bio-Rad Helios bombardment system was used for transient expression in Arabidopsis leaves for functional assays. The Helios gene gun system introduced the wheat Glu-1Dx5 glutenin gene into the rice genome (Carsono, 2007).

Microinjection

Higher plant transformation is potent for agricultural development and plant molecular biology. As a result, the field of transformation techniques continues to advance significantly. Various DNA delivery techniques have recently been developed and proven effective for transforming plant cells. Microinjection enables the introduction of DNA into specific targets, either into a specified compartment of a single cell or into whole cells within a multicellular structure under microscopic control. Microinjection methods have been applied recently in several labs and for various objectives. *Nicotiana tabacum* protoplasts were injected to transform the plants successfully. On average, less than 1–2% integrated copies per tobacco genome were found. Intranuclear injections resulted in an average integration frequency of 14% compared to 6% with cytoplasmic injections. Protoplasts of alfalfa were injected intranuclearly displayed 15–26% transformation frequency. By microinjecting various plasmids, more than 70 transformed callus lines were obtained without selection (Zhang & Yu, 2008).

Despite the extremely high transformation frequencies reported in these investigations, the actual number of transformed calli was never particularly great, and plantlets were never regenerated. A significant barrier was the lack of a suitable culture system allowing the injection of more cells with an effective and manageable regeneration system (Schnorf et al., 1991). The issues were twofold. On the one hand, it was challenging to immobilize the plant cells, which made the entire process cumbersome and ineffective. On the other hand, the absence of a selection system required that injected protoplasts be cultured separately from those not, leading to laborious and complicated culture systems such as hanging drops or single-cell culture (Crossway et al., 1986).

Much time and effort is required to deliver nucleic acids to protoplasts or whole cells using microinjection, which also requires specialized capillary needles, an inverted microscope, pumps, a light source for illumination, micromanipulators for positioning the needles, microinjectors for generating negative pressure to support the suspended cells or positive pressure for injecting materials and an injection chamber for maintaining the cell survival environment (Zhang & Yu, 2008). When DNA was introduced into the cytoplasm, the efficiency declined to 1%. Because of handling during injection, the bulk of delivered DNA is incorporated as incomplete fragments. Rapeseed was the first plant in which microinjection was shown to produce stable transgenic plants (Neuhaus et al., 1987).

Numerous laboratories have provided reports on the effective transformation of plants using microinjection techniques due to the commercial availability of improved microinjection setups, plant regeneration systems, and microculture systems. Protoplasts attach to plastic or glass surfaces, but their bond is insufficient to hold them in place for microinjection. Protoplasts are embedded in Na alginate or agarose for immobilization. These systems need ways to distinguish injected protoplasts from noninjected protoplasts. Coinjected video recording or fluorescent dyes are effective (Steinbiss & Stabel, 1983).

The holding capillary system enables intranuclear microinjection by allowing accurate orientation of protoplasts, cells, or tiny multicellular structures. Additionally, the capillary tip could be modified for various cell types. A holding capillary for microinjection can handle and develop selected protoplasts in microculture systems. Individually chosen protoplasts can be cultured and manipulated in microculture systems for microinjection by holding capillaries (Morikawa & Yamada, 1985).

The tools used for microinjecting plant cells are essentially the same as those used for injecting animal cells. One or two micromanipulators may be utilized, depending on the immobilization technique. Recovering transformants after microinjection requires high plating efficiency and regeneration frequencies. Unfortunately, many plant species are currently not meeting this condition. The observed rates of altered cells ranged from 100 to 200 cells/h. This technique offers single cells that are highly competent for both synchronous development and somatic embryogenesis. Recently, it was discovered that embryogenic carrot suspension cells could be used to microinject intact single cells with cell walls (Crossway et al., 1986).

Silicon Carbide Whisker-Mediated Transformation

Technically, whisker transformation is a simple method with no costly equipment or supplies. In Eppendorf tubes, plasmid DNA, cells, and whiskers are combined and mixed on an oscillating mixer or vortex (Greenwood & Earnshaw, 2012). Cell penetration occurs when whiskers have an average diameter of <1 μm and are combined with plasmid DNA and plant cells, but the precise procedure of whisker transformation is unknown. The premixing of whiskers and DNA in the procedure reported by Kaeppler et al. (1990) was performed before adding the cells, but it did not affect the transport of DNA. We have shown that DNA delivery is diminished as a result of prolonged storage of whiskers in aqueous suspension; thus, this must be avoided. These conditions probably limit the interaction between whiskers and DNA (Kaeppler et al., 1992). Most whisker preparations are quite heterogeneous, containing whiskers ranging from 5 to 500 μm.

There is scope to produce a whisker type that is more consistent and, for instance, has different surface features, which would allow for more effective DNA transport to plant cells. Transient gene expression has been used to determine DNA delivery with whiskers in various plants (Asano et al., 1991). A comparable frequency (65%) of coexpression of bar- and gus-related genes was observed in maize cells undergoing microprojectile bombardment. A comparable frequency (65%) of coexpression of the related genes bar and gus was seen in maize cells subjected to microprojectile bombardment (Kaeppler et al., 1992).

Fertile transgenic maize plants carrying the gus and bar genes were created via the whisker-based transformation approach of embryogenic suspension cells. Transient gene expression allows for the adjustment and evaluation of numerous parameters, ensuring that effective DNA transport is accomplished without compromising cell life. For instance, mixing time and speed can be easily regulated, affecting how effectively a transformation occurs (Frame et al., 1994). Before whisker mixing, pretreatment with combined mannitol/sorbitol increases transient expression by approximately three to fivefold (Vain et al., 1993).

Plasmid DNA delivered by whiskers in an unprecipitated state allows us to manage the DNA concentration present for transformation. In BMS cells, stable transformation was achieved with only 0.1 μg of input plasmid DNA; the transformation efficiency was 50% higher than that with 25 μg (Thompson et al., 1995).

Multiple gene copies are frequently introduced into cells via direct gene transfer, and these copies frequently exhibit complicated integration patterns. While it is yet unclear whether these integration characteristics are functionally significant, the ability to regulate the amount of input DNA by whisker transformation may make it possible to affect copy numbers. The production of stably transformed callus lines resulting from whisker treatment of embryogenic calli of a superior inbred maize variety suggests the possibility of DNA transfer to better-organized tissue (Spencer et al., 1992).

Laser-Mediated Transformation

For over 20 years, laser microbeam irradiation has been used successfully to perform subcellular microsurgery, modify subcellular structures, promote cell fusion, and microdissect chromosomes (Berns & Rounds, 1970). Weber reported that lasers could be used in plant cells to produce holes in the membrane and cell wall of chloroplasts and plant cells in hypertonic buffer. It is possible to incorporate exogenous DNA into organelles and plant cells (Weber et al., 1990).

The cell wall and cellular membrane are the two primary barriers plant cells use to naturally defend themselves from external factors. Although small molecules can freely pass through the cell wall, macromolecule and nanoparticle penetration may be restricted. The precise cutoff values rely on several factors, such as the type of plant, the molecule (e.g., 3D structure charge), the tissue, and environmental variables, such as temperature and pH. The size exclusion limit (SEL) for particles, which is frequently taken into account in nanotoxicity, has been reported to be between 5 and 20 nm, while the SEL for macromolecules has been defined in the range of 40–60 kDa (Tepfer & Taylor, 1981; Rukmana et al., 2019).

Molecules below this range can diffuse through the membrane passively by single photoporation. The cell wall must be permeable before molecules above the SEL pass through. It can be assisted by carrying out the photo injection in two steps or by puncturing the membrane and cell wall where they meet to permeabilize both barriers with a single light (Weber et al., 1988a).

The turgor, or internal pressure, of such intact plant cells will give the cell mechanical stability. When the membrane opens due to this pressure, the cytoplasmic content of the cell will extrude, preventing any inward transport. Therefore, one effective way to get around this is to pretreat with a hyperosmotic buffer that causes cell plasmolysis. It has been demonstrated that the photoporation-induced uptake of external liquid facilitates the molecular uptake and restoration of the initial cell volume (Mitchell et al., 2013).

Furthermore, it appears that plasmolysis increases post-photoporation cell viability. Therefore, hyperosmolar treatment is considered a prerequisite for cell photoporation although a properly balanced ratio between toxicity and plasmolysis from osmolar shock must be achieved. Another strategy is to increase cell wall permeability through enzymatic breakdown, releasing mostly cell wall-free protoplasts. The protoplast reacts quickly to osmolarity changes. As a result, it is difficult to establish active volume uptake such as the one in the later sentence without affecting cell viability. Despite the absence of cell walls that make protoplasts an appealing target for photoporation, reports on the process have been inconsistent until now, which may be connected to the absence of plasmolysis-driven uptake and higher susceptibility to treatments (Mitchell et al., 2013).

Laser microbeams have many characteristics that make them a perfect tool for conquering the obstructions offered by the cell walls of plants. The laser can measure less than 1 μm (Berns et al., 1981). Furthermore, manipulating organelles or nuclei can be carried out without opening the plasma membrane by focusing the beam inside the cell. The cell wall components were removed using a focused beam (Hahne & Hoffmann, 1984). Laser pulses fused the plant protoplasts and animal cells (Wiegand et al., 1987). The cells were placed in a hypertonic buffer (10 mM Tris, 0.6 M sorbitol, pH 7.2) to create differential pressure inside and outside the cell to produce holes in the cell wall. This differential pressure causes the movement of materials outside and inside the cell after applying the laser beam, and small holes are temporarily formed in the cellular membrane. Swelling of the cell indicates that the buffer has entered the cells through holes, which are resealed within 5 s after the laser pulse. Treated cells continue their growth. Hypertonic buffer reduces the volume of cells by up to 80%. Absorption of materials through the laser holes would restore the original volume of the cell. Fluorescent dyes such as bisbenzimide, absorbed by >80% of the laser-irradiated cells, are added to the buffer for visualization (Cesarone et al., 1979). DNA integration into the cells by a laser beam was verified by suspending the cells in hypertonic solutions that contained bisbenzimide-labeled DNA. The individual cells were laser irradiated, and the cells that took up the DNA showed fluorescence (Weber et al., 1988b).

A laser microbeam can puncture chloroplasts within the cells. Temporary small holes in the envelope of the isolated chloroplast were developed, and fluorescently labeled DNA was inserted into the organelles. After irradiation, the organelle membrane integrity was restored. There are two strategies to transform chloroplasts. The isolated organelles are irradiated in the capillary with DNA and then transferred to the recipient; however, this manipulation is very slow. Another method for introducing DNA into chloroplasts involves injecting DNA into a cell's cytoplasm and then puncturing the chloroplast membrane. After 1.2 s, laser holes created in chloroplasts are closed, giving enough time for DNA uptake (Weber et al., 1989). Fluorescently labeled DNA was microinjected into protoplasts. The laser was focused on specific chloroplasts, and then a single laser pulse was emitted at every organelle. The irradiated chloroplasts had fluorescence associated with them, and the cells survived the treatment. DNA appears to have entered the organelles by the holes created by the laser. Nevertheless, only temporary expression was noticed when DNA carrying the gene for herbicide resistance was used (Hirschberg & McIntosh, 1983).

Cell walls and membranes can be punctured with holes of certain diameters using a UV-laser microbeam. Target disruption is minimized by a quick laser pulse of approximately 15 ns or less in duration. In hypertonic buffer, cells and organelles assimilate exogenously supplied DNA by transient holes created through a laser. Organelles and cells that have undergone treatment express genes. Individual cells can be exposed to UV radiation in 1 h. More than 10^6 cells can be treated at a time. It is now possible to quickly manipulate the genetic makeup of plant tissue, individual organelles, and cells using a UV-laser microbeam. Crop plant cells could be given altered genes (Weber et al., 1990).

Foreign DNA was introduced into the embryonic calli of rice (*Oryza sativa*). In hypertonic buffer, calli cells were placed to remove some water, placed in a less negative osmotic potential medium with foreign DNA, and irradiated with a laser microbeam ($\lambda = 355$ nm) to produce pores in the cell wall and membranes. DNA entered the cells through laser holes due to differences in the potential gradient. Cells immersed in a fluid containing the fluorescent chemical calcein showed bright yellow-green fluorescence. β-Glucuronidase (GUS) genes were inserted successfully into rice cells, as evidenced by gene expression in post-treated cells and plantlets produced from kanamycin-resistant calli that had undergone this treatment (Ijaz & Haq, 2019).

PEG-Mediated Transformation

Polyethylene glycol (PEG) is a polyether substance that can bind with DNA and transfer the DNA into the target by piercing the membrane when combined with Ca^{2+}. The procedure is simple, reproducible, efficient, and requires no specialized tools. This method produces many transformants with high survival rates (Mathur & Koncz, 1998). This procedure is most effective, producing more than 120 transient transformants per microgram of DNA (Wu et al., 2020).

Since the target of this method is protoplasts, the method begins by preparing protoplasts. Typically, protoplasts are prepared by leaf mesophylls. Protease and cellulose enzymes break the leaf disc cell walls. Mixtures of enzymes have osmoticum for maintaining tonicity, which is important for protoplast preparation. Mannitol solution was used for the purification of protoplasts. The prepared protoplasts were suspended in the PEG/Ca^{2+} and DNA solutions at a 60 μm concentration (Sahab et al., 2019). The DNA solution and protoplasts were incubated briefly after gentle shaking. The final step is to propagate the protoplasts treated with DNA over an appropriate medium, isolate protoplasts bearing the transferred DNA using a marker, and then regenerate the entire plant from the protoplast to produce the transgenic plant. Although the method is easy, regenerating plants from protoplasts is difficult. PEG-mediated transfection was performed in canola plants using mesophyll protoplasts. Hygromycin resistance genes were introduced into the cultivars of rice plants by polyethylene glycol (25% (w/v) (Liu & Vidali, 2011).

PEG-mediated transformation in *Arabidopsis thaliana* was performed using protoplasts derived from roots, leaves, and cell suspensions. After being embedded in alginate, leaf mesophyll protoplasts can regenerate, but their production is comparatively low. Cell suspensions, in contrast, provide quickly dividing protoplasts that may be generated in 2–3 h and exhibit transitory expression of the inserted gene in 24 h. Protoplasts from roots also have higher regeneration and division capabilities. When using light-emitting enzymes as reporter proteins in nondestructive in vivo gene expression studies, such as GFP or luciferases, cell suspensions of low autofluorescence and root-derived protoplasts are of special interest. The red fluorescence of chloroplasts in leaf protoplasts prevents accurate monitoring of GFP reporter gene activity (Yang et al., 2022).

Ultrasonic/Sonication-Mediated Transformation

The human ear cannot hear ultrasound, a high-frequency sound exceeding 20 kHz. The effect of ultrasound waves is mainly due to acoustic cavitation at low energies. There are two strategies for acoustic cavitation: transient cavitation and stable cavitation. In collapse or transient cavitation, bubbles form and grow until they implode. They generate high temperatures and pressure during collapse. Due to the location of bubbles, high-pressure waves are produced, which cause damage or mechanical effects on surrounding materials such as macromolecules or cells. Stable cavitation causes rapid and large oscillations in the size of bubbles. Violent flow is generated in the surrounding medium of bubbles due to oscillation, known as microstreaming. Sheer stresses result from high microstream velocities that cause damage to cells and macromolecules (Frizzell, 1988).

Ultrasound waves have been reported to transfer genes into plant cell suspensions, protoplasts, and whole tissues. Ultrasonic-mediated gene transfer is a simple method that does not depend on the type of plant material. The plant material was suspended in a microcentrifuge tube containing the sonication medium. After rapid mixing, the DNA was added to the sample and was ready for sonication. The ultrasound pulses are then delivered from machines to homogenize tissues (20 kHz Digital Sonifier S-250 D). Ultrasound waves improve the efficiency of in vitro tissues, animal cells, and protoplasts with temporal and spatial specificity (Liu et al., 2006).

Ultrasonic transformation is also called sonication and is based on sonoporation (the breakdown of membranes using acoustic waves). It is an effective method for introducing foreign DNA into cells through acoustic cavitation, which temporarily causes changes in the permeability of membranes (Rivera et al., 2014). Critical characteristics are the time of exposure, intensity, type of application (pulsed or continuous), central frequency, duty cycle, and pulse repetition frequency (Miller et al., 2002).

Plasmid DNA can be transferred into the protoplasts of tobacco and sugar beet by exposure to ultrasound at 20 kHz, 500–900 ms, and acoustic power of 0.5–1.5 W/cm^2. The transient expression of protoplasts in sugar beets was 7–15 times higher than in electroporation (Joersbo & Brunstedt, 1990). Most changes can be seen at peak temporal and spatial intensities between 0.3 and 30 W/cm^2, frequencies 0.5–10 MHz, and on-times 0.1–1000 s (Miller, 1983).

Liposome-Mediated Gene Transfer

Liposomes have been widely used to encapsulate many molecules, such as enzymes and nucleic acids, and to transfer such molecules into cells by fusion with the plasma membrane or endocytosis. The mode by which DNA enters cells by liposomes is not fully understood. Liposomes adhere to the surface of protoplasts in the presence of PEG and are subsequently endocytosed, as determined by electron microscopy (EM) (Nagata, 1984). It is also hypothesized that removing cholesterol from the liposome membrane increased the percentage of liposomes fused to release their contents. With the ability to create pH-sensitive liposomes, delivery via endocytosis may offer a benefit. Thus, at neutral pH, liposomes are stable, but when they enter lysosomes, they pour their contents into the cytoplasm after fusion with the lysosomal membrane (Yatvin et al., 1980).

The proper composition of lipids is important when liposomes enter through a fusion mechanism to deliver their contents. Neutral liposomes hardly interact with cells; such interactions are neither chemically nor physically induced (McIntosh & Heath, 1982). Positive liposomes will be preferred because the cell wall and biological membranes are negatively charged. The interaction between cells and positively charged cells is strong, but cells have less viability. This interaction may be due to the basic hydrocarbons of liposomes, which act as detergents (Fraley et al., 1982). Negatively charged liposomes have mostly cardiolipin or phosphatidylserine as charged components and are commonly used to transfer materials into cells. The encapsulation volume is greater in the case of negatively charged liposomes, and no interaction occurs with encapsulated nucleic acids compared to positively charged liposomes (Gad et al., 1986). Cholesterol or phosphatidyl-ethanolamine (PE) is also added to control the fusion process. PE promotes membrane fusion due to a smaller layer of hydration (Portis et al., 1979). Genetic material enclosed with negatively charged liposomes was transferred into the cells by adding fusogen to the reaction mixture.

Liposome fusion was promoted by adding divalent cations such as Ca^{2+}. Nucleic acids have been introduced into the pollen tube of watermelon by liposome-mediated transformation (Gad et al., 1988). The chloramphenicol acetyltransferase (CAT) gene was transferred into the protoplast of tobacco by liposome-mediated transformation. Polylysine in an incubation buffer enhances the absorption of liposomes into protoplasts (Rosenberg et al., 1988). Mannitol changed the solution's electrostatic equilibrium, boosting the effects of monovalent salts and lowering the quantity of calcium or polylysine needed to encourage fusion (Gad et al., 1986). The most suitable targets for receiving foreign DNA appear to be protoplasts. The lack of a cell wall and the direct interaction of liposomes with the plasma membrane make protoplasts ideal for endocytosis and fusion with liposomes. Positively charged liposomes are used for transferring nucleic acids into the pollen tube of peas during germination (AHOKAS, 1987).

Possible advantages of the use of liposomes for gene transfer include improved delivery of the trapped DNA (or RNA), protection of the DNA from nuclease activity in the tissue or culture media, or both, transport of nucleic acids into a range of cells other than protoplasts by the cell wall through plasmodesmata, targeting to particular cells and enhanced interaction through modified liposomes, transport of intact tiny organelles (such as mitochondria), and low toxicity to cells (Gad et al., 1990). This technique is inefficient, as DNA larger than 10 kb cannot be easily encapsulated into liposomes (Caboche, 1990).

Pollen Tube-Mediated Gene Transfer

Cotton (*Gossypium hirsutum* L.) was the first plant on which the pollen tube-mediated gene transfer method (PTT), which involves transferring foreign DNA into the embryo of the plant, was reported (Zhou et al., 1983). In this method, the stigma of the recipient plants is removed after pollination and fertilization, and foreign DNA solution is applied to the severed style of recipient plants (Luo & Wu, 1988). The recipient plant's ovary receives foreign donor DNA through pollen tube development, fusing with fertilized but undivided recipient zygote(s). As a result, PTT is successful, and exogenous genes are inserted into the recipient's genome during the embryonic stage and are thus found in transformed seeds. Therefore, the PTT method excludes plant regeneration procedures, cell culture, or protoplast manipulations (Su et al., 2023).

In cotton, kanamycin and hygromycin resistance genes were transferred using the PTT method 2013 (Bibi et al., 2013). However, the PTT method has recently been used for gene transfer in several plant species, such as wheat, onion, papaya, watermelon, and maize. Experimental factors are important for successful PTT, such as the timing of PTT, foreign DNA concentration to be applied, and specific species DNA construction used. Additionally, depending on the timing of the application of foreign DNA after pollination, chimerism may occur in transformants because that DNA may enter the genome of the multicellular embryo. Therefore, the species initiation timing of PTT is important for increasing the transformation efficiency and reducing chimerism (Peffley et al., 2003). Specific methodological and biological aspects of species also influence the success of successful PTT. In soybeans, transformants have been yielded through PTT (Li et al., 2002). The unsuccessful soybean transformation may be due to the morphology or size of the flower in contrast to other species where transformation is successful. Smaller soybean flowers have thinner, more delicate staminal columns and are less amenable to strong physical manipulation (Shou et al., 2002; Liu et al., 2009).

The ovary drip method was used in soybeans for successful transformation. In this method, the ovary is cut and opened before foreign DNA is applied to the ovule. This modification enhanced the frequency of transformation by approximately 3.0%. However, this method is manipulated carefully during the exposure of ovary ovules and should not be damaged mechanically (Shi et al., 2012). The modified ovary drip method is more successful than the PTT method. In the modified PTT method, exogenous DNA moves through a shorter distance to reach the ovule, and as a result, there is less embryo sac degradation compared to the PTT method. In maize, recovery of transformants was greater (~6.47%) in the ovary drip method (Yang et al., 2009).

In rice, gene transformation success was increased from 2.5% vs. 6.7% using the binary vector with the T-DNA border. These improvements were also reported in *Zea mays*, where PTT was accomplished (5%) successfully utilizing a linear green fluorescent protein (Ubi Promoter-GFPnos Terminator) cassette bordered by 25 bp T-DNA boundaries (Yang et al., 2009). The pollen tube transformation method is comparatively simple, and regeneration is unnecessary (Zhang et al., 2005). Success with pollen tube transformation in numerous commercially significant crop species indicates that it might offer a practical substitute for transformation in some resistant plant species (Ali et al., 2015).

The most important prerequisite in tissue culture systems for plant transformation is a significant number of regenerable cells that are accessible for gene transfer treatment and that will maintain their capacity for regeneration throughout the required target preparation, proliferation, and selection treatments. It is not always necessary for a micropropagation system with a high multiplication ratio to provide many regenerated cells for genes. Transgenic plants are not recovered after transferring genes into regenerable cells when efficient regeneration capacity is short-lived. Additionally, tissue culture-based techniques might result in undesirable somaclonal modifications, such as changes in cytosine methylation, the generation of point mutations, and different chromosomal abnormalities (Clough, 2004). On the other hand, because these plants are extremely resistant in vitro, it has not been easy to realize whole plant transformants in many crop species (Okpe & Nkaa, 2021).

In-Planta Transformation

Although successful plant regeneration has been accomplished through the application of the coculture method, the methodology has not given regeneration in various other crop transformation protocols, which is a severe restriction on the potential of gene transfer technology. Given this significant constraint, it is important to develop transformation strategies that do not rely on tissue culture regeneration or intermediate tissue culture processes. The in-planta transformation technique provides such a chance. This method directly transfers transgenes into intact plants in naked DNA form. These techniques do not involve tissue culture procedures, rely on straightforward protocols, and take only a short time to produce fully transformed individuals. Some of the benefits of an in-plant transformation system are quick production in many uniform plants, with little labor input and fewer reagents required (Bent, 2000).

Early trials used this technique to convert Arabidopsis flowering portions using the vacuum infiltration approach and *Agrobacterium tumefaciens*. The vacuum infiltration technique involves infecting flowering plants with *Agrobacterium* cell suspension, following plant regeneration, harvesting, and selecting transgenic plants using screening media carrying marker genes such as genes for antibiotic and herbicide resistance. In another method, the floral portion of the flower bud is dipped in *Agrobacterium* culture with high optical density, and this method is referred to as the floral dip method. This process is reliable, quick, and safe from attacks of microbes, but the transformation efficiency is low, and the exogenous genome integrates randomly in the host cell (Li et al., 2010). These methods are being used for transformation in different ergonomically important crops, such as *Arabidopsis thaliana*, *Medicago truncatula*, *Arabidopsis lasiocarpa*, *Solanum lycopersicum*, *Brassica napus*, *Brassica rapa*, and *Raphanus sativus* (Jan et al., 2016).

Advantages of Transgenic Plants

Resistance to Biotic Stresses

Insect Pest Resistance

Stress resulting from other living organisms, such as viruses, bacteria, native plants, and herbivores, within the same ecosystem is known as biotic stress (Hammond-Kosack, 2000). Disease resistance genes are incorporated into crops for resistance against pathogenic diseases caused by viruses, pests, and bacteria and include tolerance against herbicides. In this way, the extensive use of herbicides and insecticides could be reduced in agriculture. Elimination of pests will result in better quality and higher yield of crops. To defend the plant against insect pests, insecticidal toxin genes from bacteria can be inserted into the target plant's genome. Examples of transgenic plants created using this technique include *Bacillus thuringiensis* (Bt) crops. The Bt (Cry) gene, obtained from the bacterium *Bacillus thuringiensis*, has effectively controlled various lepidopteran insects in numerous crops. In 1987, the Bt gene first noted insect resistance in tomatoes. Transgenic Bt tomato plants resist *Heliothis virescens* and *Spodoptera litura* (Fischhoff et al., 1987). Transgenic persimmons containing the cry1 gene were discovered to be resistant to *Monema flavescens* and *Plodia interpunctella* (Tao et al., 1997). Brinjal is one of the most popular vegetables in Asia, particularly on the Indian subcontinent, but lepidopteran insects severely damage it. The synthetic gene cry1Ab, which encodes insecticidal crystal protein (ICP), was transformed into brinjal by cultivating the cotyledons with

Agrobacterium tumefaciens (Kumar et al., 1998). In commercial cultivation for several years, Colorado potato beetle-resistant varieties achieved agronomic and technical success, enabling large reductions in insecticidal usage (Shelton et al., 2002).

Other genes, such as trypsin, lectins, and protease, introduce insect pest resistance in several species of crops. The cowpea protease trypsin inhibitor (CpTi) gene was expressed in transgenic strawberries under the control of a constitutive promoter that produced resistance to vine weevil. CpTi transgenic lines decreased the frequency of weevil pupal larvae and survival (Graham et al., 2002). Root-knot nematodes (*Meloidogyne incognita*), a different group of plant pests, severely reduce the yield of several crops. The most promising method to avoid such yield losses is genetically modifying several plant proteinase inhibitor genes. Since cysteine proteinases play a role in the digestion of root-knot worms, their activity is inhibited by proteolytic digestion when different cystatins are bound to their active sites (Moran & Camacho, 2007). Transgenic brinjal plants that express the modified rice gene cystatin using TUB-1, a root-specific promoter, showed resistance to root-knot nematodes (Papolu et al., 2016).

Disease Resistance

The second main factor for the limited yield of crops is disease, which is caused by viruses, pathogenic fungi, and bacteria. Conventional breeding has limited applications because resistance genes in a specific crop pool are unavailable. Genetic modifications help greatly to increase resistance or improve tolerance against diseases in plants and are cost-effective. Numerous genes, including glucanase, defensin, osmotin, and chitinase, are inserted into different horticultural crops worldwide to confer resistance to fungal and bacterial diseases. Numerous glycolytic enzymes within plant cells, such as glucanase, chitinase, and PR proteins, can degrade cell walls, which makes them useful for creating transgenic plants with resistance to fungi (Antony Ceasar & Ignacimuthu, 2012). The PR5 gene from tomato was expressed in sweet oranges, and regenerants displayed resistance against *Phytophthora citrophthora* (Fagoaga et al., 2001). The NPR-I gene from *Arabidopsis thaliana* was transferred to tomato. Transgenic tomatoes showed resistance against the tomato mosaic virus (Lin et al., 2004).

Chitinases are glycosyl hydrolases involved in chitin degradation that are components of bacterial and fungal cell walls and are thus involved in resistance against disease. Chitinase genes have been incorporated into many crops for fungal resistance. The ChiC gene from the carrot was transferred to the carrot. Transgenic carrots resisted *Botrytis cinerea* (Punja & Raharjo, 1996). Rice's chitinase gene (RCC2) in transgenic strawberry plants displayed increased *Sphaerotheca humuli* resistance (Asao et al., 1997).

The glucanase gene's ability to confer disease resistance may be linked to its ability to solubilize elicitors from the cell walls of fungi, which trigger the formation of antifungal phytoalexins. Glucanase has a role in the expression of phenylalanine ammonia-lyase, a plant defense gene against fungal infection. Additionally, the hydrolytic enzyme glucanase degrades the glucan component of several necrotrophic fungal infections. The higher glucanase activity in transgenic potato plants that express the soybean glucanase gene has been linked to enhanced resistance against *Phytophthora infestans* (Yoshikawa et al., 1993).

Adopting different antimicrobial protein-coding genes, such as defensins, has been recommended to eradicate an extensive range of bacterial and fungal diseases. Defensins are a group of antimicrobial peptides that significantly protect against protozoa, bacteria, and fungi while nontoxic to plant and human cells. Defensin causes pores in the membranes of fungal hyphae, which disrupts ion influx-outflux and kills fungal pathogens (Kaniewski & Thomas, 2004). Transgenic tomatoes expressing the defensin gene *Capsicum annum* displayed increased resistance to numerous fungal pathogens (Zainal et al., 2009). Incorporating viral coat proteins into plants can also result in the development of virus-resistant plants. The coat protein-encoding gene strawberry mild yellow edge potexvirus was transferred to strawberries. Transgenic strawberry plants displayed resistance against viruses (Husaini & Abdin, 2008;).

Abiotic Stresses

Abiotic stresses are environmental factors that affect the productivity and growth of crops by negatively affecting the reproductive and vegetative phases of growth. These stresses cause several molecular, biochemical, and physiological changes to crops that lead to cellular machinery damage. These alterations include oxidative stressors, ion distribution, and the disturbance of osmotic balance, which results in defective homeostasis and denaturation of essential proteins in plants. Plants react to these stresses using different mechanisms that activate transcriptional controls and cell signaling processes and produce stress condition-related tolerant proteins, osmotic solutes, and oxidants (Rai et al., 2011). The osMYB-4 cold-inducible gene was transferred to apples from rice and encoded a transcription factor that belongs to the Myb family (Pasquali et al., 2008). Transgenic apple plants overexpressing the DREB1b TF gene showed tolerance against cold and drought stresses (Jin et al., 2009).

The main abiotic stresses are salt stress and salinity, which adversely affect the quantity and quality of plants. Salinity impacts over 20% of agricultural land worldwide (R Core Team, 2013). The overexpression of the osmotin gene in transgenic strawberries and chili from tobacco showed tolerance against salt stress (Sopalun et al., 2010).

During heat stress conditions, reactive oxygen species (ROS), such as superoxide and hydrogen peroxide, are formed in the cells of plants, causing physiological changes that impact the growth and yield of crops. ROS denatures enzymes, causing cellular component damage inside plant cells. The greater ability of plants to eliminate ROS is directly connected with their tolerance to heat stress (Chaitanya et al., 2002). Cytosolic ascorbate peroxidase (cAPX) overexpression increased tolerance against heat stresses in transgenic apples and tomatoes (Wisniewski et al., 2002; Wang et al., 2006).

The herbicide glyphosate significantly inhibits 5-enolpyruvylshikimate-3-phosphate synthase (EPSP). The overexpression of EPSP in transgenic petunia cells displayed tolerance to glyphosate herbicides. Bialaphos was ineffective on transgenic pineapple carrying the bar gene (Sripaoraya et al., 2006).

Disadvantages of Transgenic Plants

For a long time, the use of transgenic crops remained controversial. Numerous issues have been brought up. There are concerns about the potential effects of genetically engineered materials on human health. For instance, it has been proposed that transgenic crops may cause allergies in some people, but it is unclear whether they are the origin of this reaction. Furthermore, it has been hypothesized that the antibiotic-resistance genes inserted in these crops will result in antibiotic resistance, creating superbugs that antibiotic treatments cannot kill. People are cautious about consuming DNA from a virus or bacteria (Sripaoraya et al., 2006). There is uncertainty regarding whether transgenic crops harm the environment. One illustration is pollen from transgenic maize, which can destroy the larvae of the Monarch butterfly. It has been demonstrated that hybrid maize has a bacterial poison in its pollen, and the wind spreads over a distance of 60 m. In this area, maize pollen is left on nearby plants, which might be consumed by nontarget animals such as the monarch butterfly, resulting in their death (Jhansi Rani & Usha, 2013).

Application of Transgenic Plants

Biopharming

Plant genetic engineering has made it possible in recent years to employ genetically modified plants to produce therapeutic recombinant proteins, particularly plant-based vaccines, also known as molecular pharming (Ma et al., 2003). In plants, generating such proteins has attracted interest due to issues related to current bioreactor systems. Mammalian cell systems are costly and difficult to grow, while bacterial systems can be grown easily. However, improper processing of recombinant proteins commonly leads to the precipitation of inactive proteins inside cells. On the other hand, plant systems can be developed and enable the purification of large amounts of proteins at the industrial scale. In some conditions, the purification step is unnecessary since plant material containing the recombinant enzymes can be added directly to the feed of animals. Both humans and livestock can benefit from this plant-based approach (Pascual, 2007).

In addition, for storing vaccine antigens, certain plant tissues are excellent sites for a long time without extensive processing. Certain tissues may be appropriate for oral administration, thus reducing the labor and costs of injectable vaccines. The economics of plant protein synthesis are complex. The cost depends on various factors: transport, plant growth, protein processing, and purification. Plant proteins could be considerably less expensive than conventional techniques (Streatfield, 2006).

Two main approaches are used for plant protein synthesis: plant viruses as transient vectors and a stable integration approach. In the stable gene expression method, a constitutive and strong promoter (e.g., CaMV S35) is used for the regulation of the transgene, and this approach is best suited for producing large numbers of soluble proteins although the yield is low (Streatfield, 2007). One method involves inserting protein or epitope-coding sequences into the virus genome's coat protein-encoding gene. Building viral vectors for manufacturing recombinant proteins that are intended to be processed by the endoplasmic reticulum is another strategy that has been applied. The virus can spread throughout the host plant and produce enough protein by serial passaging (Ahmad et al., 2012).

The manufacturing of vaccines using transgenic plants is becoming increasingly popular since their antigens are also edible. The first report and patent on the manufacture of edible vaccines in transgenic tobacco was published in 1990. Some vegetables, such as carrots, tomatoes, and potatoes, are used for vaccine expression. Potato has been used as a model for oral protein production (Polkinghorne et al., 2005). In tomatoes, the antigen-encoding genes HIVgag, HBsAg, and rabies capsid protein have been integrated successfully (Sala et al., 2003). Recombinant proteins have been demonstrated to be expressed by proplastids of cultivated carrot cells, and edible carrots maintained the target protein's structural integrity (Muller et al., 2003). Other plants utilized in manufacturing vaccines include cauliflower, cabbage, lettuce, and celery, but they have a low level of vaccine candidate expression (Ahmad et al., 2012).

Improving Food Quality and Shelf Life

Excessive softness is the fundamental factor limiting fruit storage and shelf life. By changing the activity of enzymes of the cell wall, such as polygalacturonases (PG), which contribute to tissue softening and degradation, fruit ripening is altered. The postponement in fruit ripening has also been possible to prevent the manufacture of ethylene, the hormone that causes fruit ripening. The transgenic tomato was the first commercialized plant with an extended shelf life and was developed by suppressing the gene polygalacturonase (PG) using the antisense approach. The PG gene codes for the enzyme polygalacturonase, which is involved in the degradation of pectin, a main cell wall component of fruits. The nutritional quality of fruits has been increased by recombinant DNA technology. In tomato fruits, RNAi technology was used to suppress the mitochondrial APX gene to improve the accumulation of vitamin C and make them healthier. The reduction of l-ascorbic acid in the ascorbate metabolic pathway is oxidized by ascorbate peroxidase (APX), which lowers the vitamin C concentration (Zhang et al., 2011).

Safety Assessment

Various legal systems may have varying requirements for the scientific data that must be presented in GM crop safety evaluations. The safety evaluation of GM crops typically includes a thorough molecular characterization of foreign insertion, the creation of tracing and tracking procedures to assure environmental studies, traceability, and legality to support coexistence frameworks. By a case-by-case methodology, other investigations have been carried out in toxicology, allergenicity, nutrition, and horizontal transfer while considering newly developed scientific information and technologies. Using genetically modified crops as feed may reduce human safety concerns while highlighting additional benefits, including nutritional equivalence and feeding value (Flachowsky et al., 2002). Pre- and post-marketing concerns comprise the two categories of the regulatory bodies' needed assessment from an industrial standpoint. Regulatory oversight concerns GM labeling and traceability as part of post-marketing issues (Giraldo et al., 2019).

References

Abu-Arish, A., et al. (2004). Three-dimensional reconstruction of Agrobacterium VirE2 protein with single-stranded DNA. *The Journal of Biological Chemistry, 279*, 25359–25363.

Ahmad, P., et al. (2012). Role of transgenic plants in agriculture and biopharming. *Biotechnology Advances, 30*, 524–540.

Ahokas, H. (1987). Transfection by DNA-associated liposomes evidenced at pea pollination. *Hereditas, 106*, 129–138.

Ali, A., Bang, S. W., Chung, S.-M., & Staub, J. E. (2015). Plant transformation via pollen tube-mediated gene transfer. *Plant Molecular Biology Reporter, 33*, 742–747.

Antony Ceasar, S., & Ignacimuthu, S. (2012). Genetic engineering of crop plants for fungal resistance: role of antifungal genes. *Biotechnology Letters, 34*, 995–1002.

Asano, Y., Otsuki, Y., & Ugaki, M. (1991). Electroporation-mediated and silicon carbide fiber-mediated DNA delivery in Agrostis alba L. (Redtop). *Plant Science, 79*, 247–252.

Asao, H., et al. (1997). Enhanced resistance against a fungal pathogen Sphaerotheca humuli in transgenic strawberry expressing a rice chitinase gene. *Plant Biotechnology, 14*, 145–149.

Aune, T. E. V., & Aachmann, F. L. (2010). Methodologies to increase the transformation efficiencies and the range of bacteria that can be transformed. *Applied Microbiology and Biotechnology, 85*, 1301–1313.

Babaoglu, M., Davey, M. R., & Power, J. B. (2000). Genetic engineering of grain legumes: Key transformation events. *AgBiotechNet, 2*.

Bent, A. F. (2000). Arabidopsis in planta transformation. Uses, mechanisms, and prospects for transformation of other species. *Plant Physiology, 124*, 1540–1547.

Berns, M. W., & Rounds, D. E. (1970). Cell surgery by laser. *Scientific American, 222*, 98–111.

Berns, M. W., et al. (1981). Laser microsurgery in cell and developmental biology. *Science, 213*, 505–513.

Bibi, N., et al. (2013). An efficient and highly reproducible approach for the selection of upland transgenic cotton produced by pollen tube pathway method. *Australian Journal of Crop Science, 7*, 1714–1722.
Biłas, R., Szafran, K., Hnatuszko-Konka, K., & Kononowicz, A. K. (2016). Cis-regulatory elements used to control gene expression in plants. *Plant Cell, Tissue and Organ Culture, 127*, 269–287.
Caboche, M. (1990). Liposome-mediated transfer of nucleic acids in plant protoplasts. *Physiologia Plantarum, 79*, 173–176.
Carsono, N. (2007). The establishment of in vitro culture and genetic transformation of the wheat Glu-1Dx5 gene to rice plants by gene gun bombardment. *Plant Production Science, 9*, 65–70.
Cesarone, C. F., Bolognesi, C., & Santi, L. (1979). Improved microfluorometric DNA determination in biological material using 33258 Hoechst. *Analytical Biochemistry, 100*, 188–197.
Chaitanya, K. V., Sundar, D., Masilamani, S., & Ramachandra Reddy, A. (2002). Variation in heat stress-induced antioxidant enzyme activities among three mulberry cultivars. *Plant Growth Regulation, 36*, 175–180.
Chen, C., Smye, S. W., Robinson, M. P., & Evans, J. A. (2006). Membrane electroporation theories: A review. *Medical & Biological Engineering & Computing, 44*, 5–14.
Christie, P. J. (2004). The Agrobacterium Ti plasmids. In *Plasmid biology* (pp. 455–472). ASM Press.
Clark, D. P., Pazdernik, N. J., & McGehee, M. R. (2019). Chapter 17—Regulation of transcription in eukaryotes. In D. P. Clark, N. J. Pazdernik, & M. R. McGehee (Eds.), *Molecular biology* (3rd ed., pp. 560–580). Academic Press. https://doi.org/10.1016/B978-0-12-813288-3.00017-3
Clough, S. J. (2004). Floral dip: Agrobacterium-mediated germ line transformation. In *Transgenic plants: methods and protocols* (pp. 91–101). Humana Press.
Crossway, A., et al. (1986). Integration of foreign DNA following microinjection of tobacco mesophyll protoplasts. *Molecular and General Genetics MGG, 202*, 179–185.
Danilova, S. A. (2007). The technologies for genetic transformation of cereals. *Russian Journal of Plant Physiology, 54*, 569–581.
De Jeu, M. J. (1998). In vitro techniques for ornamental breeding. In *XIX International Symposium on Improvement of Ornamental Plants* (Vol. 508, pp. 55–60).
Debnath, M., Prasad, G. B. K. S., & Bisen, P. S. (2010). *Molecular diagnostics: Promises and possibilities*. Springer Science & Business Media.
Demirer, G. S., & Landry, M. P. (2017). Delivering genes to plants. *Chemical Engineering Progress, 113*, 40–45.
Desfeux, C., Clough, S. J., & Bent, A. F. (2000). Female reproductive tissues are the primary target of Agrobacterium-mediated transformation by the Arabidopsis floral-dip method. *Plant Physiology, 123*, 895–904.
Dzul, S. P., et al. (2011). Contribution of the Klebsiella pneumoniae capsule to bacterial aggregate and biofilm microstructures. *Applied and Environmental Microbiology, 77*, 1777–1782.
Escoffre, J.-M., et al. (2009). What is (still not) known of the mechanism by which electroporation mediates gene transfer and expression in cells and tissues. *Molecular Biotechnology, 41*, 286–295.
Fagoaga, C., et al. (2001). Increased tolerance to Phytophthora citrophthora in transgenic orange plants constitutively expressing a tomato pathogenesis related protein PR-5. *Molecular Breeding, 7*, 175–185.
Fischhoff, D. A., et al. (1987). Insect tolerant transgenic tomato plants. *Bio/Technology, 5*, 807–813.
Flachowsky, G., Aulrich, K., Berk, A., Daenicke, R., & Reuter, T. (2002). Nutritional assessment of feeds from genetically modified (GM) crops. *Proceedings of the Society of Nutrition Physiology, 11*, 183–186.
Fraley, R. T., Dellaporta, S. L., & Papahadjopoulos, D. (1982). Liposome-mediated delivery of tobacco mosaic virus RNA into tobacco protoplasts: A sensitive assay for monitoring liposome-protoplast interactions. *Proceedings of the National Academy of Sciences, 79*, 1859–1863.

Frame, B. R., et al. (1994). Production of fertile transgenic maize plants by silicon carbide whisker-mediated transformation. *The Plant Journal, 6*, 941–948.

Frizzell, L. A. (1988). Biological effects of acoustic cavitation. In K. S. Suslick (Ed.), *Ultrasound: Its chemical, physical and biological effects*. VCH Publishers.

Gad, A. E., Elyashiv, G., & Rosenberg, N. (1986). The induction of large unilamellar vesicle fusion by cationic polypeptides: The effects of mannitol, size, charge density and hydrophobicity of the cationic polypeptides. *Biochimica et Biophysica Acta (BBA)-Biomembranes, 860*, 314–324.

Gad, A. E., Zeewi, B., & Altman, A. (1988). Fusion of germinating watermelon pollen tubes with liposomes. *Plant Science, 55*, 69–75.

Gad, A. E., Rosenberg, N., & Altman, A. (1990). Liposome-mediated gene delivery into plant cells. *Physiologia Plantarum, 79*, 177–183.

Giraldo, P. A., Shinozuka, H., Spangenberg, G. C., Cogan, N. O. I., & Smith, K. F. (2019). Safety assessment of genetically modified feed: Is there any difference from food? *Frontiers in Plant Science, 10*, 1592.

Graham, J., Gordon, S. C., Smith, K., McNicol, R. J., & McNicol, J. W. (2002). The effect of the cowpea trypsin inhibitor in strawberry on damage by vine weevil under field conditions. *The Journal of Horticultural Science and Biotechnology, 77*, 33–40.

Greenwood, N. N., & Earnshaw, A. (2012). *Chemistry of the elements*. Elsevier.

Guo, M., Ye, J., Gao, D., Xu, N., & Yang, J. (2019). Agrobacterium-mediated horizontal gene transfer: Mechanism, biotechnological application, potential risk and forestalling strategy. *Biotechnology Advances, 37*, 259–270.

Hahne, G., & Hoffmann, F. (1984). The effect of laser microsurgery on cytoplasmic strands and cytoplasmic streaming in isolated plant protoplasts. *European Journal of Cell Biology, 33*, 175–179.

Hammond-Kosack, K. (2000). Responses to plant pathogens. In *Biochemistry & molecular biology of plants* (pp. 1102–1109). I.K. International.

Hernandez-Garcia, C. M., & Finer, J. J. (2014). Identification and validation of promoters and cis-acting regulatory elements. *Plant Science, 217*, 109–119.

Hirschberg, J., & McIntosh, L. (1983). Molecular basis of herbicide resistance in plants in Amaranthus hybridus. *Science, 222*, 1346–1349.

Hoekema, A., Hirsch, P. R., Hooykaas, P. J. J., & Schilperoort, R. A. (1983). A binary plant vector strategy based on separation of vir- and T-region of the Agrobacterium tumefaciens Ti-plasmid. *Nature, 303*, 179–180.

Hossain, M. M., et al. (2013). The application of biotechnology to orchids. *Critical Reviews in Plant Sciences, 32*, 69–139.

Huang, T.-K., & McDonald, K. A. (2009). Bioreactor engineering for recombinant protein production in plant cell suspension cultures. *Biochemical Engineering Journal, 45*, 168–184.

Husaini, A. M., & Abdin, M. Z. (2008). Overexpression of tobacco osmotin gene leads to salt stress tolerance in strawberry (Fragaria × ananassa Duch.) plants. *Indian Journal of Biotechnology, 7*, 465–471.

Hwang, H.-H., & Gelvin, S. B. (2004). Plant proteins that interact with VirB2, the Agrobacterium tumefaciens pilin protein, mediate plant transformation. *Plant Cell, 16*, 3148–3167.

Ijaz, S., & Haq, I.U. (2019). Recombinant DNA technology. Cambridge Scholars Publishing.

Jan, S. A., et al. (2016). In-planta transformation: Recent advances. *Romanian Biotechnology Letters, 21*, 11085–11091.

Jhansi Rani, S., & Usha, R. (2013). Transgenic plants: Types, benefits, public concerns and future. *Journal of Pharmacy Research, 6*, 879–883.

Jin, W., et al. (2009). Improved cold-resistant performance in transgenic grape (Vitis vinifera L.) overexpressing cold-inducible transcription factors AtDREB1b. *HortScience, 44*, 35–39.

Joersbo, M., & Brunstedt, J. (1990). Direct gene transfer to plant protoplasts by mild sonication. *Plant Cell Reports, 9*, 207–210.

Kaeppler, H. F., Gu, W., Somers, D. A., Rines, H. W., & Cockburn, A. F. (1990). Silicon carbide fiber-mediated DNA delivery into plant cells. *Plant Cell Reports, 9*, 415–418.

Kaeppler, H. F., Somers, D. A., Rines, H. W., & Cockburn, A. F. (1992). Silicon carbide fiber-mediated stable transformation of plant cells. *Theor Appl Genet, 84*, 560.
Kaniewski, W. K., & Thomas, P. E. (2004). The potato story. *AgBioforum, 7*, 41–46.
Kavipriya, C., Yuvaraja, A., Senthil, K., & Menaka, C. (2019). Genetic transformation methods for crop improvement: A brief review. *Agricultural Reviews, 40*, 281–288.
Komari, T., et al. (2006). Binary vectors and super-binary vectors. In *Agrobacterium protocols* (pp. 15–42). Humana Press.
Koo, J., et al. (2007). A GUS/luciferase fusion reporter for plant gene trapping and for assay of promoter activity with luciferin-dependent control of the reporter protein stability. *Plant & Cell Physiology, 48*, 1121–1131.
Kotnik, T. (2017). Lightning-triggered electroporation as a mechanism for horizontal gene transfer. In *Handbook of electroporation* (pp. 369–385). Springer.
Kotnik, T., & Weaver, J. C. (2016). Abiotic gene transfer: Rare or rampant? *The Journal of Membrane Biology, 249*, 623–631.
Kotnik, T., Kramar, P., Pucihar, G., Miklavcic, D., & Tarek, M. (2012). Cell membrane electroporation-part 1: The phenomenon. *IEEE Electrical Insulation Magazine, 28*, 14–23.
Kotnik, T., Rems, L., Tarek, M., & Miklavčič, D. (2019). Membrane electroporation and electropermeabilization: Mechanisms and models. *Annual Review of Biophysics, 48*, 63–91.
Kumar, P. A., et al. (1998). Insect-resistant transgenic brinjal plants. *Molecular Breeding, 4*, 33–37.
Kumar, P., Nagarajan, A., & Uchil, P. D. (2019). Electroporation. *Cold Spring Harbor Protocols.* https://doi.org/10.1101/pdb.prot095471
Kunik, T., et al. (2001). Genetic transformation of HeLa cells by Agrobacterium. *Proceedings of the National Academy of Sciences, 98*, 1871–1876.
Lacroix, B., & Citovsky, V. (2018). Beyond Agrobacterium-mediated transformation: Horizontal gene transfer from bacteria to eukaryotes. In *Agrobacterium biology: From basic science to biotechnology* (pp. 443–462). Springer.
Li, Z., Nelson, R. L., Widholm, J. M., & Bent, A. (2002). Soybean transformation via the pollen tube pathway. *Soybean Genetics Newsletter, 29*, 1–11.
Li, J., Tan, X., Zhu, F., & Guo, J. (2010). A rapid and simple method for Brassica napus floral-dip transformation and selection of transgenic plantlets. *International Journal of Biology, 2*, 127.
Lin, W.-C., et al. (2004). Transgenic tomato plants expressing the Arabidopsis NPR1 gene display enhanced resistance to a spectrum of fungal and bacterial diseases. *Transgenic Research, 13*, 567–581.
Liu, Y. C., & Vidali, L. (2011). Efficient polyethylene glycol (PEG) mediated transformation of the moss Physcomitrella patens. *Journal of Visualized Experiments, 50*, 2560. https://doi.org/10.3791/2560
Liu, Y., Yang, H., & Sakanishi, A. (2006). Ultrasound: Mechanical gene transfer into plant cells by sonoporation. *Biotechnology Advances, 24*, 1–16.
Liu, J., Su, Q., An, L., & Yang, A. (2009). Transfer of a minimal linear marker-free and vector-free smGFP cassette into soybean via ovary-drip transformation. *Biotechnology Letters, 31*, 295–303.
Low, L.-Y., et al. (2018). Transgenic plants: Gene constructs, vector and transformation method. In *New visions in plant science* (pp. 41–61). IntechOpen.
Luo, Z., & Wu, R. (1988). A simple method for the transformation of rice via the pollen-tube pathway. *Plant Molecular Biology Reporter, 6*, 165–174.
Ma, J. K. C., Drake, P. M. W., & Christou, P. (2003). The production of recombinant pharmaceutical proteins in plants. *Nature Reviews Genetics, 4*, 794–805.
Mahnič-Kalamiza, S., Kotnik, T., & Miklavčič, D. (2012). Educational application for visualization and analysis of electric field strength in multiple electrode electroporation. *BMC Medical Education, 12*, 1–13.
Markelc, B., Cemazar, M., & Sersa, G. (2017). Effects of reversible and irreversible electroporation on endothelial cells and tissue blood flow. In *Handbook of electroporation* (pp. 607–620). Springer.

Mathur, J., & Koncz, C. (1998). PEG-mediated protoplast transformation with naked DNA. In *Arabidopsis protocols* (pp. 267–276). Humana Press.
McIntosh, D. P., & Heath, T. D. (1982). Liposome-mediated delivery of ribosome inactivating proteins to cells in vitro. *Biochimica et Biophysica Acta (BBA)-Biomembranes, 690*, 224–230.
Meijerink, M. R., Scheffer, H. J., & Narayanan, G. (2018). *Irreversible electroporation in clinical practice*. Springer.
Meyer, T., et al. (2018). Regulation of hydroxycinnamic acid degradation drives Agrobacterium fabrum lifestyles. *Molecular Plant-Microbe Interactions, 31*, 814–822.
Miller, D. L. (1983). The botanical effects of ultrasound: A review. *Environmental and Experimental Botany, 23*, 1–27.
Miller, D. L., Pislaru, S. V., & Greenleaf, J. F. (2002). Sonoporation: Mechanical DNA delivery by ultrasonic cavitation. *Somatic Cell and Molecular Genetics, 27*, 115–134.
Mitchell, C. A., et al. (2013). Femtosecond optoinjection of intact tobacco BY-2 cells using a reconfigurable photoporation platform. *PLoS One, 8*, e79235.
Mohammed, A., & Abalaka, M. E. (2011). Agrobacterium transformation: A boost to agricultural biotechnology. *Journal of Medical Genetics and Genomics, 3*, 126–130.
Moran, O., & Camacho, M. (2007). Leishmania amazonensis: Anionic currents expressed in oocytes upon microinjection of mRNA from the parasite. *Experimental Parasitology, 116*, 163–170.
Morikawa, H., & Yamada, Y. (1985). Capillary microinjection into protoplasts and intranuclear localization of injected materials. *Plant & Cell Physiology, 26*, 229–236.
Muller, C. P., Fack, F., Damien, B., & Bouche, F. B. (2003). Immunogenic measles antigens expressed in plants: Role as an edible vaccine for adults. *Vaccine, 21*, 816–819.
Nagata, T. (1984). Interaction of liposomes and protoplasts as a model system of protoplast fusion. In *Cell fusion: Gene transfer and transformation* (pp. 217–226). Raven.
Nester, E. W. (2015). Agrobacterium: nature's genetic engineer. *Frontiers in Plant Science, 5*, 730.
Neuhaus, G., Spangenberg, G., Mittelsten Scheid, O., & Schweiger, H.-G. (1987). Transgenic rapeseed plants obtained by the microinjection of DNA into microspore-derived embryoids. *Theoretical and Applied Genetics, 75*, 30–36.
Odell, J. T., Nagy, F., & Chua, N.-H. (1985). Identification of DNA sequences required for activity of the cauliflower mosaic virus 35S promoter. *Nature, 313*, 810–812.
Okpe, A. O., & Nkaa, F. A. (2021). Comparative review of plant transformation techniques. *Journal of Advances in Biology & Biotechnology, 24*, 1.
Ong, C.-T., & Corces, V. G. (2011). Enhancer function: New insights into the regulation of tissue-specific gene expression. *Nature Reviews Genetics, 12*, 283–293.
Papolu, P. K., et al. (2016). Expression of a cystatin transgene in eggplant provides resistance to root-knot nematode, Meloidogyne incognita. *Frontiers in Plant Science, 7*, 1122.
Park, S.-H., et al. (2010). Analysis of five novel putative constitutive gene promoters in transgenic rice plants. *Journal of Experimental Botany, 61*, 2459–2467.
Pascual, D. W. (2007). Vaccines are for dinner. *Proceedings of the National Academy of Sciences, 104*, 10757–10758.
Pasquali, G., Biricolti, S., Locatelli, F., Baldoni, E., & Mattana, M. (2008). Osmyb4 expression improves adaptive responses to drought and cold stress in transgenic apples. *Plant Cell Reports, 27*, 1677–1686.
Peffley, E. B., Allen, R., Song, P. & Shang, X. (2003) Direct transformation of higher plants through pollen tube pathway.
Polkinghorne, I., Hamerli, D., Cowan, P., & Duckworth, J. (2005). Plant-based immunocontraceptive control of wildlife—"potentials, limitations, and possums". *Vaccine, 23*, 1847–1850.
Portis, A., Newton, C., Pangborn, W., & Papahadjopoulos, D. (1979). Studies on the mechanism of membrane fusion: Evidence for an intermembrane calcium (2+) ion-phospholipid complex, synergism with magnesium (2+) ion, and inhibition by spectrin. *Biochemistry, 18*, 780–790.
Porto, M. S., et al. (2014). Plant promoters: An approach of structure and function. *Molecular Biotechnology, 56*, 38–49.

Potenza, C., Aleman, L., & Sengupta-Gopalan, C. (2004). Targeting transgene expression in research, agricultural, and environmental applications: Promoters used in plant transformation. *In Vitro Cellular & Developmental Biology, 40*, 1–22.
Prado, J. R., et al. (2014). Genetically engineered crops: From idea to product. *Annual Review of Plant Biology, 65*, 769–790.
Punja, Z. K., & Raharjo, S. H. T. (1996). Response of transgenic cucumber and carrot plants expressing different chitinase enzymes to inoculation with fungal pathogens. *Plant Disease, 80*, 999–1005.
R Core Team. (2013). R: A language and environment for statistical computing.
Rai, M. K., Kalia, R. K., Singh, R., Gangola, M. P., & Dhawan, A. K. (2011). Developing stress tolerant plants through in vitro selection—An overview of the recent progress. *Environmental and Experimental Botany, 71*, 89–98.
Reberšek, M., et al. (2007). Electroporator with automatic change of electric field direction improves gene electrotransfer in-vitro. *Biomedical Engineering Online, 6*, 1–11.
Rivera, A. L., Gómez-Lim, M., Fernández, F., & Loske, A. M. (2012). Physical methods for genetic plant transformation. *Physics of Life Reviews, 9*, 308–345.
Rivera, A. L., Magana-Ortiz, D., Gomez-Lim, M., Fernandez, F., & Loske, A. M. (2014). Physical methods for genetic transformation of fungi and yeast. *Physics of Life Reviews, 11*, 184–203.
Rosenberg, N., Gad, A. E., Altman, A., Navot, N., & Czosnek, H. (1988). Liposome-mediated introduction of the chloramphenicol acetyl transferase (CAT) gene and its expression in tobacco protoplasts. *Plant Molecular Biology, 10*, 185–191.
Rukmana, T. I., et al. (2019). Enzyme-assisted photoinjection of megadalton molecules into intact plant cells using femtosecond laser amplifier. *Scientific Reports, 9*, 1–9.
Sahab, S., Hayden, M. J., Mason, J., & Spangenberg, G. (2019). Mesophyll protoplasts and PEG-mediated transfections: Transient assays and generation of stable transgenic canola plants: Methods and protocols. In *Transgenic plants* (pp. 131–152). Academic Press.
Sala, F., et al. (2003). Vaccine antigen production in transgenic plants: Strategies, gene constructs and perspectives. *Vaccine, 21*, 803–808.
Santi, C., et al. (2003). Choosing a reporter for gene expression studies in transgenic actinorhizal plants of the Casuarinaceae family. *Plant and Soil, 254*, 229–237.
Schnorf, M., et al. (1991). An improved approach for transformation of plant cells by microinjection: Molecular and genetic analysis. *Transgenic Research, 1*, 23–30.
Scholthof, H. B., Scholthof, K.-B. G., & Jackson, A. O. (1996). Plant virus gene vectors for transient expression of foreign proteins in plants. *Annual Review of Phytopathology, 34*, 299–323.
Seymour, P. A., & Sander, M. (2007). Immunohistochemical detection of β-galactosidase or green fluorescent protein on tissue sections. In *Reporter genes: a practical guide* (pp. 13–23). Humana Press.
Sharma, A. K., & Sharma, M. K. (2009). Plants as bioreactors: Recent developments and emerging opportunities. *Biotechnology Advances, 27*, 811–832.
Shelton, A. M., Zhao, J.-Z., & Roush, R. T. (2002). Economic, ecological, food safety, and social consequences of the deployment of Bt transgenic plants. *Annual Review of Entomology, 47*, 845–881.
Shi, X., Wang, X., & Pei, D. (2012). Studies on gene transformation via pollen tube pathway in walnut. *Acta Horticulturae Sinica, 39*, 1243–1252.
Shou, H., Palmer, R. G., & Wang, K. (2002). Irreproducibility of the soybean pollen-tube pathway transformation procedure. *Plant Molecular Biology Reporter, 20*, 325–334.
Siddiqui, I. A., et al. (2016). Induction of rapid, reproducible hepatic ablations using next-generation, high frequency irreversible electroporation (H-FIRE) in vivo. *HPB, 18*, 726–734.
Siemens, J., & Schieder, O. (1996). Transgenic plants: Genetic transformation-recent developments and the state of the art. *Plant Tissue Culture & Biotechnology, 2*, 66–75.
Sood, P., Bhattacharya, A., & Sood, A. (2011). Problems and possibilities of monocot transformation. *Biologia Plantarum, 55*, 1–15.

Sopalun, K., Thammasiri, K., & Ishikawa, K. (2010). Micropropagation of the Thai orchid Grammatophyllum speciosum blume. *Plant Cell, Tissue and Organ Culture, 101*, 143–150.

Spencer, T. M., et al. (1992). Segregation of transgenes in maize. *Plant Molecular Biology, 18*, 201–210.

Sripaoraya, S., et al. (2006). Genetically manipulated pineapple: Transgene stability, gene expression and herbicide tolerance under field conditions. *Plant Breeding, 125*, 411–413.

Steinbiss, H.-H., & Stabel, P. (1983). Protoplast derived tobacco cells can survive capillary microinjection of the fluorescent dye Lucifer Yellow. *Protoplasma, 116*, 223–227.

Streatfield, S. J. (2006). Engineered chloroplasts as vaccine factories to combat bioterrorism. *Trends in Biotechnology, 24*, 339–342.

Streatfield, S. J. (2007). Approaches to achieve high-level heterologous protein production in plants. *Plant Biotechnology Journal, 5*, 2–15.

Su, W., Xu, M., Radani, Y., & Yang, L. (2023). Technological development and application of plant genetic transformation. *International Journal of Molecular Sciences, 24*(13), 10646. https://doi.org/10.3390/ijms241310646

Tao, R., Dandekar, A. M., Uratsu, S. L., Vail, P. V., & Tebbets, J. S. (1997). Engineering genetic resistance against insects in Japanese persimmon using the cryIA (c) gene of Bacillus thuringiensis. *Journal of the American Society for Horticultural Science, 122*, 764–771.

Tepfer, M., & Taylor, I. E. P. (1981). The permeability of plant cell walls as measured by gel filtration chromatography. *Science, 213*, 761–763.

Thompson, J. A., Drayton, P. R., Frame, B. R., Wang, K., & Dunwell, J. M. (1995). Maize transformation utilizing silicon carbide whiskers: a review. In *The Methodology of Plant Genetic Manipulation: Criteria for Decision Making: Proceedings of the Eucarpia Plant Genetic Manipulation Section Meeting held at Cork, Ireland from September 11 to 14, 1994* (pp. 75–80). Springer.

Tilman, D., Balzer, C., Hill, J., & Befort, B. L. (2011). Global food demand and the sustainable intensification of agriculture. *Proceedings of the National Academy of Sciences, 108*, 20260–20264.

Tóth, S., et al. (2006). Agrobacterium-mediated genetic transformation of the desiccation tolerant resurrection plant Ramonda myconi (L.) Rchb. *Plant Cell Reports, 25*, 442–449.

Uchida, M., Li, X. W., Mertens, P., & Alpar, H. O. (2009). Transfection by particle bombardment: Delivery of plasmid DNA into mammalian cells using gene gun. *Biochimica et Biophysica Acta (BBA)—General Subjects, 1790*, 754–764.

Ueki, S., Lacroix, B., Krichevsky, A., Lazarowitz, S. G., & Citovsky, V. (2009). Functional transient genetic transformation of Arabidopsis leaves by biolistic bombardment. *Nature Protocols, 4*, 71–77.

Vain, P., McMullen, M. D., & Finer, J. J. (1993). Osmotic treatment enhances particle bombardment-mediated transient and stable transformation of maize. *Plant Cell Reports, 12*, 84–88.

Vain, P., et al. (2004). The effect of additional virulence genes on transformation efficiency, transgene integration and expression in rice plants using the pGreen/pSoup dual binary vector system. *Transgenic Research, 13*, 593.

Van Montagu, M., & Zambryski, P. (2013). Agrobacterium and Ti plasmids. In *Brenner's encyclopedia of genetics* (Vol. 1, pp. 55–57). Elsevier Science.

Vaucheret, H., & Fagard, M. (2001). Transcriptional gene silencing in plants: Targets, inducers and regulators. *Trends in Genetics, 17*, 29–35.

Vergunst, A. C., et al. (2000). VirB/D4-dependent protein translocation from Agrobacterium into plant cells. *Science, 290*, 979–982.

Vorobiev, E., & Lebovka, N. (2017). Pulsed electric energy-assisted biorefinery of oil crops and residues. In *Handbook of electroporation* (pp. 2863–2881). Springer.

Walden, R., & Wingender, R. (1995). Gene-transfer and plant-regeneration (techniques). *Trends in Biotechnology, 13*, 324–331.

Wang, Y., Wisniewski, M., Meilan, R., Cui, M., & Fuchigami, L. (2006). *Transgenic tomato (Lycopersicon esculentum) overexpressing cAPX exhibits enhanced tolerance to UV-B and heat stress* (Vol. 8, p. 87).

Weaver, J. C., Smith, K. C., Esser, A. T., Son, R. S., & Gowrishankar, T. R. (2012). A brief overview of electroporation pulse strength–duration space: A region where additional intracellular effects are expected. *Bioelectrochemistry, 87*, 236–243.

Weber, G., Monajembashi, S., Greulich, K. O., & Wolfrum, J. (1988a). Genetic manipulation of plant cells and organelles with a laser microbeam. In *Progress in Plant Protoplast Research: Proceedings of the 7th International Protoplast Symposium, Wageningen, the Netherlands, December 6–11, 1987* (pp. 319–322). Springer.

Weber, G., Monajembashi, S., Greulich, K. O., & Wolfrum, J. (1988b). Microperforation of plant tissue with a UV laser microbeam and injection of DNA into cells. *Naturwissenschaften, 75*, 35–36.

Weber, G., Monajembashi, S., Wolfrum, J., & Greulich, K. (1989). A laser microbeam as a tool to introduce genes into cells and organelles of higher plants. *Berichte der Bunsengesellschaft für physikalische Chemie, 93*, 252–254.

Weber, G., Monajembashi, S., Wolfrum, J., & Greulich, K. (1990). Genetic changes induced in higher plant cells by a laser microbeam. *Physiologia Plantarum, 79*, 190–193.

Wiegand, R., et al. (1987). Laser-induced fusion of mammalian cells and plant protoplasts. *Journal of Cell Science, 88*, 145–149.

Wisniewski, M., Fuchigami, L., Wang, Y., Srinivasan, C., & Norilli, J. (2002). Overexpression of a cytosolic ascorbate peroxidase gene in apple improves resistance to heat stress. In *XXXVI International Horticultural Congress and Exhibition (Abstr.)* (p. 147).

Wu, S., Zhu, H., Liu, J., et al. (2020). Establishment of a PEG-mediated protoplast transformation system based on DNA and CRISPR/Cas9 ribonucleoprotein complexes for banana. *BMC Plant Biology, 20*, 425. https://doi.org/10.1186/s12870-020-02609-8

Yang, A., Su, Q., & An, L. (2009). Ovary-drip transformation: A simple method for directly generating vector-and marker-free transgenic maize (Zea mays L.) with a linear GFP cassette transformation. *Planta, 229*, 793–801.

Yang, D., Zhao, Y., Liu, Y., Han, F., & Li, Z. (2022). A high-efficiency PEG-Ca2+-mediated transient transformation system for broccoli protoplasts. *Frontiers in Plant Science, 13*, 1081321.

Yarmush, M. L., Golberg, A., Serša, G., Kotnik, T., & Miklavčič, D. (2014). Electroporation-based technologies for medicine: Principles, applications, and challenges. *Annual Review of Biomedical Engineering, 16*, 295–320.

Yatvin, M. B., Kreutz, W., Horwitz, B. A., & Shinitzky, M. (1980). pH-sensitive liposomes: possible clinical implications. *Science, 210*, 1253–1255.

Yoshikawa, M., Tsuda, M., & Takeuchi, Y. (1993). Resistance to fungal diseases in transgenic tobacco plants expressing the phytoalexin elicitor-releasing factor, β-1,3-endoglucanase, from soybean. *Naturwissenschaften, 80*, 417–420.

Zainal, Z., Marouf, E., Ismail, I., & Fei, C. K. (2009). Expression of the Capsicuum annum (chili) defensin gene in transgenic tomatoes confers enhanced resistance to fungal pathogens. *American Journal of Plant Physiology, 4*, 70–79.

Zhang, Y., & Yu, L.-C. (2008). Microinjection as a tool of mechanical delivery. *Current Opinion in Biotechnology, 19*, 506–510.

Zhang, Y., Yin, X., Yang, A., Li, G., & Zhang, J. (2005). Stability of inheritance of transgenes in maize (Zea mays L.) lines produced using different transformation methods. *Euphytica, 144*, 11–22.

Zhang, X., Henriques, R., Lin, S.-S., Niu, Q.-W., & Chua, N.-H. (2006). Agrobacterium-mediated transformation of Arabidopsis thaliana using the floral dip method. *Nature Protocols, 1*, 641–646.

Zhang, Y., Li, H., Shu, W., Zhang, C., & Ye, Z. (2011). RNA interference of a mitochondrial APX gene improves vitamin C accumulation in tomato fruit. *Scientia Horticulturae (Amsterdam), 129*, 220–226.

Zhou, G.-Y., et al. (1983). Chapter 29: Introduction of exogenous DNA into cotton embryos. In *Methods in enzymology* (Vol. 101, pp. 433–481). Elsevier.

Zuo, J., Niu, Q., & Chua, N. (2000). An estrogen receptor-based transactivator XVE mediates highly inducible gene expression in transgenic plants. *The Plant Journal, 24*, 265–273.

Chapter 3
Genomics

Siddra Ijaz, Imran Ul Haq, Zakia Habib, and Hayssam Mohamed Ali

Abstract Plant breeding research has historically progressed due to unintentional selection following crop development and the desire for increased food availability. The progress in achieving this objective unveiled the constituents of plant genomes that control the whole plant life. Plant genomics seeks to create high-throughput genome-wide techniques, tools, and approaches to unravel the fundamentals of genetic traits, genetic diversity, and by-product output; to comprehend phenotypic development across the plant developmental stages along with its genetics by ecological factors; to map significant loci in the genome; and subsequently to hasten crop improvement. The accessibility of cost-effective, high-throughput DNA sequencing systems has directed the complete sequencing of hundreds of plant genomes, which has broad implications for all aspects of plant biology research and its applications. Constantly increased efforts have been put into plant genomics studies over the last 30 years. However, these technologies have offered several unanticipated challenges. This chapter briefly reviews developments in plant genomics research over the previous decades, plant genome sequencing initiatives, and major challenges in the genomics era.

Keywords Genome · Genome organization · Sequencing · Molecular markers · Functional genomics

S. Ijaz (✉) · Z. Habib
Centre of Agricultural Biochemistry and Biotechnology (CABB), University of Agriculture Faisalabad, Faisalabad, Pakistan

I. U. Haq
Department of Plant Pathology, University of Agriculture Faisalabad, Faisalabad, Pakistan

H. M. Ali
Department of Botany and Microbiology, College of Science, King Saud University, Riyadh, Saudi Arabia

S. Ijaz et al. (eds.), *Trends in Plant Biotechnology*,
https://doi.org/10.1007/978-981-97-0814-7_3

Introduction

The plant kingdom is the foundation of our planet's food supply chain. Early human societies domesticated plants, and through agronomic practices, purposeful plant breeding, and other processes, high-yielding crop species were produced that offered food and other necessities for all life forms, particularly humans (Kossmann, 2012; Schatz et al., 2012). Approximately 300,000 plant species worldwide are incredibly diverse (Li et al., 2015b). Plant genomics strives to evaluate, sequence, distinguish, and examine the whole plant's genetic makeup, composition, and function, as well as their relationships and networks (Abdurakhmonov, 2016). Until recently, the single gene level was frequently the focus of molecular analyses of plants. This perspective has evolved due to recent technical advancements, which make it possible to analyze organisms in the context of how their genomes are organized, expressed, and interact. Genomics explores how genes and their genetic information are arranged in the genome, the mechanisms for acquiring and examining these data, and how this arrangement affects and functions biologically (Varma & Shrivastava, 2009). Genomic techniques influence every facet of plant biology, and because they depend on data encoded in DNA, they enable molecular investigations to be conducted at the level of several species. The traditional concept of revealing genes underlying biological functions has been reversed by plant genomics, which now focuses on determining the biological functions underlying genes. Additionally, it narrows the genetic gap between phenotype and genotype and facilitates an understanding of how a gene's activity can be modulated by its genetic background, the genetic systems it interacts with, and the gene's isolated effects (Campos-de Quiroz, 2002). This chapter reviews developments in plant genomics research over the previous decades, plant genome sequencing initiatives, and major challenges in the plant genomics era.

Genome Evolution

Genome evolution is how a genome alters its sequence or size over time. Double-stranded DNA makes up the genomes of all cellular creatures. Thus, a crucial subject for our comprehension of early life evolution is the genesis of DNA and the processes that control DNA replication. Therefore, the evolutionary origins must be considered to completely comprehend the activities in modern living cells (Alberts et al., 2002a).

Pre-RNA world: It is doubtful that RNA was the first type of molecule to serve as the foundation for a self-replicating group of biochemical catalysts although it looks ideally prepared to do so. The early formation of lengthy RNA molecules by entirely nonenzymatic methods is challenging to imagine from a chemical perspective. For example, it is challenging to synthesize ribonucleotides, the building blocks of RNA, without the aid of enzymes. In addition, a lengthy series of 3′–5′

phosphodiester links must develop in the presence of several competing processes, such as hydrolysis, 2′–5′ linkages, and 5′–5′ linkages, for RNA to be formed. These issues have led to the hypothesis that the first molecules capable of catalysis and information storage may have been polymers similar to RNA but chemically simpler. These substances do not remain in cells today and do not leave any traces in the form of fossils. Nevertheless, due to their simplicity, these "RNA-like polymers" are more likely than RNA itself to be the first biopolymers on Earth capable of information storage and catalysis. The synthesis of RNA employing one of these less complex chemicals as both a template and a catalyst would have marked the change from the pre-RNA world toward the RNA world (Alberts et al., 2002b).

RNA World

It is believed that a simpler evolutionary stage, when just RNA molecules were used to store and transfer genetic information, led to the transition from DNA to RNA and protein. This perception, referred to as the "RNA world hypothesis," contends that the ribosome, an RNA enzyme that can duplicate RNA molecules, developed from an early-evolutionary RNA enzyme or "ribozyme" that could replicate RNA molecules (Gilbert, 1986; Razvan & Unrau, 2017).

Shift from RNA to DNA Genomes

The creation of protein enzymes, which complemented and eventually replaced most of the catalytic functions of ribozymes, was perhaps the first significant change in the evolution from the RNA to the DNA world (Alberts et al., 2002a). Furthermore, DNA is a modified form of RNA in which "regular" ribose sugar found in RNA is reduced to deoxyribose sugar in DNA although nucleotide uracil is methylated into thymidine. A key justification for RNA coming before DNA in evolution is the creation of building blocks of DNA from RNA precursors (Forterre et al., 2004).

Mechanism of Acquisition of New Genes

All species experience evolutionary innovation due to the emergence of novel genes. New genes can be attained via gene duplication, the acquisition of genes from other species, transposable elements (TEs), lateral gene transfer, gene fusion and fission, and de novo gene origination (Alberts et al., 2002a).

Gene Duplication

Gene duplication was the earliest method for gene creation, and it seems that this method is still the most prevalent. The crucial step for creating new genes throughout genome evolution is probably the duplication of already existing genes. A duplication may involve the entire genome, substantial portions of the genome, a single gene, a single exon, or even a specific region of an exon. Duplications are often categorized based on the length of the duplicated genome (Betrán & Long, 2002). Diverse pathways contribute to the production of duplicate genes, and new information about these mechanisms is always being found. These mechanisms include transpositions involving transposable elements, tandem duplications originating from unequal crossover, unequal exchange of sister chromatids, replication slippage, whole-genome duplications originating from nondisjunction, retroposition originating from the retrotranscription of an RNA intermediate, and retroposition originating from the retrotranslation of an intermediate (Morgante et al., 2005).

Transposable Elements

Transposable elements are characterized as jumping elements and can jump from one position to another across the genome. These DNA segments encode particular proteins. These elements can copy themselves across the genome. These elements are categorized into two forms: DNA transposons and retrotransposons. The foremost ones can excise themselves from the genome and then insert them into another position in the genome (cut and paste mechanism). The latter makes a copy of itself through an RNA intermediate followed by insertion into a new position (copy and paste mechanism). TE insertions in the genome induce mutations and add to the evolution of the genome's size, much like viral insertions, although they often do not truly encode cellular proteins (Fedoroff, 2012). Genome evolution as a whole is impacted in various ways by transposable elements. The potential of transposons to start recombination events that result in genomic rearrangements is the most fundamental of these. This refers to the fact that various copies of the same element contain similar sequences and might, as a result, start recombination between two sections of the same chromosome or between separate chromosomes (Alberts et al., 2002a). Deleting critical genes will often make the ensuing rearrangement deleterious, but there have been some recorded examples in which it has been advantageous. The duplication of the β-globin gene that gave rise to the Gy and Ay constituents of this family of genes occurred approximately 35 million years ago due to recombination between LINE-1 elements (Maeda & Smithies, 1986).

Furthermore, it is interesting that employing transposable element proteins as cellular proteins is another approach to gaining additional genes. These scenarios are referred to as TE protein domestication. It was discovered that the activity of the vertebrate immune system and plant light perception are two surprising processes in which TE-domesticated proteins are implicated (Feschotte & Pritham, 2007).

Drosophila has also been used as an example of domestication in several cases (Casola et al., 2007).

Lateral Gene Transfer

Lateral gene transfer refers to gene origination from an unrelated genome rather than a vertical origin (vertical inheritance from parents to the next generation). Lateral gene transfer plays a substantial part in the evolution of prokaryotic genomes, according to comparisons of bacterial and archaeal genome sequences (Alberts et al., 2002a). This type of transfer between bacteria is well known, and it has also occurred between the nuclear and organelle genomes (chloroplasts and mitochondria) (Roger, 1999). Nevertheless, the latest transfer activities involving organelles or endosymbiotic bacteria persist (Hotopp et al., 2007). For instance, extensive sequencing investigation has shown that the genome of the intracellular endosymbiotic bacterium Wolbachia was largely incorporated into the Drosophila genome (Hotopp et al., 2007).

Existing Gene Fusion and Fission

Existing genes could also undergo fission, in which a transcript splits into several distinct transcripts, or fusion, in which two or several genes join together to produce a new gene. The alcohol dehydrogenase gene (Drosophila) is an example of a chimeric fusion gene that occasionally involves duplicates of the same gene. When this occurs, the resultant genes go through parallel evolution and diverge from the roles of their parent genes (Jones & Begun, 2005).

De Novo Gene Origination

Noncoding DNA regions can act as a de novo source for new genes. Numerous novel Drosophila genes originating from noncoding DNA have been identified. No homologs from any other species for these recently evolved Drosophila genes are likely capable of coding proteins (Begun et al., 2007).

Genome Organization

Genome organization refers to the complex structure of an individual's genetic material into three-dimensional structures known as chromosomes (Sood & Misteli, 2022). The "genome" is referred to as the entire set of DNA (genetic material) containing all genes in an organism (Goldman & Landweber, 2016).

Genome Organization in Prokaryotes

Prokaryotic cells typically comprise single cells that lack a distinct cell nucleus. Instead, they have a nucleoid region comprising a single and circular double-stranded DNA molecule as their genetic material. Prokaryotes can be generally categorized into two domains: bacteria and archaea. Prokaryotic genomes are much smaller than eukaryotic genomes. In addition to this single chromosome, prokaryotic cells have an autonomous, extrachromosomal DNA molecule that can be circular or linear. The genomic size of *E. coli* is approximately 4639 kb, consisting of 4405 genes, much smaller than the eukaryotic genome, i.e., yeast. DNA is a dynamic genetic material that organizes into a compact structure to fit into a prokaryotic cell (Brown, 2002). Prokaryotic genome organization is achieved by coiling and supercoiling DNA molecules around itself.

Furthermore, the winding of DNA molecules around itself is known as supercoiling. Supercoiling can be either positive or negative winding either in a similar direction of helical or in the opposite direction, respectively. Twist and writhe are other parameters that help to coil DNA into a compact structure in the prokaryotic nucleoid region (Gilbert & Allan, 2014). In prokaryotes, HU proteins are specific DNA-binding proteins that help in DNA organization. At the lowest level of organization, negatively supercoiled DNA develops plectonemic loops with the help of gyrase, topoisomerase I and NAPs (proteins associated with the nucleoid). These proteins help bend and wrap DNA (Dillon & Dorman, 2010).

Genome Organization in Eukaryotes

Prokaryotes have a fraction of the DNA that eukaryotes do. An *E. coli* bacterium, for instance, has 4.6 million bps of DNA, whereas a human has approximately 3 billion bps. The genome of eukaryotes, which includes animals and humans, comprises multiple double-stranded linear DNA molecules placed within a membrane-bound nucleus. In addition to the nuclear genome, eukaryotes have organelle genomes, such as the mitochondrial genome (Clark & Pazdernik, 2012). Additionally, all eukaryotes have smaller, typically circular, mitochondrial genomes. The smallest genomes of eukaryotes are less than 10 Mb long, whereas the highest seems to be over 100,000 Mb. The human chromosome averages slightly under 5 cm of DNA although chromosomes are significantly shorter than the DNA molecules they carry. A DNA molecule must consequently be packaged in a highly ordered manner to fit inside a chromosome. In the 1970s, biochemical studies and electron microscopy were used to make significant discoveries about DNA packaging. Although the relationship between nuclear DNA and the DNA-binding proteins known as histones is already known, its precise nature has not yet been defined. In the first level of organization, histone proteins interact with DNA and convert it into a structure known as chromatin (Misteli, 2007; Clark & Pazdernik, 2012).

Nucleosome Assembly

The nucleosome is the structural and functional unit of chromatin structure molecules. Eight proteins, two of H2A and H2B, H4A, and H4B, create a core octamer, approximately 1140–150 bp of DNA (Khorasanizadeh, 2004). Each nucleosome is alienated by 50–70 bp DNA stretches, known as linker DNA, resulting in 190–220 bp DNA segments. The nucleosome is represented as beads on a string structure.

In next-level organization, there are more histone proteins in the core octamer, known as H5-linker histones (Zhou et al., 1998; Travers, 1999). Therefore, a linker histone and a nucleosome form another structure known as the chromatosome. These linker histones serve as clamps and prevent the removal of coiled DNA from nucleosomes. This structure becomes more condensed into 30 nm fibers with widths of 30 nm at a further level of organization. These 30 nm fibers then convert into highly organized three-dimensional structures known as chromosomes (Clark & Pazdernik, 2012).

Endosymbiotic Theory

The endosymbiotic idea is one of the hypotheses that persists currently. Endosymbiotic interactions are thought to have occurred among the first life beings. In this sort of symbiosis, a larger cell known as the host is joined by a smaller cell defined as an endosymbiont. According to the endosymbiotic hypothesis, the larger cell "took in" or devoured the smaller cell. Today's eukaryotic cells are represented by larger cells, while prokaryotic cells are represented by smaller cells. Eukaryotes are known for having nuclei. Theories of nucleus evolution usually emphasize one of the following: (1) prokaryotic cell membrane invagination, (2) engulfing of archaeon into the eubacterial host cell, or (3) autogenous origination of new membranous system containing nuclear envelope within host cell from archaeal source subsequently attaining mitochondria. Mereschkowsky is credited with developing the endosymbiotic theory of nucleus genesis. He proposed that a prokaryote called mycoplasma, which was absorbed from an amoeboid cell similar to eukaryotic cytoplasm, gave rise to the nucleus (Martin et al., 2015). In an endosymbiotic theory, the in-foldings of the cell membrane from prokaryotic cells subsequently pinched off, resulting in an endomembrane compartment enclosing a nucleoid region known as the membrane-bound nucleus and considered the first eukaryotic compartment.

Furthermore, aerobic bacteria (proteobacterium) parasitize eukaryotic cells, enter them as parasites or prey, and lead to an endosymbiont living in another's cell by skipping digestion. Furthermore, the aerobic ability of these bacteria becomes an asset for energy production, thus serving as an energy-rich environment. These proteobacteria are then further assimilated and converted into organelles, mitochondria. Another bacteria, cyanobacteria, can photosynthesize in the host cell and convert it into the chloroplast (Zimorski et al., 2014; Sato, 2020).

Genome Sequencing

The term "genome sequencing" is defined as the process of comprehensively analyzing the genome to determine all of its bases and their positions. It entails determining the sequence in which the nucleotide bases that make up the entire set of chromosomes in the genome are arranged. Genome sequencing seems intuitively very simple because DNA sequencing is extremely common. If 1000 base pairs (bps) can be fully sequenced in a single run, a genome of 1 billion bps will further require 1 million lanes, and just because present automated sequencers can load 96 samples per run, it would have seemed that only 10,416 runs are needed to sequence the entire genome. Even though these basic figures are accurate for a single genome, fully sequencing the entire genome is trickier than it seems (Liu, 2008).

Whole-Genome Sequencing

WGS is an approach for ascertaining complete or almost all DNA sequencing of an individual's genome in one spell. It includes sequencing of chromosomal DNA of the whole organism and DNA found in mitochondria and chloroplasts (in plants). There are two major shotgun-based whole-genome sequencing approaches (Liu, 2008).

Whole-Genome Shotgun

Shotgun sequencing and hierarchical sequencing (clone-by-clone) methods were initially employed to sequence an entire genome (Kaiser et al., 2003). With WGS sequencing, the genomic DNA sequence is deliberately fragmented into tiny bits and then cloned and inserted into plasmids, followed by sequencing on both strands. The sequences are collected, and then, using bioinformatics methods, they are aligned and put together into draft sequences (Staden, 1979; Anderson, 1981). Shotgun sequencing libraries featuring various insert sizes are typically created for WGS sequencing. BAC, plasmid, and cosmid libraries are usually used to support WGS. BAC libraries frequently feature inserts that are 100–200 kb in size. Cosmid libraries can store intermediary inserts up to 40 kb in size. Plasmid libraries used for shotgun sequencing range in size from 2 to 8 kb but are typically 4 kb or smaller. Sequences produced from BAC libraries act as substantial anchoring islands for sequence assembly and clone alignment; sequences produced from cosmid libraries act as more manageable anchoring islands, and sequences produced from plasmid libraries offer high-genome coverage for whole-genome sequencing. Utilizing BAC and cosmid insert libraries for genome assembly is very beneficial, particularly for regions with high sequence repetition. Thus, the physical map created by BAC-based contig creation employing restriction fingerprinting can also be utilized to discover the BAC clones used to generate the DNA sequences. The physical map information would act as anchoring points for the DNA sequences' location to prevent incorrect assembly based solely on sequence overlaps inside the repeating region. The shotgun method is quicker and less expensive than hierarchical shotgun sequencing. Due to incorrect assembly of the entire genome sequence, it has been increasingly vulnerable to errors. Therefore, the WGS technique is better suited to sequencing relatively short genomes and those with straightforward repetitive structures. However, it enables a higher degree of genome coverage while using the same economic means because of its quickness and comparatively lower prices (Liu, 2008). The most popular method for sequencing genomes of microorganisms is WGS sequencing. This approach has been employed for sequencing several organisms, such as *Archaeoglobus fulgidus*, *Helicobacter pylori*, and *H. influenzae*, as well as *Thermotoga maritima* (Nelson et al., 1999).

Hierarchical Shotgun Sequencing

An alternative that is better suited for larger, more complicated genomes, such as those with many repetitions, is the hierarchical shotgun technique (Green, 2002). A divide-and-conquer method for genome sequencing known as "hierarchical shotgun sequencing" has produced high-quality genomic sequences as references for a variety of plant as well as animal species, including *Saccharomyces cerevisiae*, *Arabidopsis thaliana*, maize, and rice (Goffeau et al., 1996; Arabidopsis Genome Initiative 2000; Sasaki, 2005; Schnable et al., 2009). This method involves breaking down a genome into larger pieces, typically cosmids or BACs. These are then arranged using time-consuming hybridization or PCR-based procedures into a minimum tiling path. Following that, a shotgun method is used to individually sequence each of the chosen groups of clones in a minimal path (Kaiser et al., 2003; Beier et al., 2016). Currently, bacterial artificial chromosomes make up the majority of larger clone libraries based on inserts. Physical maps are created by measuring size or directly sequencing restricted short fragments (van Oeveren et al., 2011). Although next-generation sequencing tools have been employed for higher level clone sequencing (Choulet et al., 2014; Li et al., 2015a), minimum tiling path sequencing has traditionally been executed via automated Sanger sequencing (Howe et al., 2013). The upside of the hierarchical shotgun technique is that fewer errors are made when the shotgun sequences are combined into contigs to create the genomic sequence. The justification is that each BAC's chromosomal position is known according to the physical map, and there are generally fewer disparate parts to put together inside every BAC.

Additionally, compared to the entire genome, the diversity of repetitive regions within a particular BAC would be substantially smaller. This method's flaws include its high price and slow sequence generation speed. However, the genomic sequences produced via the hierarchical shotgun method are highly qualitative after they have been produced (Kaiser et al., 2003).

Next-Generation Sequencing

The next-generation sequencing approach between 2004 and 2006 revolutionized biomedical research, resulting in an exceptional increase in sequencing information (Mardis, 2013). Beyond the capabilities of the conventional DNA sequencing technologies created in the late 1970s, they offer a highly effective, quick, and affordable DNA sequencing platform. They are constantly being enhanced to become quicker, more effective, and less expensive. Since 2004, they have been functional in various biological disciplines (Barba et al., 2014). NGS includes two approaches: second-generation and third-generation sequencing.

Second-Generation Sequencing

Second-generation sequencing is the predominant sequencing technology after traditional Sanger sequencing (first-generation sequencing). Short-read NGS first includes library preparation, sequencing, and data analysis (Quail et al., 2012). Furthermore, sequencing by synthesis (SBS) and sequencing by hybridization (SBH) are the two foremost categories of second-generation sequencing approaches.

Sequencing by Hybridization

Sanger approaches have been improved upon by SBH sequencing, which does not use di-deoxy terminators and combines several synthesis cycles with imaging and techniques to add new nucleotides to the developing chain. The DNA oligonucleotides of reference sequences arranged on filters and hybridized to tagged pieces of the DNA that needed to be sequenced were the basis for the invention of this technique in the early 1980s. Constructing more contiguous sequence-based data on the overlap from probe hybridization locations was probable. To find disease-linked SNPs (single nucleotide polymorphisms) in particular genes or to detect substantial chromosomal anomalies, SBH is primarily used in technologies that rely on utilizing specialized probes to interrogate sequences (rearrangements, indels, and copy number variants) (Church, 2006; Qin et al., 2012).

Sequencing by Synthesis (SBS)

The "second generation" techniques often employ solid support with tiny wells where the sequencing reactions occur. Some technologies use "reversible" terminators, which permit the incorporation reactions for nucleotides to move ahead ordinarily while illustrating the integrated nucleotides and eliminating the synthesis-blocking replacements on the characterized nucleotides to allow the next base in the sequence to be added. However, most new SBS techniques do not use di-deoxy terminators (Fuller et al., 2009). Modern SBS techniques rely on substantially shorter reads (up to 300–500 bp) than the original Sanger sequencing strategy offered. Furthermore, they frequently have higher chances of errors than Sanger sequencing.

Furthermore, it relies on massively parallel sequencing of trillions of shorter DNA sequence reads (approximately 50–300 bp) to deliver a precise reading based on a consensus sequence. Nevertheless, some methods experience sequence issues that cannot be fixed by adding more reads (Quail et al., 2012). Now that the trend has reversed from long-read to short-read technologies, greater primary read lengths are generated while upholding massively parallel aspects of this approach. Most SBS approaches use a technique where the distinct DNA molecules that have to sequence are dispersed throughout the wells or attached to certain spots on the solid surface of a specific substrate. PCR/rolling circle amplification methods amplify DNA molecules. The DNA molecules are then exposed to DNA synthesis processes, which allow labeled nucleotides/chemical reactions based on particular nucleotide additions to be captured or recognized (Slatko et al., 2018).

Pyrosequencing

454 Pyrosequencing is a novel method based on the identification of pyrophosphate, which is a by-product as a result of the incorporation of nucleotides, to validate whether a base was added to a developing nucleotide chain (Ronaghi et al., 1996). Pyrosequencing is a leading approach in SGS technology to create reads with differential characteristics (Schatz et al., 2010). A primed DNA template was attached to a microbead (emulsion beads) and underwent amplification via emulsion PCR in Roche 454 pyrosequencing. DNA pieces can link straightly to beads, preferably a single segment for each bead, because the DNA sequences within the beads and adaptors are complementary. In a picoliter size device (picotiter plate) where pyrophosphate loss is detected, DNA synthesis is subsequently followed by chemical reaction-based identification of DNA synthesis. In this picotiter plate, beads are incubated with sequencing reagents such as DNA polymerase and certain enzymes such as apyrase, ATP sulfurylase, and luciferase alongside their substrates (APS and luciferin). In this reaction, pyrophosphate as a by-product is eliminated upon addition to nucleotides by the DNA polymerase enzyme in a growing chain. This by-product is then converted into the ATP form in the presence of an ASP substrate. This ATP interacts with the luciferase enzyme that transforms luciferin into oxyluciferin, resulting in a light-emitting reaction. Furthermore, light intensity is unswervingly linked to the number of nucleotides added to a chain (Ronaghi et al., 1998; Nyrén, 2007). The apyrase enzyme eliminates all unutilized nucleotides and ATP molecules. This reaction is repeated several times unless a particular DNA fragment

is extended. The chains of nucleotides that make up the DNA sequence are created by analyzing the fluorescent light produced when nucleotides are added, which is imaged by a camera. Roche 454 initially produced reads with lengths of approximately 100 bp. However, technology has advanced such that read lengths have become equivalent to those generated by Sanger sequencing (approximately 800 bp), generating approximately 700 Mb from approximately 1 million reads (Slatko et al., 2018). Due to its ability to analyze longer reads, it is a frequently used approach for either transcriptomic or genomic sequencing and de novo genome assembly (Lai et al., 2012).

Ion Torrent

Ion torrent is distinct from other NGS systems in sequencing detection and depends on pH fluctuations due to H^+ ion loss following nucleotide assimilation rather than fluorescent dyes. This technology is based on directly converting the sequence information of nucleotides into numerical data on a semiconductor device chip. Furthermore, the device can identify which nucleotide has been integrated into the developing strand by adding nucleotides in a specific order, followed by the generation of H^+ ions. This alters the pH of the solution, which a pH meter-like ion sensor may measure as a voltage shift. Since no nucleotide is added, there is no voltage surge. Voltage change occurs when the right nucleotide is inserted by periodically overflowing and wiping out the sequencing chamber alongside sequencing reagents containing just one of four nucleotides (Rothberg et al., 2011).

Furthermore, two hydrogen atoms are generated, and the voltage increases when two neighboring nucleotides integrate the same nucleotide. As a result, it is possible to identify "runs" of individual nucleotides. It might be challenging to distinguish lengthy homopolymer chains of similar nucleotides (Egan et al., 2012; Slatko et al., 2018).

Currently, two systems are employed for this technology: the Ion PGM, which is used in laboratories, and the new Ion Proton, which has a better throughput upon the discharge of the following hardware chip, which is anticipated to contain 660 million sensors. The novel proton system is expected to have approximately 165 million sensors equivalent to a 250 bp read length. Therefore, every hardware chip advancement raises the throughput for both systems (Egan et al., 2012). This system presently focuses on shotgun genome sequencing of several microbial organisms (Howden et al., 2011). This technology is quite efficient and less laborious and gives accurate results. Furthermore, the system's key benefit is the lack of a camera, illuminator, and scanner; instead, nucleotide integration is transmuted straight into voltage and recorded directly, considerably accelerating the practice (Slatko et al., 2018). Numerous applications are endorsed, including miRNA sequencing, small RNA sequencing, ChIP-seq, transcriptome and microbial sequencing, copy number variation identification, targeted sequencing, and de novo DNA and RNA sequencing (Furey, 2012).

Third-Generation Sequencing

Third-generation sequencing is important for sequencing larger DNA or RNA fragments, unlike other techniques. Pacific Biosciences (PacBio), which has developed two systems of sequencing, the initial RSII model and SequelT, is the current market leader in this field. PacBio sequencing can sequence prolonged fragments, ranging from 30 to 50 kb or greater. Attaching a specified DNA polymerase to the center of a well (ZMW) containing attached DNA that requires sequencing is part of the SMRT technique (Slatko et al., 2018). The ability of this process to measure pace for every nucleotide inclusion in production is a significant benefit (Zhang et al., 2015). Most importantly, a recent sequencing system in third-generation sequencing is nanopore technology.

Nanopore Sequencing

Nanopore sequencing is an innovative method for analyzing an individual's DNA and produces longer reads (~1000 or 10,000 bp) than second-generation sequencing without the prerequisite of amplification for library preparation (Branton et al., 2008; Leggett & Clark, 2017). A DNA/RNA strand molecule can be immediately evaluated as it goes through a nanopore via the intrinsic electrical or chemical characteristics of every nucleotide to discern order, which is the attractiveness of this technology. Currently, two techniques can be employed for sequencing DNA molecules: exonuclease sequencing, in which nucleotide bases are fragmented once at a moment and then moved via nanopore in the same orientation as they were cleaved (Lieberman et al., 2010), and strand sequencing, in which a particular enzyme is synthesized to pass a molecule of single-stranded DNA via nanopore for identification (Clarke et al., 2009). The promptness at which DNA molecules travel by nanopores, which may enable extraordinarily high DNA sequencing throughput, is a possible benefit of nanopore sequencing. GridION's system is capable of producing sequences at higher rates. This method's limitations prevent it from using this potential benefit because decelerating the pace at which DNA moves via a nanopore is necessary to improve reading precision. Despite several obstacles, nanopore sequencing has a bright future (Egan et al., 2012).

Metagenomic Sequencing

Metagenomics mainly refers to the application of sequencing approaches to evaluate the total genomic information present in an organism (Roumpeka et al., 2017). Shotgun metagenomics randomly tests a sample's microbial genomes (hence the term "shotgun"). In addition to retrieving entire genome sequences, it can be used to profile microbial populations' taxonomic makeup and functional potential.

Sequence data can be examined without assembly; however, for the majority of investigations, larger, longer contiguous sequences are preferable (contigs) (Quince et al., 2017). Whole-genome shotgun (WGS) metagenomics has recently become a dominant approach to exploring microbial communities utilizing high-throughput sequencing. The objective of WGS metagenomics is to sequence every genome present in a biological sample to evaluate the functional abilities and biodiversity of the microbial population. It is feasible to assess the entire biodiversity of a habitat, including archaea, bacteria, viruses, plasmids, and eukaryotes, along with its gene content when the complete genetic information of a sample is retrieved (Pérez-Cobas et al., 2020). The foremost gain of WGS metagenomics is that it provides an opportunity to define the genetic and genomic diversity of the observed community and the prospective and unique functionalities present. Additionally, it is conceivable to assemble whole genomes from metagenome data using appropriate sequencing to better understand microbial ecosystem genomic diversity and obtain draft genomes for uncultured microorganisms (Wilkins et al., 2019). Furthermore, depending on the characteristics of genomic regions, the sequencing depth and the chosen sequencing method can uncover some specific chromosomal regions while sequencing metagenomes (Ross et al., 2013).

Personal Genomics

Personal genomics has revolutionized DNA sequencing approaches and made it possible to identify the genomic sequences of an individual (McGuire et al., 2007). Cell and tissue genome sequences in healthy and diseased conditions have been recognized. Currently, high-throughput DNA sequencing methods produce fewer base pairs long (35–450 bp), making it difficult to de novo assemble (i.e., create the genome sequence entirely from sequencing reads without any additional knowledge) a genome. De novo assembly may soon be practical because of improvements in computing and sequencing methods (Li et al., 2010). Using high-throughput DNA sequencing methods, the genome of an individual is sequenced. The generated massive sequence data are then compared to reference sequences (haploid genome) to locate changes. SNPs and small indels are often easy to identify after correcting for sequencing oversights and assuming they can be discovered in a single sequence read (Li et al., 2009). However, there are some DNA sequencing errors in the reference human genome sequence (0.01%), and not all variations in the targeted genome from the reference genome are inevitably SNPs (Lander, 2001; Snyder et al., 2010).

Genome Assembly

Genome assembly is a computational-based procedure that practices a higher number of short sequences called reads produced from several positions in target DNA as input to discern the sequence structure of the genetic information (DNA) within an organism's cell (Rice & Green, 2019; Thrash et al., 2020). The entire DNA molecules in an organism's cell are called the genome. However, genome sequencing has become significantly more affordable and quicker since the human genome was available in 2001. Assembly is the in silico-based reconstruction of the original genomic sequences from the sequenced and smaller fragments. One can conduct a "comparative assembly," in which the assembly is guided by an already sequenced, strongly related organism, or one can undertake de novo assembly, combining sequenced fragments to build contigs without employing a sequenced reference genome. Since the second strategy is difficult, numerous heuristic-based approaches have been created to address the issue (Pérez-Cobas et al., 2020).

Since the previous 10 years, the price of genome sequencing has decreased, leading to the publication of at least 183 plant reference genomes and the emergence of the first efforts that involved de novo assemblies of many organisms from similar species (Li et al., 2014; Zhang et al., 2016). Conversely, some plant genome assemblies are based on chromosomes (Arabidopsis Genome Initiative 2000; Schnable et al., 2009). Most of them include important agricultural species, are based on short-read sequencing and are composed of 100–1000 contigs and scaffolds often not mapped to particular chromosomal locations. The assembly of plant genomes is difficult, perhaps even more difficult than vertebrate genome assembly. Extreme genome sizes (Neale et al., 2014), high repetitiveness caused by mobile genetic elements, and the polyploidy trait of several plants (Salman-Minkov et al., 2016), especially many agricultural species, provide specific difficulties (Schatz et al., 2012). In recent times, long-read sequencing (Branton et al., 2008; Eid et al., 2009) and long-range scaffolding approaches, including chromosome confirmation capture (Burton et al., 2013), optical mapping (Schwartz et al., 1993), and DNA dilution-based methods (Zheng et al., 2016), were presented to address the challenges in short-read assembly and subsequently the assembly of the whole chromosome (Koren & Phillippy, 2015).

Short-Read Genome Assembly

The fundamental steps in short-read assembly are contig assembly followed by scaffolding and subsequent gap filling (Compeau et al., 2011). The readings are combined into large consensus sequences/contigs with no gaps throughout the contig assembly process. Large-insert reads, which frequently arise through extensive DNA inserts of numerous kbps in length, are utilized to link the contigs in the following scaffolding stage. In the subsequent scaffolding stage, large-insert reads that typically come from longer DNA inserts are used to assemble the contigs. A "scaffold" refers to the connected contigs arranged into a set. If there is no overlap between contigs after being scaffolded, spaces between them are referred to as "gaps," and unexplained bases and imprecise expenses are determined from the insert length of larger insert reads (El-Metwally et al., 2013). The assembly is finished by carefully filling the gaps with additional independent reads (gap-filling phase). Scaffolding along with gap-filling stages could be carried out repeatedly to improve the assembly quality until no contigs have been scaffolded or any more gaps have been filled (Kosugi et al., 2015). The de Bruijn graph is widely utilized in many different assemblers and is the most prevalent method for short-read assembly (Nagarajan & Pop, 2013; Simpson & Pop, 2015). Typically, lengthy reads are brought together using the overlap-layout consensus (OLC) method, while short reads are assembled by converting them to k-mers via the de Bruijn graph method (Pérez-Cobas et al., 2020).

Challenges in Short-Read Assembly

The major challenges in short-read assembly are the choice of assembly approach, sequencing error, repetitive structures, scaffolding complexity, irregular read depth, and computational cost (Sohn & Nam, 2018). The most important challenge is the correction (adjustment) of the sequencing error. When repairing sequencing errors, it is important to remember that the frequency and kinds of errors differ based on the NGS platform and library preparation technique. For instance, short reads with approximately 1% unpredictable sequencing errors are typically produced by Illumina systems (Alic et al., 2016). The second most important issue is inconsistent read depth, which is brought on by copy number variations, PCR, high GC leaning, and cloning. An uneven depth frequently results in breakdowns in the assembly, which introduces gaps. It could be necessary to employ an optimal k for resolving issues (Sims et al., 2014). It is not easy to address the topological complexity of repeating regions in genomes. This problem can be remedied if reads are sufficiently extensive to highlight the repeated sequences. However, long reads are chosen to solve this issue, as Illumina sequences are often shorter and do not typically span a repeated region. The fourth issue is an algorithmic complication, which has a high cost of computing (Sohn & Nam, 2018).

Long-Read Genome Assembly

Repetitive elements, inconsistent sequencing depth, and missing gaps are considered major obstacles to de novo assembly, especially in the case of second-generation short reads, as previously addressed. However, large-insert reads can partially alleviate the issues in genomic areas with lengthy repetitions, such as LINEs and LTRs with sizes of 6–8 and 1–10 kb, respectively, but they do not fully fix them (Koito & Ikeda, 2013; Ashton et al., 2015) in long gaps or low-depth areas. HOWEVER, long SMS readings that cross the repetitions might easily fix the issue. Furthermore, because longer SMS reads have been created using PCR-independent techniques and are considered less biased to areas containing higher GC/AT contents, they are highly useful for dealing with irregular sequencing, gaps, and depth difficulties

(Gross et al., 2014; Goodwin et al., 2015). Long-read-only assemblers are sometimes categorized into hybrid and long-read-only groups. Long-read-only approaches employ longer SMS reads to resolve sequencing inaccuracies and create genome assembly, unlike hybrid methods, which use longer/shorter NGS reads to improve assembly grade and lower sequencing expenses. Scaffolding and gap-filling via longer reads could be more effective than short-read assembly (Sohn & Nam, 2018).

Genome Annotation

A genome sequence must be annotated with biologically related data to be used to its full potential. This information can vary from gene models as well as functional data, including "Kyoto Encyclopedia of Genes and Genomes" (KEGG) pathways and gene ontology (GO) pathways (Primmer et al., 2013), to microRNAs and modifications of epigenetics (The ENCODE Project Consortium, 2012). Annotation is frequently limited to transcripts or protein-coding sequences (CDS) in the domain of genetic nonmodel species. Although annotating genes for newly sequenced organisms can be challenging because they often lack current gene models, automated annotation of genes is now potentially possible for individuals (Yandell & Ence, 2012). However, comprehensive genome annotation involves significant effort and demands bioinformatics expertise. It should be highlighted that the accuracy of genome assembly significantly impacts successful annotation. Only contiguous and nearly complete genomes, interrupted only by tiny gaps, will produce satisfactory findings. Large genomes typically have lengthy genes and require many contiguous assemblies for efficient annotation (Yandell & Ence, 2012). This approach can be generally categorized into two phases: the "computational phase," which simultaneously uses several pieces of evidence from other genomes or transcriptome information based on particular species to make early predictions for genes and transcripts. All (occasionally contradictory) data are then combined into a gene annotation in a subsequent "annotation phase," according to a set of guidelines established by the annotation process (Ekblom & Wolf, 2014).

Genome Publishing

There are currently an increasing number of draft genome sequences being created. Conventional databases that offer access to genomes as well as meta-information, including ENSEMBL and Wellcome Trust Sanger Institute (from NCBI), cannot annotate and collect entirely impending genomes. As a result, NCBI currently offers the option to submit draft genome sequences and user-released annotations. All accessible raw information must be displayed alongside the assembled genome and entire related metadata, such as a BioProject on NCBI, to enable other handlers to enhance the assembly with its annotation (Ekblom & Wolf, 2014).

Molecular Markers

Molecular markers are particular sequences of nucleotides exhibiting an identified chromosomal position that is significant in controlling a specific gene or trait. The heterogeneity (polymorphism) in nucleotide sequences among individuals enables researchers to look for molecular markers. These polymorphisms are based on indels (insertions and deletions), translocations, point mutations, errors in replication, and duplications (Smýkal et al., 2012). These markers are considered strong candidates for evaluating and assessing genetic diversity, population structure, DNA fingerprinting, and phylogenetic associations within or among different species. These markers were subsequently employed in genomic plant breeding, exhibiting exceptional efficiency via marker-assisted selection (Altıntaş et al., 2008). DNA markers have significant reverence for plant breeders to accurately map several interrelated genes associated with targeted agronomic qualities, including abiotic factor resistance and insect and disease tolerance. These markers are constant and not affected by environmental factors or epistatic and pleiotropic effects (Adhikari et al., 2017). Unlike morphological and biochemical markers, these markers are potential tools for identifying high genetic differences and allow analysis at each developing stage (Dar et al., 2014). Molecular markers have revolutionized conventional breeding programs, thus enhancing affectivity and reducing costs and labor (Jiang, 2013).

Classification of Molecular Markers

Molecular markers are categorized into three based on gene action mode (dominant/codominant markers), identification (PCR-based markers and hybridization-based markers), and transmission (maternal and paternal nuclear or organelle inheritance) (Semagn et al., 2006).

Characteristics of an Ideal Molecular Marker

- High reproducibility
- Highly polymorphic
- Locus specificity
- Codominant inheritance (it can differentiate homozygosity and heterozygosity of an organism)

- It must have Mendelian inheritance
- The expression should be self-determining (it should not be dependent on the environment)
- Cost-effective
- It should not be laborious and time-consuming

Furthermore, neither of these molecular markers exhibited all traits simultaneously (Dong et al., 2018). Some of these molecular markers are codominant and thus used in breeding programs. Some markers have high polymorphism and can be used in studying genetic diversity and population structure. Hence, a researcher should consider molecular markers as per his or her requirements and potential (Merritt et al., 2015).

Hybridization-Based Molecular Markers

These molecular markers are independent of PCR and include RFLP (restriction fragment length polymorphism), RLGS (restriction landmark genomic scanning), and VNTR (variable number tandem repeat) markers (Adhikari et al., 2017). RFLP is an extensively used marker for the analysis of genetic diversity (Landry et al., 1987).

Restriction Fragment Length Polymorphism (RFLP)

RFLP is a marker-based method and is considered the leading generation of markers. This marker system detection relies on hybridization. Individuals of similar species display various polymorphisms, including point mutations, InDels, inversions, duplications, unequal crossing over, rearrangements, and translocations (Nadeem et al., 2018). The prime step in the RFLP technique is to recognize and isolate pure DNA. This purified DNA is then treated with restriction enzymes that act upon particular sequences known as restriction sites (Amiteye, 2021). The restricted fragments were then resolved by polyacrylamide gel electrophoresis, which resulted in a series of bands with different lengths. These particular banding patterns appear as a result of genetic variations. This variability results in either the presence or absence of recognition sites, resulting in fragments varying in length and polymorphisms. If the recognition site varies even in a single bp, restriction enzymes cannot restrict the fragment (Madhumati, 2014). These markers are locus specific, highly reproducible, and abundantly distributed throughout the genome, making them suitable and preferable markers for mapping studies. Moreover, they serve as prospective candidates in linkage analysis and quickly assess whether a recessive trait in an individual is linked to a homozygous or heterozygous state (Winter & Kahl, 1995). This methodology has become more sophisticated, which frequently restricts its use in plant breeding due to the need for radioactive or toxic chemicals, high-quality DNA in large amounts, and prior information on sequences for probe generation (Wünsch & Hormaza, 2002).

In previous studies, RFLP markers have been utilized to fingerprint the germplasm of *Pennisetum purpureum* for detecting genetic diversity and heterozygosity. Furthermore, linkage analyses showed that 64 markers were associated with QTLs of 26 plant characteristics. These QTL-associated markers begin the genetic analysis of crucial plant yield and quality parameters (Smith et al., 1993). In another study, the association between RFLP phenotypes and substantial agronomic traits was examined in wheat. Identifying unique alleles distinctive to these regions was possible by comparing the RFLP phenotypes of leaf rust and stem rust resistance genes. Therefore, these markers can be quite useful for identifying genes linked to particular agronomic traits (Paull et al., 1998).

Furthermore, these markers have also been used for studying the inheritance of morphological characteristics in maize. However, the employment of RFLP markers is strenuous and inefficient. These limitations have directed the generation of PCR-based molecular markers for effective genetic analysis (Zebire, 2020). In *Sorghum bicolor*, RFLPs have been used to detect QTLs associated with drought and lodging tolerance before flowering (Kebede et al., 2001).

PCR-Based Markers

PCR development has updated the area of molecular biology and played an important role in the advancement and evolution of DNA markers. Several approaches based on PCR have been extensively employed for studying genetic diversity in various plants, as they are cost-effective, highly reproducible, reliable, convenient to use, time-efficient, and require no radioactive probes (Barman & Kundu, 2019). PCR-based markers are prevailing tools for evaluating polymorphisms, diversity analysis, and plant genetic linkages (Amom & Nongdam, 2017). This approach allows the exponential amplification of a targeted sequence without the need for molecular cloning (Adhikari et al., 2017). The widely employed PCR-based markers are RAPD, microsatellites/SSR, ISSRs, AFLPs, SCARs, and CAPS (Grover & Sharma, 2016).

Random Amplified Polymorphic DNA

RAPD is a PCR-based marker technique in which genomic DNA from an organism is amplified using short fragments of oligonucleotides, known as primers. In the RAPD system, random primers are used with a size of 10–15 bases. These markers amplify anonymous DNA stretches from a complex DNA template strand using arbitrary primers. The amplification products depend on the primer size and length and their targeted DNA. PCR amplification occurs if two hybridization sites (approximately 3000 bp) are the same and in reverse directions. The PCR amplicons were resolved by agarose gels and visualized by ethidium bromide (Jones et al., 1997). As a result, polymorphisms are detected either in the targeted DNA sequence or the binding sites of the primers by validating either the presence or absence of a particular RAPD amplification (Jiang, 2013). This PCR-based system generates high polymorphism. The integrity and amount of DNA, concentration of $MgCl_2$, PCR buffer, annealing temperature, and nature of DNA polymerase are crucial aspects troubling the reproducibility of these markers. These markers are dominant and can detect multiple loci (0.5–5 kb) in an assay (Adhikari et al., 2017). No prior sequence requirement for primer synthesis, no need for radioactive probes, hybridization, simplicity, ease of use, cost, and time efficiency are promising aspects of this technique for detecting RAPD markers linked to disease resistance or various agronomic traits. However, low reproducibility, reliability, presence of false positive results, sensitivity, and the inability to discriminate between homozygous and heterozygous states of an individual are some drawbacks of this technique, making them less preferential for genome-wide analysis (Bardakci, 2001).

In preceding studies, RAPDs have been utilized to investigate genetic diversity in fungal isolates, *Alternaria alternata*, and activating leaf spot disease in potatoes (Esfahani, 2018). RAPD analysis was conducted in another study to detect markers linked with plant height. From a total of 200 primers, two RAPD primers were associated with height because these were found only in tall potato crops and did not appear in dwarf potato crops (Dhanasekar et al., 2010). Furthermore, another study studied the genetic diversity of *Sesamum indicum* using RAPDs and SSRs. They stated that RAPDs have a low PIC (polymorphic information content) value (0.186) compared to SSR markers (0.194) (Dar et al., 2014). Furthermore, they demonstrated that RAPD markers showed a higher marker index. The RAPD primer, RPI-B11, was potentially linked with evaluating genetic diversity in *Sesamum indicum*. RAPD primers were also used in the genetic diversity analysis of 54 Indian pea varieties (Thakur et al., 2018). In previous reports, RAPDs have been employed for identifying abiotic factors in various plants, such as salinity and heat tolerance genes (Damra et al., 2017; Krupa-Małkiewicz & Bienias, 2018).

Amplified Fragment Length Polymorphism

AFLP is an operative approach for PCR to identify polymorphisms in restricted-digested DNA sequences. Consequently, this method combines the efficacy of PCR technology with the RFLP technique (Lynch & Walsh, 1998). In this technique, the amplification is targeted and selective rather than random, as in RAPD, to avoid nonreproducible results. These markers are dominant, inexpensive, and efficient, and no prior sequence data are needed. The constructive aspect of this method is that even degraded or minor-quality DNA can be used (Ridout et al., 1999). It is the fastest method for genetic diversity analysis and constructing high-density genetic maps linked to targeted traits. It is a potential candidate for identifying single nucleotide variations by simultaneously arraying variations in multiple loci (Bangarwa et al., 2021). In previous studies, 14 AFLP markers were used for screening 36 alfalfa resistant and 36 alfalfa-susceptible plants to downy mildew disease. Four AFLP markers (ACACTC150, ACACTC208, ACACTC216, and ACACTC486) were significantly linked to disease resistance. These markers could be useful for crop improvement in disease resistance (Obert et al., 2000).

Simple Sequence Repeats

SSRs are repeated tandem motifs (one to six nucleotides) (Tautz, 1989). These repeats can be mono- (A), di- (GT), tri- (ATT), tetra- (ATCG), penta- (TAACC), and hexanucleotide (TGACTC) repeats and are widely distributed across the genomes of different taxa (Weber, 1990; Adal et al., 2015). These repeated motifs are found in the nuclear genome, chloroplast genome, and mitochondria (Rajendrakumar et al., 2007). These sequences are found in coding and noncoding regions and expressed sequence tags. This polymorphism appears due to discrepancies, mobile elements, double-strand DNA recombination, and misalignment of single-stranded DNA (Nadeem et al., 2018). In previous reports, SSRs have been used to screen resistant and susceptible mung bean genotypes against salt stress. Hence, this can assist in breeding selection for prior detection of stresses instead of morphological evaluation (Nirmala et al., 2016). In another study, 23 SSR primers were used for genetic diversity in 104 cucumber genotypes. This information will aid genotype detection, molecular breeding, and crop progress (Dar et al., 2017). In another preceding study, 25 yield-linked SSRs were used for the characterization of 96 genotypes of common beans. The high degree of genetic variability could be exploited to improve common bean genetics by focusing on factors that affect yield (Gupta et al., 2018).

Cleaved Amplified Polymorphic Sequences

CAPSs were previously identified as PCR-RFLP markers because they are derived from PCR and RFLP combinations (Maeda et al., 1990). This approach relies on PCR-based amplification of targeted DNA after restricting the amplified products (Jarvis et al., 1994). These products are then visualized on either agarose gels or polyacrylamide gels. The sequence data in a genomics databank, RAPD bands, or cDNA sequences synthesize the primers employed in this approach. They are quite adaptable, and their ability to detect DNA polymorphisms could be increased in combination with other markers, such as SCAR, RAPD, or AFLP. CAPSs are codominant markers that are effective in genotyping, molecular identification, and genome mapping (Spaniolas et al., 2006). Previously, a CAPS marker (NEM06) was identified as associated with resistance to root-knot nematode (RKN) in sugarbeet (Weiland & Yu, 2003). Recent studies have used CAPS markers to analyze genetic diversity and variety identification in tea germplasm (Hu et al., 2014). The only pitfall of this technique is the occurrence of mutations in the recognition sites of restriction enzymes that halt CAPS markers' ability to identify polymorphisms. However, a promising alternative known as derived-CAPS (dCAPS) that introduces SNPs (single nucleotide polymorphisms) into PCR products by creating mismatches in primers and detects the ensuing variations by restriction patterns has been developed to address this problem (Adhikari et al., 2017).

Single Nucleotide Polymorphisms

SNPs are characterized as single-base pair variations (G, C, T, and A) found in an individual's genomic sequence of a population (Brookes, 1999). It is considered the most common genetic variation present within an organism. In the human genome, SNPs are repeated after every 100–300 nucleotide sequence, and approximately 10 million SNPs have been observed in humans (Minotti et al., 2018). SNPs can be categorized as either transitions or transversions based on nucleotide changes. Single-base substitutions are insertions or deletions (InDels) typically found in mRNA. SNPs are significantly found in plants and animals, and the probability for every single SNP is after 100–300 nucleotide sequences. SNPs are widely distributed across the genome and can be found at varying rates in coding and noncoding sequences or the intergenic space between two genes (Xu, 2010). However, SNP frequency is comparatively higher in noncoding regions. SNPs in coding regions could generate synonymous or nonsynonymous changes that either modify the amino acid sequence or have no alteration (Barman & Kundu, 2019). Since a nucleotide is the smallest unit of heredity, SNPs offer the most comprehensive and simplest type of new-generation molecular marker (Bangarwa et al., 2021).

However, the discovery of numerous SNPs is substantially facilitated by improvements in next-generation sequencing and high-throughput genotyping approaches, making them desirable tools as genetic markers (Davey et al., 2011). Furthermore, several genotyping methods have been established based on allele-specific discrimination and molecular detection mechanisms. Among them, the simplest and most efficient approach is SNP-RFLP, while the CAPS marker technique can also be used for SNP identification. In this technique, if the restriction sites are present in one allele and not in another, the digested products will result in various lengths. Hence, SNPs can be identified by examining the data stored in the database (Nadeem et al., 2018). These markers could be commendably utilized in creating genetic maps and marker-assisted screening and identifying targeted genes via positional cloning. Several gel-based SNP genotyping assays are available to confirm SNP markers (Adhikari et al., 2017). SNPs are preferred DNA markers because they quickly replace other markers, such as SSRs, for use in genetics and plant breeding (Edwards & Batley, 2010). These markers are desirable because they are codominant, extremely reproducible, and cost-effective. The only drawback of this technique is the high developmental cost (Jiang, 2013; Madhumati, 2014). In previous studies, 34 SNPs have been identified linked to abiotic stress (abscisic acid stress) and ripening genes in chickpeas. These genes can be used for abiotic stress improvement in many crops (Roorkiwal et al., 2014).

Copy Number Variations (CNV)

Copy number variation is based on changes in genomic structure that comprise certain duplications or deletions (>50 bp) in a DNA segment. Its size ranges from 1 kb to 3 Mb. These variations involve genomic disorders, evolutionary adaptability, and disease development (Gordeeva et al., 2022). As a result, these variations lead to replicas of particular DNA fragments linked to homologous chromosomes (Hovhannisyan et al., 2019). Recently, these variations have been identified as highly prevalent variabilities in plant, animal, and human genomes (Bickhart & Liu, 2014; Żmieńko et al., 2014; Nowakowska, 2017). CNVs are a complex class of cytogenetic variations (also known as aneuploidy), as these are distinguishable from whole-chromosome losses or gains (Zarrei et al., 2015). Genes affected by CNVs are associated with certain agronomic traits in various crops. In previous studies, these variations are significant in the flowering period as well as vernalization in rapeseed (Schiessl et al., 2017a, 2017b) and wheat, resistance to abiotic (frost) stress in wheat (Würschum et al., 2015), and biotic stress in tobacco (Chen et al., 2018).

Functional Genomics

Large-scale genome studies have significantly altered how biology is perceived. Genomics is the study of gene organization, role, sequence, quantity, and structure throughout the genome. Functional genomics involves the analysis of gene function and regulation of gene expression during each growth stage at the cell or organism level. For that, high-throughput approaches have been used in genome-wide analysis by interpreting the expression of genes, translation, and protein-to-protein interactions. Microarrays have been the primary large-scale experimental approaches for analyzing gene expression (transcriptomics). Mass spectrometry is frequently employed for protein analysis (proteomics). Moreover, computational tools were also developed due to technological advancements to analyze their output (Zhuang et al., 2014).

Forward and Reverse Genetics

Forward genetics, or classical genetics/forward genetic screening, is a strategy to detect and characterize genes linked to an individual's phenotype. Forward genetics is based on the genetics of a particular phenotypic trait, whereas reverse genetics is based on the phenotypes that result from specific genetic sequences. Reverse genetics evaluate a particular gene function by investigating the phenotypic effects of these genes. Forward genetics begins from the mutated phenotype, followed by progeny ratio analysis to indicate the presence of that specific gene. The appropriate gene was cloned and sequenced to determine its DNA and protein sequences. In contrast to classical genetics, reverse genetics often works oppositely. Reverse genetics is a novel approach conceived by recombinant DNA technology (Aklilu, 2021).

Forward Genetics

Forward genetics typically involves the evaluation of particular phenotypes followed by the selection and screening of novel genotypes. These plants are then immediately integrated into breeding programs, and to understand the function of these genes, mutations in phenotypes can be cloned to generate a suitable genetic marker. The process of forward genetics includes three steps (Zakhrabekova et al., 2013):

1. Development of a mutagenized population (variations can be either induced or natural)
2. Selection and screening of the mutant phenotype
3. Identification of genes and associated functions

Development of the Mutagenized Population

The prime phase in forward and reverse genetics is mutant development (mutagenized populations). Selection of plant parts, mutagens, and dosage optimization is necessary for these approaches. Typically, seeds are mutagenized for seed-propagated crops (Serrat et al., 2014). For mutagenesis, individuals are assessed based on heritable traits from their nonmutagenized parental lines. Each cell in the first generation (known as M1) treated with mutagens (either chemical or physical mutagenesis) must have the ability to accumulate different mutations. Once potentially desirable traits have been found, efforts are made to stabilize the heritability of these traits by introducing novel alleles into the population. Stable plants are typically generated after 5–10 generations of self-crossing and then moved on for varietal testing before formal release. Furthermore, increasing treatment dosages, monitoring the durability of materials, and assessing the capacity to produce complete plants are all necessary for optimizing the mutagenesis process (Mba et al., 2010).

Genetic Mapping

The next step is the detection of the loci of the genes associated with the desired traits, which is achieved by physical mapping. Identifying a specific marker associated with a desired trait is the main step in genetic mapping, followed by marker coinheritance via physical linkage. The next phase is mapping population development, parental line screening for genetic marker polymorphism, and population genotyping. After that, linkage analysis was carried out to identify recombination frequency among markers.

QTL Mapping

QTL mapping refers to association-based statistical analysis of genotypes and phenotypes at particular positions throughout the genome. Genotypic and phenotypic variability are crucial factors in QTL mapping. Phenotypic variations are easily observable among the inbred lines, and experimental crosses between these lines give both genotypic and phenotypic variations. A set of markers that could discriminate among different strains employed in experimental crosses are used to quantify genetic diversity throughout the genome. However, with the advancement of technology and understanding of genetic structure, the origins of these markers have changed over time. Hence, genetic markers based on visible characteristics have been replaced by RFLPs after restriction enzymes. Following PCR advancement, many PCR-based markers have been identified, including SNPs, AFLPs, SSRs, and SSLPs. When phenotypes and genotypes have been identified, association studies between these two can be conducted (Aklilu, 2021).

Exome Sequencing

Exome sequencing is a potential technique that can be used instead of whole-genome sequencing to sequence the targeted coding regions selectively (Ng et al., 2009).

Reverse Genetics

Reverse genetics explore the effects of generated (induced) variations in a particular gene and determine gene functionality. Gene silencing and homologous recombination are two methods for targeted disruption or alternation processes, whereas chemical mutagenesis and transposon-mediated mutagenesis rely on random, nontargeted disruptions. These processes are often followed by screening an organism's library for lesions at particular locations (Aklilu, 2021). T-DNA insertion mutants for a model plant, such as Arabidopsis, have been accessible to researchers. Recently, two methods for reverse genetics have been developed: targeted gene silencing by RNAi and TILLING (Aklilu, 2021).

Gene Silencing

RNA interference is a gene silencing approach to inhibit the expression of a targeted gene either by sense RNA or antisense RNA. It operates on the principle that double-stranded sequences can detect and degenerate complementary sequences (Aklilu, 2021). Double-stranded RNA was introduced into *Caenorhabditis elegans*, an effective way to silence gene expression (Kuttenkeuler & Boutros, 2004). RNAi is a potential tool for silencing gene expression and analyzing their loss of functions in phenotypes when mutated alleles are inaccessible. This technique uses dicer (endonuclease enzyme) to cut single-stranded RNA (small interfering RNAs) from double-stranded RNA. These siRNAs then form an RISC complex and destroy the targeted complementary sequence, ensuring the silencing of the particular gene (Pattanayak et al., 2005).

Targeted-Induced Local Lesions in Genomes (TILLINGs)

TILLINGs is a novel approach that detects mutations in a target gene, including a heteroduplex, using mismatch-specific endonucleases (Cel1). Approximately all of the genes' mutant alleles will be found in a population if the mutation rate is relatively high and the population size is large enough. This method entails amplifying the target gene via PCR with fluorescently labeled primers, forming a DNA heteroduplex in wild-type and mutant alleles, and then performing endonuclease restriction that precisely cleaves at the EMS-induced mismatch. In many cases, Li-COR is used to determine amplicon cleavage fragment size (McCallum et al., 2000). Therefore, TILLINGs can be used on crops with complex genetic makeup, i.e., wheat (Slade et al., 2005). The primary disadvantage of the TILLING method is genetic engineering avoidance, which leads to GMOs, which are prohibited for use in agriculture in numerous nations.

Homologous Recombination

Homologous recombination is linked with the restoration of double-stranded breaks, just as it has been discovered that homologous recombination primarily begins with a double-strand break. The most widely accepted theories for gene targeting rely on the repair of double-strand breaks and strand annealing (Iida & Terada, 2004). Recently, a Drosophila reverse genetic system utilizing this approach was created.

Although it is potentially important, the process is laborious and entails the generation of transgenic flies (Stemple, 2004). It is now possible to target reproducible genes in rice via homologous recombination. The future of homologous recombination-based gene targeting as a common practice for engineering the rice genome and likely other plants appears effective with the addition of a site-specific recombination system (Iida & Terada, 2004).

Transposon-Mediated Mutagenesis

Transposons are mobile elements that may change their location across the genome (Hayes, 2003). Without any sequence associated with the target locus, these DNA sequences can splice into a new position in a genome. Transposon-mediated signature-tagged mutagenesis has made it possible to pinpoint crucial genes and genes involved in infection ability. In Drosophila and *C. elegans*, reverse genetics is now being conducted using individual libraries with insertions of transposable elements. Inaccurate excision in P-elements could cause mutations in the closest gene. Arabidopsis, maize, and other plants use transposon-based techniques (Stemple, 2004). The lower frequency of mutations caused by insertional mutagenesis makes it necessary to screen many individuals for polymorphisms in the studied genes (Gilchrist & Haughn, 2005).

Moreover, the addition of key genes typically results in mortality, necessitating the generation of milder mutations in these genes to comprehend their function. Insertional mutagenesis has also been applied to the T-DNA of the *A. tumefaciens* Ti plasmid, which conveys the genes for plant cell transformation. More than 70% of the genes in Arabidopsis have gene knockouts because of T-DNA insertional mutagenesis (Alonso et al., 2003); however, despite the increasing availability of high-coverage genome sequences, no equivalent resources are accessible for rice or maize. Incorporation of T-DNA into the genome of plants remains indistinguishable, in contrast to other effective gene targeting systems (notably Physcomitrella, mouse, and Drosophila) (Iida & Terada, 2004). The small number of allele types constrains similar RNAi approaches and insertional mutagenesis (McCallum et al., 2000). For many filamentous fungi, gene substitution by transformation is a frequently employed method (Fang et al., 2002; Lalucque & Silar, 2004).

Chemical Mutagenesis

Ethyl methanesulfonate (EMS) and ethylnitrosourea (ENU) are extensively utilized mutagens for chemical-related mutagenesis research. Guanine bases are alkylated by the chemical mutagen known as EMS. A G/C-A/T transition will eventually occur due to the alkylated guanine pairing with thymine rather than the desired cytosine base. The most often employed mutagen in plants is EMS. Approximately 50% of EMS-induced mutations in specific genes in Arabidopsis change the amino acid pattern of the encoded protein, whereas 5% cause gene products to terminate

(Gilchrist & Haughn, 2005) prematurely. In addition to causing point mutations, ENU is a stronger mutagen than EMS. Chemical mutagenesis produces induced point mutations that give rise to various alleles for genetic research, making it appealing for reverse genetics. It causes many randomly dispersed recessive mutations in each genome (Gilchrist & Haughn, 2005). It is useful for reverse genetics, as it is already broadly utilized in several individuals for genetic selection (Henikoff et al., 2004).

Transcriptome Sequencing and Hybridization-Based Approaches

Transcriptomics analyzes gene expression for counting every mRNA transcript in cells (Wang et al., 2009). The entire collection of transcripts in a cell at a specific time and stage of development is known as the transcriptome. Transcriptome sequencing, sometimes referred to as RNA-seq or mRNA-seq, involves enriching single-stranded mRNA from total RNA, followed by reverse transcription into double-stranded cDNA, which is utilized to perform high-throughput sequencing and correlation analysis (Qiang-Long et al., 2014). Several approaches, including hybridization and sequence-based methods, have been developed to determine and quantify the transcriptome. For hybridization-based methods, fluorescently labeled cDNA is commonly incubated with premade or high-density oligo microarrays (Clark et al., 2002).

Furthermore, the development of genomic tiling microarrays that represent the genome with a great density allows for the high-resolution mapping of transcribed regions, from a few bp to approximately 100 bp (David et al., 2006). In addition to high-resolution tiling arrays that query large genomes, hybridization-based techniques are cost-effective and high throughput. However, these techniques have several drawbacks, such as dependence on already known genome sequence information and the high background levels imposed by cross-hybridization (Bertone et al., 2004). Sequence-based methods directly identify the cDNA sequence as opposed to microarray techniques. Primarily, cDNA/EST libraries are subjected to sequencing using the Sanger method; however, this method is often nonquantitative and costly (Gerhard et al., 2004).

Microarray: High-Throughput Transcriptomic Technique

High-throughput analyses of differential expression of genes are effective tools in finding new genes and learning more about specific biological processes at the genome level. Microarray-based transcriptional profiling has emerged as the most important technique used for functional genomics, and it has persuasively shown how raw sequencing data can be transformed into a thorough indulgence of gene activity (Holtorf et al., 2002). The initial stage in making proteins is mRNA synthesis, and variations in mRNA abundance correspond to protein levels. Changing the rate at which genes are transcribed is the main way to alter mRNA steady-state levels. Such variations in relative mRNA levels reflect responses to various stimuli or environmental changes and developmental programs (Holtorf et al., 2002; Royce et al., 2007). The benefit of microarray-based expression is the ability to simultaneously monitor many different genes, whether they have an anonymous or specified sequence, utilizing a minute quantity of biological samples.

RNA-Seq: Transcriptome Sequencing

RNA-Seq is a novel high-throughput sequencing technology for mapping and quantifying the transcriptome. A cDNA library and adaptors linked to either single or both ends are often created from an RNA population (total or partitioned, such as poly(A)+). After that, each molecule is high-throughput sequenced with or without amplification to obtain tiny sequences from single and both ends, recognized as single-end sequencing and paired-end sequencing, respectively. According to the DNA sequencing method, the reads are often 30–400 bp (Holt & Jones, 2008). Additionally, RNA-Seq 30-bp short reads highlight the association between two exons, but pair-end shorter or prolonged reads must indicate connectivity between several exons. Because of these aspects, RNA-Seq is excellent for analyzing complicated transcriptomes. Furthermore, RNA-Seq can detect sequence differences in transcribed areas, such as SNPs (Morin et al., 2008).

Although DNA sequences are reliably mapped to distinct sections of the genome, RNA-Seq offers a second benefit over DNA microarrays: it has very little to no background signal or none. The number of sequences collected coincides with the highest limit of quantification for RNA-Seq, which does not exist. As a result, a high dimensionality range of transcript expression levels on a particular transcript may be identified: a study that examined 16 million mapped reads in *Saccharomyces cerevisiae* suggested that this range was higher than 9000-fold (Nagalakshmi et al., 2008). In contrast, DNA microarrays have a very limited dynamic range due to their lack of sensitivity toward genes produced at lower or higher levels (~100–1000-fold) (Holtorf et al., 2002).

ChIP-Seq

ChIP-Seq is a sequencing approach for the genome-wide assessment of DNA-binding proteins, nucleosomes, and histone modifications. Understanding the gene regulatory networks that enable many biological processes requires a precise map of the DNA-binding sites for transcription factors, other DNA-binding proteins, and core transcriptional components (Farnham, 2009). The location of nucleosomes significantly influences gene regulation in combination with dynamic modifications to DNA and histones (Li et al., 2007; Jiang & Pugh, 2009). ChIP, a primary technique that concentrates DNA segments to which a particular protein or category of nucleosomes is associated, is the primary tool for researching these mechanisms (Solomon et al., 1988). For some applications, such as de novo genome assembly, these short reads of 35 bp produced by NGS platforms present significant challenges, yet they are suitable for ChIP-Seq. ChIP-Seq-improved mapping of protein-binding sites enables the identification of sequence motifs and a more precise database of targets for particular enhancers and other transcription factors (Johnson et al., 2007; Visel et al., 2009). Since then, ChIP-Seq has become an essential technique for researching gene regulation and epigenetic processes as sequencing costs have dropped (Park, 2009).

References

Abdurakhmonov, I. Y. (2016). Genomics era for plants and crop species—Advances made and needed tasks ahead. In *Plant genomics*. IntechOpen.

Adal, A. M., Demissie, Z. A., & Mahmoud, S. S. (2015). Identification, validation and cross-species transferability of novel Lavandula EST-SSRs. *Planta, 241*, 987–1004.

Adhikari, S., Saha, S., Biswas, A., Rana, T. S., Bandyopadhyay, T. K., & Ghosh, P. (2017). Application of molecular markers in plant genome analysis: A review. *Nucleus, 60*, 283–297.

Aklilu, E. (2021). Review on forward and reverse genetics in plant breeding. *All Life, 14*, 127–135.

Alberts, B., Johnson, A., Lewis, J., Raff, M., Roberts, K., & Walter, P. (2002a). How genomes evolve. In *Molecular biology of the cell* (4th ed.). Garland Science.

Alberts, B., Johnson, A., Lewis, J., Raff, M., Roberts, K., & Walter, P. (2002b). The RNA world and the origins of life. In *Molecular biology of the cell* (4th ed.). Garland Science.

Alic, A. S., Ruzafa, D., Dopazo, J., & Blanquer, I. (2016). Objective review of de novo stand-alone error correction methods for NGS data. *Wiley Interdisciplinary Reviews: Computational Molecular Science, 6*, 111–146.

Alonso, J. M., Stepanova, A. N., Leisse, T. J., Kim, C. J., Chen, H., Shinn, P., et al. (2003). Genome-wide insertional mutagenesis of Arabidopsis thaliana. *Science, 301*, 653–657.

Altıntaş, S., Toklu, F., Kafkas, S., Kilian, B., Brandolini, A., & Özkan, H. (2008). Estimating genetic diversity in durum and bread wheat cultivars from Turkey using AFLP and SAMPL markers. *Plant Breeding, 127*, 9–14.

Amiteye, S. (2021). Basic concepts and methodologies of DNA marker systems in plant molecular breeding. *Heliyon, 7*, e08093.

Amom, T., & Nongdam, P. (2017). The use of molecular marker methods in plants: A review. *International Journal of Current Research and Review, 9*, 1–7.

Anderson, S. (1981). Shotgun DNA sequencing using cloned DNase I-generated fragments. *Nucleic Acids Research, 9*, 3015–3027.

Arabidopsis Genome Initiative. (2000). Analysis of the genome sequence of the flowering plant Arabidopsis thaliana. *Nature, 408*, 796–815.

Ashton, P. M., Nair, S., Dallman, T., Rubino, S., Rabsch, W., Mwaigwisya, S., et al. (2015). MinION nanopore sequencing identifies the position and structure of a bacterial antibiotic resistance Island. *Nature Biotechnology, 33*, 296–300.

Bangarwa, S. K., Balwan, K. L. S., Choudhary, M., & Choudhary, M. K. (2021). An introduction to DNA-markers and their role in crop improvement. *The Pharma Innovation Journal, 10*, 638–643.

Barba, M., Czosnek, H., & Hadidi, A. (2014). Historical perspective, development and applications of next-generation sequencing in plant virology. *Viruses, 6*, 106–136.

Bardakci, F. (2001). Random amplified polymorphic DNA (RAPD) markers. *Turkish Journal of Biology, 25*, 185–196.

Barman, M., & Kundu, S. (2019). Molecular markers and a new vista in plant breeding: A review. *International Journal of Current Microbiology and Applied Sciences, 8*, 1921–1929.

Begun, D. J., Lindfors, H. A., Kern, A. D., & Jones, C. D. (2007). Evidence for de novo evolution of testis-expressed genes in the Drosophila yakuba/Drosophila erecta clade. *Genetics, 176*, 1131–1137.
Beier, S., Himmelbach, A., Schmutzer, T., Felder, M., Taudien, S., Mayer, K. F. X., et al. (2016). Multiplex sequencing of bacterial artificial chromosomes for assembling complex plant genomes. *Plant Biotechnology Journal, 14*, 1511–1522.
Bertone, P., Stolc, V., Royce, T. E., Rozowsky, J. S., Urban, A. E., Zhu, X., et al. (2004). Global identification of human transcribed sequences with genome tiling arrays. *Science, 306*, 2242–2246.
Betrán, E., & Long, M. (2002). Expansion of genome coding regions by acquisition of new genes. *Genetica, 115*, 65–80.
Bickhart, D. M., & Liu, G. E. (2014). The challenges and importance of structural variation detection in livestock. *Frontiers in Genetics, 5*, 37.
Branton, D., Deamer, D. W., Marziali, A., Bayley, H., Benner, S. A., Butler, T., et al. (2008). The potential and challenges of nanopore sequencing. *Nature Biotechnology, 26*, 1146–1153.
Brookes, A. J. (1999). The essence of SNPs. *Gene, 234*, 177–186.
Brown, T. A. (2002). Genome anatomies. In *Genomes* (2nd ed.). Wiley-Liss.
Burton, J. N., Adey, A., Patwardhan, R. P., Qiu, R., Kitzman, J. O., & Shendure, J. (2013). Chromosome-scale scaffolding of de novo genome assemblies based on chromatin interactions. *Nature Biotechnology, 31*, 1119–1125.
Campos-de Quiroz, H. (2002). Plant genomics: An overview. *Biological Research, 35*, 385–399.
Casola, C., Lawing, A. M., Betrán, E., & Feschotte, C. (2007). PIF-like transposons are common in drosophila and have been repeatedly domesticated to generate new host genes. *Molecular Biology and Evolution, 24*, 1872–1888.
Chen, S., Ren, F., Zhang, L., Liu, Y., Chen, X., Li, Y., et al. (2018). Unstable allotetraploid tobacco genome due to frequent homologous recombination, segmental deletion, and chromosome loss. *Molecular Plant, 11*, 914–927.
Choulet, F., Alberti, A., Theil, S., Glover, N., Barbe, V., Daron, J., et al. (2014). Structural and functional partitioning of bread wheat chromosome 3B. *Science, 345*, 1249721.
Church, G. M. (2006). Genomes for all. *Scientific American, 294*, 46–55.
Clark, D. P., & Pazdernik, N. (2012). *Molecular biology*. Elsevier.
Clark, T. A., Sugnet, C. W., & Ares, M., Jr. (2002). Genomewide analysis of mRNA processing in yeast using splicing-specific microarrays. *Science, 296*, 907–910.
Clarke, J., Wu, H.-C., Jayasinghe, L., Patel, A., Reid, S., & Bayley, H. (2009). Continuous base identification for single-molecule nanopore DNA sequencing. *Nature Nanotechnology, 4*, 265–270.
Compeau, P. E. C., Pevzner, P. A., & Tesler, G. (2011). How to apply de Bruijn graphs to genome assembly. *Nature Biotechnology, 29*, 987–991.
Damra, E. M., Kasrawi, M., & Akash, M. W. (2017). Development of scar marker linked to heat stress tolerance in tomato. In *Proceedings of the 65th ISERD International Conference, Mecca, Saudi Arabia* (pp. 23–24).
Dar, A. A., Choudhury, A. R., & Arumugam, N. (2014). A study on seed protein profile of Indian cultivars of Sesamum indicum L. *International Journal of Current Biotechnology, 2*, 10–17.
Dar, A. A., Mahajan, R., Lay, P., & Sharma, S. (2017). Genetic diversity and population structure of Cucumis sativus L. by using SSR markers. *3 Biotech, 7*, 1–12.
Davey, J. W., Hohenlohe, P. A., Etter, P. D., Boone, J. Q., Catchen, J. M., & Blaxter, M. L. (2011). Genome-wide genetic marker discovery and genotyping using next-generation sequencing. *Nature Reviews Genetics, 12*, 499–510.
David, L., Huber, W., Granovskaia, M., Toedling, J., Palm, C. J., Bofkin, L., et al. (2006). A high-resolution map of transcription in the yeast genome. *Proceedings of the National Academy of Sciences, 103*, 5320–5325.
Dhanasekar, P., Dhumal, K. N., and Reddy, K. S. (2010). Identification of RAPD markers linked to plant type gene in pigeonpea.

Dillon, S. C., & Dorman, C. J. (2010). Bacterial nucleoid-associated proteins, nucleoid structure and gene expression. *Nature Reviews Microbiology, 8*, 185–195.

Dong, M., Wang, Z., He, Q., Zhao, J., Fan, Z., & Zhang, J. (2018). Development of EST-SSR markers in Larix principis-rupprechtii Mayr and evaluation of their polymorphism and cross-species amplification. *Trees, 32*, 1559–1571.

Edwards, D., & Batley, J. (2010). Plant genome sequencing: Applications for crop improvement. *Plant Biotechnology Journal, 8*, 2–9.

Egan, A. N., Schlueter, J., & Spooner, D. M. (2012). Applications of next-generation sequencing in plant biology. *American Journal of Botany, 99*, 175–185.

Eid, J., Fehr, A., Gray, J., Luong, K., Lyle, J., Otto, G., et al. (2009). Real-time DNA sequencing from single polymerase molecules. *Science, 323*, 133–138.

Ekblom, R., & Wolf, J. B. W. (2014). A field guide to whole-genome sequencing, assembly and annotation. *Evolutionary Applications, 7*, 1026–1042.

El-Metwally, S., Hamza, T., Zakaria, M., & Helmy, M. (2013). Next-generation sequence assembly: Four stages of data processing and computational challenges. *PLoS Computational Biology, 9*, e1003345.

Esfahani, M. N. (2018). Analysis of virulence and genetic variability of Alternaria alternata associated with leaf spot disease in potato plants in Iran. *Acta Mycologica, 53*. https://doi.org/10.5586/am.1105

Fang, G.-C., Hanau, R. M., & Vaillancourt, L. J. (2002). The SOD2 gene, encoding a manganese-type superoxide dismutase, is up-regulated during conidiogenesis in the plant-pathogenic fungus Colletotrichum graminicola. *Fungal Genetics and Biology, 36*, 155–165.

Farnham, P. J. (2009). Insights from genomic profiling of transcription factors. *Nature Reviews Genetics, 10*, 605–616.

Fedoroff, N. V. (2012). Transposable elements, epigenetics, and genome evolution. *Science, 338*, 758–767.

Feschotte, C., & Pritham, E. J. (2007). DNA transposons and the evolution of eukaryotic genomes. *Annual Review of Genetics, 41*, 331–368.

Forterre, P., Filée, J., & Myllykallio, H. (2004). Origin and evolution of DNA and DNA replication machineries. In *The genetic code and the origin of life* (pp. 145–168). Springer.

Fuller, C. W., Middendorf, L. R., Benner, S. A., Church, G. M., Harris, T., Huang, X., et al. (2009). The challenges of sequencing by synthesis. *Nature Biotechnology, 27*, 1013–1023.

Furey, T. S. (2012). ChIP–seq and beyond: New and improved methodologies to detect and characterize protein–DNA interactions. *Nature Reviews Genetics, 13*, 840–852.

Gerhard, D. S., Wagner, L., Feingold, E. A., Shenmen, C. M., Grouse, L. H., Schuler, G., et al. (2004). The status, quality, and expansion of the NIH full-length cDNA project: The Mammalian Gene Collection (MGC). *Genome Research, 14*, 2121–2127.

Gilbert, W. (1986). Origin of life: The RNA world. *Nature, 319*, 618.

Gilbert, N., & Allan, J. (2014). Supercoiling in DNA and chromatin. *Current Opinion in Genetics & Development, 25*, 15–21.

Gilchrist, E. J., & Haughn, G. W. (2005). TILLING without a plough: A new method with applications for reverse genetics. *Current Opinion in Plant Biology, 8*, 211–215.

Goffeau, A., Barrell, B. G., Bussey, H., Davis, R. W., Dujon, B., Feldmann, H., et al. (1996). Life with 6000 genes. *Science, 274*, 546–567.

Goldman, A. D., & Landweber, L. F. (2016). What is a genome? *PLoS Genetics, 12*, e1006181.

Goodwin, S., Gurtowski, J., Ethe-Sayers, S., Deshpande, P., Schatz, M. C., & McCombie, W. R. (2015). Oxford Nanopore sequencing, hybrid error correction, and de novo assembly of a eukaryotic genome. *Genome Research, 25*, 1750–1756.

Gordeeva, V., Sharova, E., & Arapidi, G. (2022). Progress in methods for copy number variation profiling. *International Journal of Molecular Sciences, 23*, 2143.

Green, P. (2002). Whole-genome disassembly. *Proceedings of the National Academy of Sciences, 99*, 4143–4144.

Gross, D. C., Lichens-Park, A., & Kole, C. (2014). *Genomics of plant-associated bacteria*. Springer.
Grover, A., & Sharma, P. C. (2016). Development and use of molecular markers: Past and present. *Critical Reviews in Biotechnology, 36*, 290–302.
Gupta, N., Zargar, S. M., Mahajan, R., Sharma, S., Nehvi, F. A., & Salgotra, R. K. (2018). Candidate gene-based characterization of common bean genotypes. *Indian Journal of Genetics and Plant Breeding, 78*, 389–392.
Hayes, F. (2003). Transposon-based strategies for microbial functional genomics and proteomics. *Annual Review of Genetics, 37*, 3.
Henikoff, S., Till, B. J., & Comai, L. (2004). TILLING. Traditional mutagenesis meets functional genomics. *Plant Physiology, 135*, 630–636.
Holt, R. A., & Jones, S. J. M. (2008). The new paradigm of flow cell sequencing. *Genome Research, 18*, 839–846.
Holtorf, H., Guitton, M.-C., & Reski, R. (2002). Plant functional genomics. *Naturwissenschaften, 89*, 235–249.
Hotopp, J. C. D., Clark, M. E., Oliveira, D. C. S. G., Foster, J. M., Fischer, P., Torres, M. C. M., et al. (2007). Widespread lateral gene transfer from intracellular bacteria to multicellular eukaryotes. *Science, 317*, 1753–1756.
Hovhannisyan, G., Harutyunyan, T., Aroutiounian, R., & Liehr, T. (2019). DNA copy number variations as markers of mutagenic impact. *International Journal of Molecular Sciences, 20*, 4723.
Howden, B. P., McEvoy, C. R. E., Allen, D. L., Chua, K., Gao, W., Harrison, P. F., et al. (2011). Evolution of multidrug resistance during Staphylococcus aureus infection involves mutation of the essential two component regulator WalKR. *PLoS Pathogens, 7*, e1002359.
Howe, K., Clark, M. D., Torroja, C. F., Torrance, J., Berthelot, C., Muffato, M., et al. (2013). The zebrafish reference genome sequence and its relationship to the human genome. *Nature, 496*, 498–503.
Hu, C.-Y., Tsai, Y.-Z., & Lin, S.-F. (2014). Development of STS and CAPS markers for variety identification and genetic diversity analysis of tea germplasm in Taiwan. *Botanical Studies, 55*, 1–15.
Iida, S., & Terada, R. (2004). A tale of two integrations, transgene and T-DNA: Gene targeting by homologous recombination in rice. *Current Opinion in Biotechnology, 15*, 132–138.
Jarvis, P., Lister, C., Szabo, V., & Dean, C. (1994). Integration of CAPS markers into the RFLP map generated using recombinant inbred lines of Arabidopsis thaliana. *Plant Molecular Biology, 24*, 685–687.
Jiang, G.-L. (2013). Molecular markers and marker-assisted breeding in plants. In *Plant breeding from laboratories to fields* (Vol. 3, pp. 45–83). IntechOpen.
Jiang, C., & Pugh, B. F. (2009). Nucleosome positioning and gene regulation: Advances through genomics. *Nature Reviews Genetics, 10*, 161–172.
Johnson, D. S., Mortazavi, A., Myers, R. M., & Wold, B. (2007). Genome-wide mapping of in vivo protein-DNA interactions. *Science, 316*, 1497–1502.
Jones, C. D., & Begun, D. J. (2005). Parallel evolution of chimeric fusion genes. *Proceedings of the National Academy of Sciences, 102*, 11373–11378.
Jones, C. J., Edwards, K. J., Castaglione, S., Winfield, M. O., Sala, F., Van de Wiel, C., et al. (1997). Reproducibility testing of RAPD, AFLP and SSR markers in plants by a network of European laboratories. *Molecular Breeding, 3*, 381–390.
Kaiser, O., Bartels, D., Bekel, T., Goesmann, A., Kespohl, S., Pühler, A., et al. (2003). Whole genome shotgun sequencing guided by bioinformatics pipelines—An optimized approach for an established technique. *Journal of Biotechnology, 106*, 121–133.
Kebede, H., Subudhi, P. K., Rosenow, D. T., & Nguyen, H. T. (2001). Quantitative trait loci influencing drought tolerance in grain sorghum (Sorghum bicolor L. Moench). *Theoretical and Applied Genetics, 103*, 266–276.
Khorasanizadeh, S. (2004). The nucleosome: From genomic organization to genomic regulation. *Cell, 116*, 259–272.

Koito, A., & Ikeda, T. (2013). Intrinsic immunity against retrotransposons by APOBEC cytidine deaminases. *Frontiers in Microbiology, 4*, 28.

Koren, S., & Phillippy, A. M. (2015). One chromosome, one contig: Complete microbial genomes from long-read sequencing and assembly. *Current Opinion in Microbiology, 23*, 110–120.

Kossmann, J. (2012). Grand challenges in plant biotechnology. *Frontiers in Plant Science, 3*, 61.

Kosugi, S., Hirakawa, H., & Tabata, S. (2015). GMcloser: Closing gaps in assemblies accurately with a likelihood-based selection of contig or long-read alignments. *Bioinformatics, 31*, 3733–3741.

Krupa-Małkiewicz, M., & Bienias, A. (2018). BSA and molecular markers screening for salt stress tolerant mutant of Petunia obtained in in vitro culture. *Ciência Rural, 48*, e20170042.

Kuttenkeuler, D., & Boutros, M. (2004). Genome-wide RNAi as a route to gene function in Drosophila. *Briefings in Functional Genomics, 3*, 168–176.

Lai, Z., Kane, N. C., Kozik, A., Hodgins, K. A., Dlugosch, K. M., Barker, M. S., et al. (2012). Genomics of compositae weeds: EST libraries, microarrays, and evidence of introgression. *American Journal of Botany, 99*, 209–218.

Lalucque, H., & Silar, P. (2004). Incomplete penetrance and variable expressivity of a growth defect as a consequence of knocking out two K+ transporters in the euascomycete fungus Podospora anserina. *Genetics, 166*, 125–133.

Lander, E. S., Linton, L. M., Birren, B., Nusbaum, C., Zody, M. C., Baldwin, J., Devon, K., Dewar, K., Doyle, M., FitzHugh, W., Funke, R., Gage, D., Harris, K., Heaford, A., Howland, J., Kann, L., Lehoczky, J., LeVine, R., McEwan, P., McKernan, K., Meldrim, J., Mesirov, J. P., Miranda, C., Morris, W., Naylor, J., & Raymond, C. (2001). Initial sequencing and analysis of the human genome. *Nature, 409*, 860–921.

Landry, B. S., Kesseli, R. V., Farrara, B., & Michelmore, R. W. (1987). A genetic map of lettuce (Lactuca sativa L.) with restriction fragment length polymorphism, isozyme, disease resistance and morphological markers. *Genetics, 116*, 331–337.

Leggett, R. M., & Clark, M. D. (2017). A world of opportunities with nanopore sequencing. *Journal of Experimental Botany, 68*, 5419–5429.

Li, B., Carey, M., & Workman, J. L. (2007). The role of chromatin during transcription. *Cell, 128*, 707–719.

Li, R., Li, Y., Fang, X., Yang, H., Wang, J., Kristiansen, K., et al. (2009). SNP detection for massively parallel whole-genome resequencing. *Genome Research, 19*, 1124–1132.

Li, R., Fan, W., Tian, G., Zhu, H., He, L., Cai, J., et al. (2010). The sequence and de novo assembly of the giant panda genome. *Nature, 463*, 311–317.

Li, Y., Zhou, G., Ma, J., Jiang, W., Jin, L., Zhang, Z., et al. (2014). De novo assembly of soybean wild relatives for pan-genome analysis of diversity and agronomic traits. *Nature Biotechnology, 32*, 1045–1052.

Li, F., Fan, G., Lu, C., Xiao, G., Zou, C., Kohel, R. J., et al. (2015a). Genome sequence of cultivated Upland cotton (Gossypium hirsutum TM-1) provides insights into genome evolution. *Nature Biotechnology, 33*, 524–530.

Li, X., Yang, Y., Henry, R. J., Rossetto, M., Wang, Y., & Chen, S. (2015b). Plant DNA barcoding: From gene to genome. *Biological Reviews, 90*, 157–166.

Lieberman, K. R., Cherf, G. M., Doody, M. J., Olasagasti, F., Kolodji, Y., & Akeson, M. (2010). Processive replication of single DNA molecules in a nanopore catalyzed by phi29 DNA polymerase. *Journal of the American Chemical Society, 132*, 17961–17972.

Liu, Z. J. (2008). *Aquaculture genome technologies*. John Wiley & Sons.

Lynch, M., & Walsh, B. (1998). *Genetics and analysis of quantitative traits*. Sinauer.

Madhumati, B. (2014). Potential and application of molecular markers techniques for plant genome analysis. *International Journal of Pure & Applied Bioscience, 2*, 169–188.

Maeda, N., & Smithies, O. (1986). The evolution of multigene families: Human haptoglobin genes. *Annual Review of Genetics, 20*, 81–108.

Maeda, M., Uryu, N., Murayama, N., Ishii, H., Ota, M., Tsuji, K., et al. (1990). A simple and rapid method for HLA-DP genotyping by digestion of PCR-amplified DNA with allele-specific restriction endonucleases. *Human Immunology, 27*, 111–121.

Mardis, E. R. (2013). Next-generation sequencing platforms. *Annual Review of Analytical Chemistry, 6*, 287–303.

Martin, W. F., Garg, S., & Zimorski, V. (2015). Endosymbiotic theories for eukaryote origin. *Philosophical Transactions of the Royal Society of London Series B, Biological Sciences, 370*, 20140330.

Mba, C., Afza, R., Bado, S., & Jain, S. M. (2010). Induced mutagenesis in plants using physical and chemical agents. In *Plant cell culture: essential methods* (Vol. 20, pp. 111–130). Wiley.

McCallum, C. M., Comai, L., Greene, E. A., & Henikoff, S. (2000). Targeting Induced Local Lesions IN Genomes (TILLING) for plant functional genomics. *Plant Physiology, 123*, 439–442.

McGuire, A. L., Cho, M. K., McGuire, S. E., & Caulfield, T. (2007). The future of personal genomics. *Science, 317*, 1687.

Merritt, B. J., Culley, T. M., Avanesyan, A., Stokes, R., & Brzyski, J. (2015). An empirical review: Characteristics of plant microsatellite markers that confer higher levels of genetic variation. *Applications in Plant Sciences, 3*, 1500025.

Minotti, L., Agnoletto, C., Baldassari, F., Corrà, F., & Volinia, S. (2018). SNPs and somatic mutation on long non-coding RNA: New frontier in the cancer studies? *High Throughput, 7*, 34.

Misteli, T. (2007). Beyond the sequence: Cellular organization of genome function. *Cell, 128*, 787–800.

Morgante, M., Brunner, S., Pea, G., Fengler, K., Zuccolo, A., & Rafalski, A. (2005). Gene duplication and exon shuffling by helitron-like transposons generate intraspecies diversity in maize. *Nature Genetics, 37*, 997–1002.

Morin, R. D., Bainbridge, M., Fejes, A., Hirst, M., Krzywinski, M., Pugh, T. J., et al. (2008). Profiling the HeLa S3 transcriptome using randomly primed cDNA and massively parallel short-read sequencing. *BioTechniques, 45*, 81–94.

Nadeem, M. A., Nawaz, M. A., Shahid, M. Q., Doğan, Y., Comertpay, G., Yıldız, M., et al. (2018). DNA molecular markers in plant breeding: Current status and recent advancements in genomic selection and genome editing. *Biotechnology and Biotechnological Equipment, 32*, 261–285.

Nagalakshmi, U., Wang, Z., Waern, K., Shou, C., Raha, D., Gerstein, M., et al. (2008). The transcriptional landscape of the yeast genome defined by RNA sequencing. *Science, 320*, 1344–1349.

Nagarajan, N., & Pop, M. (2013). Sequence assembly demystified. *Nature Reviews Genetics, 14*, 157–167.

Neale, D. B., Wegrzyn, J. L., Stevens, K. A., Zimin, A. V., Puiu, D., Crepeau, M. W., et al. (2014). Decoding the massive genome of loblolly pine using haploid DNA and novel assembly strategies. *Genome Biology, 15*, 1–13.

Nelson, K. E., Clayton, R. A., Gill, S. R., Gwinn, M. L., Dodson, R. J., Haft, D. H., et al. (1999). Evidence for lateral gene transfer between archaea and bacteria from genome sequence of Thermotoga maritima. *Nature, 399*, 323–329.

Ng, S. B., Turner, E. H., Robertson, P. D., Flygare, S. D., Bigham, A. W., Lee, C., et al. (2009). Targeted capture and massively parallel sequencing of 12 human exomes. *Nature, 461*, 272–276.

Nirmala, S., Mukesh, Y., Venkataraman, B. K., Kumar, S. R., & Kumar, J. P. (2016). Hybridization between salt resistant and salt susceptible genotypes of mungbean (Vigna radiata L. Wilczek) and purity testing of the hybrids using SSRs markers. *Journal of Integrative Agriculture, 15*, 521–527.

Nowakowska, B. (2017). Clinical interpretation of copy number variants in the human genome. *Journal of Applied Genetics, 58*, 449–457.

Nyrén, P. (2007). The history of pyrosequencing®. In *Pyrosequencing® protocols* (pp. 1–13). Springer.

Obert, D. E., Skinner, D. Z., & Stuteville, D. L. (2000). Association of AFLP markers with downy mildew resistance in autotetraploid alfalfa. *Molecular Breeding, 6*, 287–294.

Park, P. J. (2009). ChIP–seq: Advantages and challenges of a maturing technology. *Nature Reviews Genetics, 10*, 669–680.

Pattanayak, D., Agarwal, S., Sumathi, S., Chakrabarti, S. K., Naik, P. S., and Khurana, S. M. (2005). Small but mighty RNA-mediated interference in plants.

Paull, J. G., Chalmers, K. J., Karakousis, A., Kretschmer, J. M., Manning, S., & Langridge, P. (1998). Genetic diversity in Australian wheat varieties and breeding material based on RFLP data. *Theoretical and Applied Genetics, 96*, 435–446.

Pérez-Cobas, A. E., Gomez-Valero, L., & Buchrieser, C. (2020). Metagenomic approaches in microbial ecology: An update on whole-genome and marker gene sequencing analyses. *Microbial Genomics, 6*, mgen000409.

Primmer, C. R., Papakostas, S., Leder, E. H., Davis, M. J., & Ragan, M. A. (2013). Annotated genes and nonannotated genomes: Cross-species use of gene ontology in ecology and evolution research. *Molecular Ecology, 22*, 3216–3241.

Qiang-Long, Z., Shi, L., Peng, G., & Fei-Shi, L. (2014). High-throughput sequencing technology and its application. *Journal of Northeast Agricultural University (English Edition), 21*, 84–96.

Qin, Y., Schneider, T. M., & Brenner, M. P. (2012). Sequencing by hybridization of long targets. *PLoS One, 7*, e35819.

Quail, M. A., Smith, M., Coupland, P., Otto, T. D., Harris, S. R., Connor, T. R., et al. (2012). A tale of three next generation sequencing platforms: Comparison of ion torrent, Pacific biosciences and illumina MiSeq sequencers. *BMC Genomics, 13*, 1–13.

Quince, C., Walker, A. W., Simpson, J. T., Loman, N. J., & Segata, N. (2017). Shotgun metagenomics, from sampling to analysis. *Nature Biotechnology, 35*, 833–844.

Rajendrakumar, P., Biswal, A. K., Balachandran, S. M., Srinivasarao, K., & Sundaram, R. M. (2007). Simple sequence repeats in organellar genomes of rice: Frequency and distribution in genic and intergenic regions. *Bioinformatics, 23*, 1–4.

Razvan, C., & Unrau, P. J. (2017). Transitioning to DNA genomes in an RNA world. *eLife, 6*, e32330.

Rice, E. S., & Green, R. E. (2019). New approaches for genome assembly and scaffolding. *Annual Review of Animal Biosciences, 7*, 17–40.

Ridout, C. J., Donini, P., Ridout, C. J., & Donini, P. (1999). Use of AFLP in cereals research. *Trends in Plant Science, 4*, 76–79.

Roger, A. J. (1999). Reconstructing early events in eukaryotic evolution. *The American Naturalist, 154*, S146–S163.

Ronaghi, M., Karamohamed, S., Pettersson, B., Uhlén, M., & Nyrén, P. (1996). Real-time DNA sequencing using detection of pyrophosphate release. *Analytical Biochemistry, 242*, 84–89.

Ronaghi, M., Uhlén, M., & Nyrén, P. (1998). A sequencing method based on real-time pyrophosphate. *Science, 281*, 363–365.

Roorkiwal, M., Nayak, S. N., Thudi, M., Upadhyaya, H. D., Brunel, D., Mournet, P., et al. (2014). Allele diversity for abiotic stress responsive candidate genes in chickpea reference set using gene based SNP markers. *Frontiers in Plant Science, 5*, 248.

Ross, M. G., Russ, C., Costello, M., Hollinger, A., Lennon, N. J., Hegarty, R., et al. (2013). Characterizing and measuring bias in sequence data. *Genome Biology, 14*, 1–20.

Rothberg, J. M., Hinz, W., Rearick, T. M., Schultz, J., Mileski, W., Davey, M., et al. (2011). An integrated semiconductor device enabling non-optical genome sequencing. *Nature, 475*, 348–352.

Roumpeka, D. D., Wallace, R. J., Escalettes, F., Fotheringham, I., & Watson, M. (2017). A review of bioinformatics tools for bio-prospecting from metagenomic sequence data. *Frontiers in Genetics, 8*, 23.

Royce, T. E., Rozowsky, J. S., & Gerstein, M. B. (2007). Toward a universal microarray: Prediction of gene expression through nearest-neighbor probe sequence identification. *Nucleic Acids Research, 35*, e99.

Salman-Minkov, A., Sabath, N., & Mayrose, I. (2016). Whole-genome duplication as a key factor in crop domestication. *Nature Plants, 2*, 1–4.
Sasaki, T. (2005). The map-based sequence of the rice genome. *Nature, 436*, 793–800.
Sato, N. (2020). *Endosymbiotic theories of organelles revisited: Retrospects and prospects*. Springer.
Schatz, M. C., Delcher, A. L., & Salzberg, S. L. (2010). Assembly of large genomes using second-generation sequencing. *Genome Research, 20*, 1165–1173.
Schatz, M. C., Witkowski, J., & McCombie, W. R. (2012). Current challenges in de novo plant genome sequencing and assembly. *Genome Biology, 13*, 1–7.
Schiessl, S., Huettel, B., Kuehn, D., Reinhardt, R., & Snowdon, R. (2017a). Post-polyploidisation morphotype diversification associates with gene copy number variation. *Scientific Reports, 7*, 1–18.
Schiessl, S., Huettel, B., Kuehn, D., Reinhardt, R., & Snowdon, R. J. (2017b). Targeted deep sequencing of flowering regulators in Brassica napus reveals extensive copy number variation. *Scientific Data, 4*, 1–10.
Schnable, P. S., Ware, D., Fulton, R. S., Stein, J. C., Wei, F., Pasternak, S., et al. (2009). The B73 maize genome: Complexity, diversity, and dynamics. *Science, 326*, 1112–1115.
Schwartz, D. C., Li, X., Hernandez, L. I., Ramnarain, S. P., Huff, E. J., & Wang, Y.-K. (1993). Ordered restriction maps of Saccharomyces cerevisiae chromosomes constructed by optical mapping. *Science, 262*, 110–114.
Semagn, K., Bjørnstad, Å., & Ndjiondjop, M. N. (2006). An overview of molecular marker methods for plants. *African. Journal of Biotechnology, 5*.
Serrat, X., Esteban, R., Guibourt, N., Moysset, L., Nogués, S., & Lalanne, E. (2014). EMS mutagenesis in mature seed-derived rice calli as a new method for rapidly obtaining TILLING mutant populations. *Plant Methods, 10*, 1–14.
Simpson, J. T., & Pop, M. (2015). The theory and practice of genome sequence assembly. *Annual Review of Genomics and Human Genetics, 16*, 153–172.
Sims, D., Sudbery, I., Ilott, N. E., Heger, A., & Ponting, C. P. (2014). Sequencing depth and coverage: Key considerations in genomic analyses. *Nature Reviews Genetics, 15*, 121–132.
Slade, A. J., Fuerstenberg, S. I., Loeffler, D., Steine, M. N., & Facciotti, D. (2005). A reverse genetic, nontransgenic approach to wheat crop improvement by TILLING. *Nature Biotechnology, 23*, 75–81.
Slatko, B. E., Gardner, A. F., & Ausubel, F. M. (2018). Overview of next-generation sequencing technologies. *Current Protocols in Molecular Biology, 122*, e59.
Smith, R. L., Schweder, M. E., Chowdhury, M. K. U., Seib, J. C., & Schank, S. C. (1993). Development and application of RFLP and RAPD DNA markers in genetic improvement of Pennisetum for biomass and forage production. *Biomass and Bioenergy, 5*, 51–62.
Smýkal, P., Aubert, G., Burstin, J., Coyne, C. J., Ellis, N. T. H., Flavell, A. J., et al. (2012). Pea (Pisum sativum L.) in the genomic era. *Agronomy, 2*, 74–115.
Snyder, M., Du, J., & Gerstein, M. (2010). Personal genome sequencing: Current approaches and challenges. *Genes & Development, 24*, 423–431.
Sohn, J., & Nam, J.-W. (2018). The present and future of de novo whole-genome assembly. *Briefings in Bioinformatics, 19*, 23–40.
Solomon, M. J., Larsen, P. L., & Varshavsky, A. (1988). Mapping protein DNA interactions in vivo with formaldehyde: Evidence that histone H4 is retained on a highly transcribed gene. *Cell, 53*, 937–947.
Sood, V., & Misteli, T. (2022). The stochastic nature of genome organization and function. *Current Opinion in Genetics & Development, 72*, 45–52.
Spaniolas, S., May, S. T., Bennett, M. J., & Tucker, G. A. (2006). Authentication of coffee by means of PCR-RFLP analysis and lab-on-a-chip capillary electrophoresis. *Journal of Agricultural and Food Chemistry, 54*, 7466–7470.
Staden, R. (1979). A strategy of DNA sequencing employing computer programs. *Nucleic Acids Research, 6*, 2601–2610.

Stemple, D. L. (2004). TILLING—A high-throughput harvest for functional genomics. *Nature Reviews Genetics, 5*, 145–150.

Tautz, D. (1989). Hypervariability of simple sequences as a general source for polymorphic DNA markers. *Nucleic Acids Research, 17*, 6463–6471.

Thakur, B., Sharma, S., Sharma, I., Sharma, P., & Zargar, S. M. (2018). Diversity analysis of pea genotypes using RAPD markers. *Legume Research an International Journal, 41*, 196–201.

The ENCODE Project Consortium. (2012). An integrated encyclopedia of DNA elements in the human genome. *Nature, 489*, 57.

Thrash, A., Hoffmann, F., & Perkins, A. (2020). Toward a more holistic method of genome assembly assessment. *BMC Bioinformatics, 21*, 1–8.

Travers, A. (1999). The location of the linker histone on the nucleosome. *Trends in Biochemical Sciences, 24*, 4–7.

van Oeveren, J., de Ruiter, M., Jesse, T., van der Poel, H., Tang, J., Yalcin, F., et al. (2011). Sequence-based physical mapping of complex genomes by whole genome profiling. *Genome Research, 21*, 618–625.

Varma, A., & Shrivastava, N. (2009). The role of plant genomics in biotechnology. In *Biotechnology* (Fundamentals of biotechnology) (Vol. VIII, p. 317).

Visel, A., Blow, M. J., Li, Z., Zhang, T., Akiyama, J. A., Holt, A., et al. (2009). ChIP-seq accurately predicts tissue-specific activity of enhancers. *Nature, 457*, 854–858.

Wang, Z., Gerstein, M., & Snyder, M. (2009). RNA-Seq: A revolutionary tool for transcriptomics. *Nature Reviews Genetics, 10*, 57–63.

Weber, J. L. (1990). Informativeness of human (dC-dA) n·(dG-dT) n polymorphisms. *Genomics, 7*, 524–530.

Weiland, J. J., & Yu, M. H. (2003). A cleaved amplified polymorphic sequence (CAPS) marker associated with root-knot nematode resistance in sugarbeet. *Crop Science, 43*, 1814–1818.

Wilkins, L. G. E., Ettinger, C. L., Jospin, G., & Eisen, J. A. (2019). Metagenome-assembled genomes provide new insight into the microbial diversity of two thermal pools in Kamchatka, Russia. *Scientific Reports, 9*, 1–15.

Winter, P., & Kahl, G. (1995). Molecular marker technologies for plant improvement. *World Journal of Microbiology and Biotechnology, 11*, 438–448.

Wünsch, A., & Hormaza, J. I. (2002). Molecular characterisation of sweet cherry (Prunus avium L.) genotypes using peach [Prunus persica (L.) Batsch] SSR sequences. *Heredity (Edinburgh), 89*, 56–63.

Würschum, T., Boeven, P. H. G., Langer, S. M., Longin, C. F. H., & Leiser, W. L. (2015). Multiply to conquer: Copy number variations at Ppd-B1 and Vrn-A1 facilitate global adaptation in wheat. *BMC Genetics, 16*, 1–8.

Xu, Y. (2010). *Molecular plant breeding*. CABI.

Yandell, M., & Ence, D. (2012). A beginner's guide to eukaryotic genome annotation. *Nature Reviews Genetics, 13*, 329–342.

Zakhrabekova, S., Gough, S. P., Lundh, L., & Hansson, M. (2013). Functional genomics and forward and reverse genetics approaches for identification of important QTLs in plants. *Proceedings of the Azerbaijan National Academy of Science, 68*, 23–28.

Zarrei, M., MacDonald, J. R., Merico, D., & Scherer, S. W. (2015). A copy number variation map of the human genome. *Nature Reviews Genetics, 16*, 172–183.

Zebire, D. A. (2020). Applications of molecular markers in genetic diversity studies of maize. *Nigerian Journal of Biotechnology, 37*, 101–108.

Zhang, W., Sun, Z., Menghe, B., & Zhang, H. (2015). Single molecule, real-time sequencing technology revealed species-and strain-specific methylation patterns of 2 Lactobacillus strains. *Journal of Dairy Science, 98*, 3020–3024.

Zhang, J., Chen, L.-L., Xing, F., Kudrna, D. A., Yao, W., Copetti, D., et al. (2016). Extensive sequence divergence between the reference genomes of two elite indica rice varieties Zhenshan 97 and Minghui 63. *Proceedings of the National Academy of Sciences, 113*, E5163–E5171.

Zheng, G. X. Y., Lau, B. T., Schnall-Levin, M., Jarosz, M., Bell, J. M., Hindson, C. M., et al. (2016). Haplotyping germline and cancer genomes with high-throughput linked-read sequencing. *Nature Biotechnology, 34*, 303–311.

Zhou, Y.-B., Gerchman, S. E., Ramakrishnan, V., Travers, A., & Muyldermans, S. (1998). Position and orientation of the globular domain of linker histone H5 on the nucleosome. *Nature, 395*, 402–405.

Zhuang, J., Zhang, J., Hou, X.-L., Wang, F., & Xiong, A.-S. (2014). Transcriptomic, proteomic, metabolomic and functional genomic approaches for the study of abiotic stress in vegetable crops. *Critical Reviews in Plant Sciences, 33*, 225–237.

Zimorski, V., Ku, C., Martin, W. F., & Gould, S. B. (2014). Endosymbiotic theory for organelle origins. *Current Opinion in Microbiology, 22*, 38–48.

Żmieńko, A., Samelak, A., Kozłowski, P., & Figlerowicz, M. (2014). Copy number polymorphism in plant genomes. *Theoretical and Applied Genetics, 127*, 1–18.

Chapter 4
Transcriptomics

Askim Hediye Sekmen Cetinel, Hayssam Mohamed Ali, and Sadia Ali

Abstract Transcriptomics has become an integral approach in all areas of plant sciences. Advances in plant transcriptomics and new tools and techniques that are taking the ground to excel in this field are the emerging horizons of life sciences. An introduction to plant transcriptomics is provided here, along with a description of the fundamental procedures, tools, and techniques employed in transcriptome analysis. This chapter covers competitive transcriptome analysis, which compares how homologous genes in different species are expressed and the potential of functional genomics and gene regulatory mechanisms.

Keywords RNA-seq · Next-generation sequencing · Expressed sequence tags · Microarray

Introduction

The transcriptome consists of ribosomal RNA (rRNA) that makes up 80–90% of the whole transcriptome, and the rest constitutes 5–15% of transfer RNA, 2–4% of messenger RNA, and a negligible amount that could be 1% noncoding RNA (Lindberg & Lundeberg, 2010). In particular, repetitive-rich intragenic and intergenic regions are called junk or selfish DNA. This part of the DNA has long been thought to be genetically inactive. It has been demonstrated that junk DNA (noncoding DNA),

A. H. S. Cetinel (✉)
Department of Biology, Science Faculty, Ege University, Izmir, Turkey

H. M. Ali
Department of Botany and Microbiology, College of Science, King Saud University, Riyadh, Saudi Arabia
e-mail: hayhassan@ksu.edu.sa

S. Ali
Centre of Agricultural Biochemistry and Biotechnology (CABB), University of Agriculture, Faisalabad, Pakistan

S. Ijaz et al. (eds.), *Trends in Plant Biotechnology*,
https://doi.org/10.1007/978-981-97-0814-7_4

which makes up 98.8% of the genomes of humans and 0.25% of those of prokaryotes, increases with the complexity of an organism (Taft et al., 2007).

The majority of ubiquitous transcripts discovered thus far have only been found in a few cell types, usually mutants, under certain growth conditions and in a limited number of tissues (Jacquier, 2009). As a result, one of the most important goals for comprehending the functional components of a genome is uncovering and analyzing the intricacy of a transcriptome. By examining molecular and cellular components, it could be possible to reveal the intricacy of the gene codes of living things, leading to a better understanding of many biological concerns, including the beginning and course of the disease.

The identification, description, and cataloging of the expression of all gene transcripts in a given cell at a certain phase are the main goals of whole-transcriptome studies. It has enormous potential to assess differential transcript expression in healthy and pathological circumstances and show the right gene structure and splicing patterns (Wang et al., 2009).

Hybridization- or sequencing-based techniques are frequently used for transcriptome profiling. In hybridization-based techniques, complementary probe sequences are bound to fluorescently tagged fragments either on the solid surface or in solution (Okoniewski & Miller, 2006). However, these methods have some drawbacks; for instance, resolution may be limited, or some methods may not produce results with high specificity and high sensitivity (Mantione et al., 2014). However, these methods have significant drawbacks, including the inability to distinguish between isoforms and the consideration of partial transcript structures for genes (Livak & Schmittgen, 2001). Next-generation sequencing (NGS) has enabled RNA sequencing (RNA-Seq) with a significant ability to investigate the transcriptome. These methods for studying the transcriptome enabled the study of complicated mRNAs from particular cells or tissues, resulting in measures of gene expression at a vast scale. However, the data provided by the various approaches have varying applications. Each technology's benefits and drawbacks are now available, and none fully satisfy all experimental demands. These distinctions complement the technologies, and it is crucial to validate any ostensible variations in the quantity of the levels of mRNA using a variety of these methods along with sound experimental design and overall analysis.

In the past 25 years, advances in hybridization technology and tag sequence-based methods have given researchers a thorough understanding of this discipline. However, without a doubt, the completion of next-generation sequencing (NGS) platforms and all of the associated "Seq" applications has fundamentally changed how molecular biologists work. NGS has also been used to evaluate the transcriptomes of both plants that have been sequenced and those that have not yet been. The results have been made available in open resources. It is possible to incorporate data from marker and epigenome analysis, transcriptome and small RNA evaluation, and other analyses to comprehend the underlying mechanisms of plant growth and stress. Plants with altered gene expression transcriptomes offer information about their downstream constituents. Together with forward genetics, these sophisticated reverse genetics methods may be utilized to characterize the functional properties of individual genes.

The chapter's main objective is to deliver a comprehensive overview of the plant transcriptome, the techniques used in transcriptome profiling, its applicability from both a biology and a bioinformatics perspective, and recent developments in transcriptome profiling of plants.

Plant Transcriptome

The "plant transcriptome" refers to the whole complement of messenger RNA molecules generated by a plant cell or tissue at any time (McGettigan, 2013). Plant transcriptomes are an effective resource for estimating varying levels of gene expression and determining the regulatory networks of genes for different functions of important genes. Transcriptome data comprise biological details about the transcriptional activity of genes in a specific cell, tissue, or person (a population of cells) and a pool of samples under certain developmental conditions, environmental factors, or experimental treatments.

In molecular biology, the central idea is the transmission of genomic information from genes to the activity of cells. There are two steps to this process: A subset of RNA known as messenger RNAs, or mRNAs, is subsequently translated into protein by the RNA polymerase enzymes, which first transcribe DNA, the heritable, permanent repository of genetic information, into RNA, a short-lived information carrier. RNA is essential for directing several cellular biological processes and bridging the genome and the proteome. Not all RNAs translate into functional proteins, which is a crucial distinction to note. Other RNAs have structural functions, such as ribosome assembly-related rRNAs, while others are transporters, such as tRNAs, or some are involved in regulatory activities, such as noncoding RNAs and interfering RNAs (Botchkareva, 2017).

In the genome, every gene transcriptional landscape has been made public by researchers from the time of completion of the human genome project to explore more of the functional processes behind variation in phenotypes at a genome-wide transcriptional level. Significant improvements in high-throughput sequencing have made it feasible to quickly generate many genomic and RNA data.

Transcriptome Profiling

The whole assembly of RNA molecules in a cell, a tissue, or an organism is known as the transcriptome. Transcriptomics is most frequently used to study mRNAs, which are coding transcripts. However, it also offers valuable information on the content of noncoding RNAs found in cells, such as rRNA, tRNA, lncRNA, siRNA, and others. Transcriptome analysis uses high-throughput methods to examine the transcriptome and genome total production of the RNA transcript in the cell under specified conditions. Transcription profiling is used to identify patterns of changes in a cell's function as a whole rather than just one or a few genes in many different genres of research in different fields, including disease identification, discovery of various biomarkers, assessment of risk for experimental medicines or ecological contaminants, etc. For instance, different mechanisms in plants or other living organisms are identified as responsible for the response against different environmental stresses. These kinds of findings are made possible by transcriptomics utilizing specific methodologies.

Gene transcription, posttranscriptional alterations, and translation are only a few examples of the many stages at which the complicated process of gene expression is controlled. According to Davuluri et al. (2008), a single gene may produce several transcripts if it has many transcription start sites (TSSs). Additionally, several transcripts with various structures originating from the same gene may be produced due to alternative polyadenylation and alternate splicing of the original RNA (Chen & Manley, 2009). Every mature transcript will eventually code for a distinct protein due to various TSSs. Nonprotein coding RNAs are not translated into proteins and have crucial structural and catalytic roles. For instance, guide RNAs (gRNAs) control RNA editing, small nucleolar RNAs (snoRNAs) control rRNA splicing, tRNAs and rRNAs are essential for translation, miRNAs are engaged in translational repression, and small nuclear RNAs (snRNAs) are involved in the splicing of mRNA (Eddy, 2001). The control of gene expression patterns, alternative splicing, structure of transcript, dynamical regulatory pattern of transcripts in diverse tissues, and deep understanding of the gene regulation in both healthy and diseased states are all revealed by studying the transcriptome (Nagalakshmi et al., 2008). For a thorough knowledge of the operation of biological systems and the vast array of plant species, specific tissue domains and their interconnections must be studied. Functional features, including development, biochemistry, physiology, and growth, eventually result from basic gene expression programs regularly controlled in many organs and tissues. Every morphological level (whole organism, organ, tissue, and cell) exhibits specific molecular circumstances, making spatially defined, high resolution, and high-throughput investigations of tissues essential. This necessity led to the development of numerous techniques that may produce such spatial information.

Techniques in Transcriptomics

Transcriptomics has developed into a well-known study area since the late 1990s when technology started to advance. Early in the 1990s, the first analyses of the complete transcriptome were attempted. Technology breakthroughs have often changed the field of transcriptomics.

Before recording transcripts, RNA must first be extracted from the experimental organism for all transcriptomic techniques. RNA extraction methods involve mechanically disrupting sample cell tissues, deactivating RNase through chaotropic salts, distressing large molecules and genetic molecule complexes, and separating RNA from unwanted biochemical compounds comprising proteins and DNA (Chomczynski & Sacchi, 2006). It is possible to enrich transcripts using poly-A affinity techniques or by depleting rRNA with sequence-specific probes (Zhao et al., 2014). Degraded RNA may impact subsequent outcomes, for instance, enhancement of messenger RNA from deteriorated samples. This will lead to the loss of 5' mRNA ends and inconsistent signals over a transcript's length. Snap-freezing of tissue is routine before RNA extraction, and precautions are made to limit exposure to RNase enzymes after isolation (Chomczynski & Sacchi, 2006).

Sequencing-Based Approaches

In sequence-based methods, expressed sequence tags (ESTs) were the first used. Later,

techniques utilizing short tags, such as serial gene expression (SAGE) analysis and massively parallel signature sequencing (MPSS), were employed along with RNA-seq.

Expressed Sequence Tags (EST)

Putney and colleagues first used expressed sequence tags in early 1980 during insert sequencing from 178 clones produced from a rabbit muscle cDNA library. Later, as high-throughput DNA sequencing technology was developed, it was suggested as an additional reserve for genome sequencing initiatives. Adams and colleagues started the initiative for systematic sequencing in 1991 and developed the term expressed sequence tags (EST) (Adams et al., 1991). It produces a sequence of between 100 and 700 kb using currently available capillary-based sequencing methods. Therefore, the EST represents the associated cDNA clone's partial sequence, which a range of bioinformatic methods may distinguish. The main emphasis of conventional analysis is EST detection by homology searches in preexisting DNA and protein sequence databases. Large-scale EST sequencing has replaced plant genome sequencing as a promising substitute.

Approximately 3.8 million sequences from approximately 200 species comprise the current plant EST collections. They are a useful resource for identifying genes and examining plant metabolism. Many plant-specific EST databases already give users access to sequence information and data mining tools built on the bioinformatics platform. It is possible to preselect genes by searching EST libraries. Additional tools for breeders and researchers are ESt-based molecular markers, including SNPs, SSRs, and indels (Agnieszka & Ski, 2005). ESTs are incredibly helpful for identifying genes and validating gene predictions since they reflect the expressed section of a genome. They, therefore, serve as an affordable substitute for whole-genome sequencing. It is mainly valuable for eukaryote species, whose genomes are often larger and less densely packed with genes than those of prokaryotes.

ESTs are still used to uncover novel genes, especially for those species whose economies of scale prevent full genome sequencing. The quantity of ESTs produced has increased exponentially due to the development of more affordable and quick sequencing technology (Parkinson, 2009). The importance of EST sequencing for plant sciences, both model plants and nonmodel plants, cannot be overstated. All sequencing projects by EST provide a plethora of data that may be explored via in-depth study. The possibility of merging in silico approaches with conventional laboratory investigations and through free access to plant EST databases has sped up our knowledge of the functions performed by a single gene in addition to the general structure and physiology of plants.

Serial Analysis of Gene Expression (SAGE)

Serial analysis of gene expression, or SAGE, uses high-throughput sequencing to deliver a quantitative profile of the genetic expression pattern. In the middle of the 1990s, SAGE, the first sequencing method, was created to assess the transcriptome in terms of identifying and quantifying transcripts, including ncRNAs (Velculescu et al., 1995). Instead of assessing the level of gene regulation, the SAGE method quantifies a "tag" that signifies the transcriptome. The approach is built based on two ideas. First, a brief (10 or 11 base pairs) oligonucleotide sequence tag has sufficient data to differentiate one transcript from another. Genes are identified by utilizing these tags and the relative frequency of their transcript. The second is that a chain of short sequence tags facilitates effective serial evaluation of transcriptions, as SAGE uses consecutive dispensation, where 25–50 SAGE tags are assessed in each track of a DNA sequencer.

A genetic sequence of a certain interval located next to the 3′-most restriction position of a specific restriction enzyme, such as NlaIII, is referred to as a tag for SAGE purposes. The location and sequence of the tag are crucial for identifying the gene from which the tag was obtained. Restriction enzymes release SAGE tags from cDNAs, which are then joined, multiplied by PCR, and sequenced as concatenated fragments. A set of tags with matching numerical counts makes up the SAGE technique's data output, digitally depicting genetic expression within cells. Tentatively, given a random nucleotide distribution across the genome, a sequence segment as brief as nine base pairs can distinguish 49 transcripts, equivalent to 2,62,144 possibilities. This ability seems sufficient to distinguish between all transcripts in higher plants. Two of SAGE's key advantages are the digital output and the identification of novel genetic elements. However, the information produced by producing a short sequence tag may not be sufficient to identify a novel sequence properly. Longer tags used in a modified version of this strategy are more beneficial. Comprehensive sequencing and sequence analysis skills are needed for SAGE. Sequencing must be automated. Novel genes will also differ from known sequences, and it might be challenging to generate a full-length cDNA using a 10-bp oligonucleotide probe.

Massive Parallel Signature Sequencing (MPSS)

Massively parallel signature sequencing (MPSS) is a distinctive signature-based transcription resource for investigating mRNA and small RNA. At Lynx Therapeutics in the early 2000s, Sydney Brenner created MPSS, a bacteria-free library preparation known as mega clone technology. It is currently a standard tool in the molecular biology toolbox. This makes it easier to find gene expression information in exemplary plant species such as Arabidopsis and rice. It is an unrestricted approach that counts the number of individual mRNA molecules to analyze the expression level in a sample and examines each gene's output. Amplification of the products of tagged PCR generated from cDNA results in approximately 100,000 PCR products with distinctive tags for each associated mRNA molecule.

An MPSS dataset's signature sequences (MPSS) has typical sensitivities in the range of a few mRNA molecules per cell. Because MPSS uses longer tags, it can identify novel genes well. It can also find novel transcripts, primarily in species without a whole-genome sequence. This method has also been used to investigate small RNA expression and mRNA transcription. The public may access the MPSS database, which contains expression profiles for several significant crops, such as rice, maize, and soybean. MPSS can quantify genes with very low expression levels. All clones in the library can be sequenced simultaneously, which is called passive parallel sequencing. MPSS has several identified downsides. One is that expression levels cannot be measured for a limited subset of genes because no signatures can be given. Due to sequence quirks, certain genes and transcripts are also inappropriate for the technique.

RNA-Seq

RNA sequencing is a contemporary method for investigating genome-wide transcriptome characterization and profiling by applying high-throughput sequenced data. RNA sequencing has made great advancements in recent years and is now a crucial method for transcriptome profiling. For identifying, categorizing, and measuring diverse sets of RNA molecules synthesized in cells, RNA-seq datasets have been created. Important details about the activated genes, their degree of expression, and the regulatory triggers may be learned from this. This is especially helpful in identifying and mapping noncoding RNA species that control many biological processes. Because of its digital read-out, RNA-seq allows precise RNA quantification at better resolution and a larger dynamic range than earlier approaches, enabling the discovery of weakly expressed transcripts. RNA-seq also provides a general overview of which genome regions are transcribed. It also makes identifying novel genes, exons, and transcript isoforms possible. The functional information about the genome may be derived from a contemporaneous RNA-seq analysis by integrating it with other genomic data sources. To investigate how DNA methylation affects the transcriptome in Arabidopsis thaliana, the researchers combined RNA-seq and bisulfite sequencing.

The main advantages of RNA-seq over earlier gene expression profiling approaches include the capability to find novel genes and their transcripts and the scalability of the technology. RNA-seq has a significant influence, but its full potential is still untapped. RNA-seq has had a significant influence on research in a variety of domains, including genetics, biology, and medicine, where it has established itself as a standard research tool by serving to describe the functional genome and measure RNA across periods and in altering circumstances.

Early RNA-seq methods used Sanger sequencing technology, which was novel at the time but also expensive and low throughput. With the introduction and spread of NGS technology, it is only lately available. RNA extraction, reverse transcription into cDNA, adaptive ligation, amplification, and sequencing are among the processes in the RNA-seq method.

Next-Generation Sequencing

Next-generation sequencing (NGS) was developed in 2004. High-throughput sequencing and assembly were made possible by next-generation sequencing (NGS). Depending on the equipment, NGS may generate billions of short reads simultaneously, typically between 50 and 800 base pairs (bp). Now, it may only take a few days or even hours to create sequence data, as opposed to years in the past. NGS is also called second-generation sequencing technology (Sanger, Maxam, and Gilbert's first sequencing methods comprise the first generation). Due to the lack of a subcloning requirement, NGS is replacing traditional capillary sequencing in numerous applications. In several areas of molecular biology, the large volume of data has altered how issues are approached, opening up intriguing new possibilities in various scientific fields. In particular, this technique has significantly advanced the study of transcriptomics. All forms of nucleic acids may be successfully sequenced utilizing NGS technologies, whether using a whole genome or focused strategy, including DNA, mRNA, and small RNA sequencing, as in common studies. The development of NGS has also made it feasible to sequence certain RNA subtypes on a greater scale, including methylated DNA and short noncoding RNAs (snoRNAs) and long noncoding RNAs (lncRNAs). The rise of next-generation DNA sequencing enabled the expansion of high-throughput sequencing of the whole transcriptome. RNA sequencing (RNA-seq), an application of next-generation sequencing, was published in 2008 in a series of landmark articles spanning multiple species. Next-generation sequencing (NGS) technologies are becoming important tools in plant research. The creation of whole-genome molecular markers, the compilation of the agricultural reference genome, transcriptome sequencing for the study of gene expression, and the search for markers in genes with established functions. Some of these methods have produced successful crop breeding results.

Only a few hundred base pair long short reads are produced by next-generation sequencing. This enables more continuous restoration of genomes with high quantities of repetitive elements and extremely accurate de novo assemblies.

Third-Generation Sequencing

Third-generation sequencing (TGS) was introduced in 2011, almost immediately after NGS started (Van Dijk et al., 2018). It is now feasible to sequence nucleotide molecules (DNA or RNA) before even amplifying the template using PCR for single-molecule sequencing (SMS) and the real-time sequencing capabilities of the newly developed TGS methods. They also make it possible to examine the produced data in real time. This is a notable advancement because TGS can sequence DNA or RNA without amplifying the template, partly because it lessens the biases that PCR added while creating the library. In 2009, Helicos Bioscience introduced HeliScope, a genomic study system that performed single-molecule sequencing by utilizing the fluorescence phenomenon (Pushkarev et al., 2009). This workflow involved library preparation before sequencing, representing the first attempt at single-molecule sequencing.

In summary, this approach entails shearing the template into small DNA fragments (100–200 nt), polyadenylation of these fragments and poly(A) tail production. The oligo-d(T) primers are then mounted on a disposable glass flow cell with 25 channels for sequencing once the tailed DNA has been hybridized to them. Before sequencing, the 30 ends were treated with terminal transferase enzymes to prevent DNA extension. Finally, after including a fluorescent nucleotide, the DNA sequence identification depends on the emission of light identified using imaging.

New third-generation genomic techniques are being used to close the quality gap between pricey short-read assemblies and high-quality reference sequences. When combined with optical mapping or other long-range data, long-read assemblies, which currently outperform short-read assemblies, have the potential to assemble genomes with previously unknown completeness and quality.

This approach streamlined the process of preparing the library by removing the need for PCR amplification steps, but it also had significant downsides in the form of time-consuming sequencing, elevated expenses, significant rates of mistakes, and, finally, the generation of brief reads (32 bp).

Hybridization-Based Approaches

Northern blotting, the RNA gel blot, which involves hybridizing a tagged probe to an RNA target and measuring the subsequent band size and intensity of the signal to check and count the number of expressions, is the ancestor of many contemporary methods. The principle behind many of these techniques is focused probe-target hybridization, but recent developments in genomic technology now make it possible to examine thousands of genes simultaneously. The plant transcriptome is analyzed using various hybridization techniques, including microarray technologies, tilling arrays, and real-time quantitative PCR (qPCR).

Microarray Technology

Microarray is a technique used in laboratories to immediately quantify the expression levels of thousands of genes. High-throughput gene expression analysis and DNA microarray are two terms often used to describe tens of thousands of tiny dots on a microscope slide, each bearing a distinguishable DNA sequence or gene, which can be utilized interchangeably. They are known as DNA microarrays. Many people refer to these slides as DNA or gene chips. When identifying gene expression expressed by a group of genes, the probes are represented by the DNA molecules attached to each slide. Expression analysis has undergone a revolution thanks to DNA microarray. Data are examined for expression patterns that alter in response to various treatments or periods in time. They can be made of short oligonucleotides or entire cDNA clones. Early microarrays resembled a northern blot in that they were constructed using cDNA snippets that had been robotically griddled and fixed on slides for microscopes (Schena et al., 1995).

The profile of gene expression in the sample, such as cells or tissues, quantitatively and qualitatively at a certain time or in a known physiological or pathological state is provided by studying the full transcriptome using microarrays (or DNA chips). Since this profile reflects a sample's innate quality, it may be used for various tasks, including identifying biomarkers, investigating mechanisms of action or signaling pathways, developing novel therapeutic targets, and studying fundamental concepts.

Microarrays can provide only data on targets genuinely characterized by microarray probes, have poor signal resolution, and fluctuate significantly at low signal intensities.

Tiling Arrays

Tiling arrays, a whole-genome microarray based on oligonucleotides, are useful for functional whole-genome analysis beyond simple gene expression profiling. A normal expression microarray or a tiling array, another kind of microarray, fundamentally differs in probe design. The most well-known technique for examining tiling array expression was developed by Kampa et al. (2004). Tiling arrays look for recognizable continuous sequences, such as an unexpressed genomic region. The probe design, particularly whether the probes are gapped or overlapped, affects the high resolution of tiling arrays. Whole-genome tiling arrays may be used to analyze transcription-factor binding sites, map genomic methylation sites, and map genomic polymorphisms. These duties include analyzing genomic regions for transcription, antisense transcription, and alternative splicing (CGH). It may be used to locate a genome region where active transcription was not recognized. Overlapping probes are used in tiling arrays for SNP detection so that every base is checked for mutations. A whole-genome tiling array may use several million oligoprobes. Tiling arrays, for example, have been used to find new long noncoding RNAs and identify spatiotemporal gene expression patterns (Guttman et al., 2009).

The "transcriptional noise" generated by Tilling's ultrasensitive detection capabilities is a problem with its utilization. The method also does not provide areas of interest where the array identifies a distinct beginning or end. Finally, since arrays often provide only chromosome and location numbers, sequencing must frequently be performed separately.

Real-Time Quantitative PCR (qPCR)

In situ hybridizations may be used to analyze the quantitative and qualitative expression of genes in specific tissues. Quantitative real-time PCR (QRT-PCR), which provides swift and consistent results while encompassing a broad spectrum of variability, has yielded solid, quantitative expression data for a single gene (Bustin, 2002; Ginzinger, 2002; Klein, 2002). Using fluorescently labeled oligonucleotide primers or dyes that are unique to a certain sequence in dsDNA, fluorescence signals may be produced. The amplification process may be observed in real time, and the magnitude of the signal corresponds directly to the quantity of PCR product. The starting concentration level of a particular gene transcription in a pattern mix is determined using the amplification curve. One of its main benefits is the enormous dynamic range of QRT-PCR, which can accurately detect transcript levels covering over eight orders of magnitude. qRT–PCR may be performed with one tube per amplification target and parallel control reactions using unlabeled primers and a dye such as SYBR Green. As an alternative, several pairs of control primers, each labeled, each marked with a distinct fluorochrome to enable precise identification, are added to the sample. The next step is to create a pair of fluorescently labeled gene-specific primers, one of which is created. Both methods average the results of reproduced reactions, with the former employing SYBR Green being less costly than the latter.

One of the most fascinating new techniques for measuring single genes uses what are known as polonies or polymerase colonies (Mitra et al., 2003). The amplification results in many polonies, spherical colonies of DNA, dispersed throughout the surface because the PCR products are efficiently fixed. Every individual polony is made from a single template molecule, much like a conventional colony lifts blot, and specific genes or transcripts may be located by hybridizing tagged probes. One may obtain a quantitative estimate of gene expression by calculating the proportion of polonies coming from a certain transcript to the total (Mikkilineni et al., 2004). This method may undergo modifications to monitor RNA splicing or other processes in addition to expression analyses (Zhu et al., 2003).

Spatial Transcriptomics

Spatial transcriptomics is a rapidly expanding area that can quantify tissue structure and architecture at the single-cell or subcellular level (Dries et al., 2021). It refers to efforts to measure the mRNA expression of several genes in the anatomical environment of tissues and cells as "spatial transcriptomics." The mechanistic knowledge of numerous physiological mechanisms within a state of well-being and disease that may be gained from such information cannot be achieved using conventional technology. In extracting biological signals from raw data, computational technique development is crucial. Many strategies have been devised to overcome technological constraints, including spatial precision, gene inclusiveness, detection sensitivity, and potential technical distortions.

Spatial transcriptomics may be applied to describe transcriptional patterning and control in tissues. It may highlight the tissue neighborhoods and regional elements that influence illness.

Methodologies Used for Spatial Transcriptomics Analysis

Imaging-Based Spatial Transcriptomics Techniques

Imaging-Based In Situ RNA Hybridization

In situ RNA hybridization is the foundation of the imaging-based spatial transcriptomics method. While visualizing mRNAs in situ, a method for distinguishing the two different mRNA species must exist. One involves mRNA hybridization to fluorescently labeled gene-specific probes. This spatial transcriptomics technique is known as in situ hybridization (ISH). In imaging-based spatial transcriptomics, transcripts are quantified by in situ hybridization of mRNAs utilizing labeled complementary probes that can be seen under a microscope. Despite its primarily qualitative character, ISH was useful for analyzing the expression patterns of genes in samples taken and entire organisms (Moses & Pachter, 2022). In 1998, the quantitative method known as single-molecule fluorescence in situ hybridization (smFISH), which allows for the visual representation of each labeled transcript as a single spot under a microscope, was developed for imaging probe-tagged transcripts (Femino et al., 1998).

Imaging-Based In Situ RNA Sequencing

In situ sequencing (ISS) of increased mRNAs, a technique utilized in the imaging-based in situ RNA sequencing approach, instantaneously sequences transcripts within a tissue block or segment. Instead of gene-specific probes, ISS methods profile 1–2 nucleotides at a time of primed and amplified transcripts in the tissue. Finally, each transcript may be identified by viewing and recording each base or two-base sequence using a unique fluorophore. The rolling circle amplification (RCA) technique may amplify targeted or untargeted primers while preserving spatial localization. Rolling circle amplification could be selective, making ISH techniques more suited to mRNA detection than ISS approaches.

Sequencing-Based Techniques

Oligonucleotide Array-Indexed (or Spatially Barcoded) RNA Sequencing

On in situ arrays, transcripts from tissue segments could be captured while maintaining their spatial locations. These arrays may be made by imprinting spot barcodes, unique molecular identifiers (UMIs), and poly-T oligonucleotides on commercialized microarray slides to record polyadenylated transcripts, similar to spatial transcriptomics (ST) and Visium technologies. Arrayed oligos with spatially precise indices, such as spotted, self-assembled beads, and in situ produced arrays, have been used as primers in spatially barcoding RNA sequencing techniques. The complementary DNAs (cDNAs) made in situ from tissue sections placed on top of the array include these oligos. The locations of the RNA transcripts are then determined by sequencing the indexed cDNAs.

Contrary to imaging approaches, spatial barcoding techniques are not limited by the diffraction-limited sample of tightly packed molecules in cells. They also do not require specialized end-user expertise or equipment. The limitations of currently accessible oligo arrays, however, allow the suggested approaches to come up short of the tissue dissociative, single-cell RNA sequencing's single-cell resolution and high RNA grab efficiency. A good oligo array should have a few feature-to-feature gaps and tiny features, approximately the size of camera sensor pixels. A homogenous oligo allocation that tissue transcripts may readily obtain (a continuous distribution from now on).

Single-Cell RNA Sequencing (siRNA-Seq)

With single-cell RNA-seq, sometimes called scRNA-seq or transcriptome sequencing, each cell in a sample is counted for the number of RNA molecules it contains. This information, known as the transcriptome, provides a snapshot of the genes actively transcribed when the cells were taken. The most cutting-edge method for examining the variety and complexity of RNA transcripts within individual cells is single-cell RNA sequencing (scRNA-seq) technology. It also displays the characteristics of many cell types and how those cells contribute to the complex organization of tissues, organs, and living things. Since its first discovery in 2009 (Tang et al., 2009), studies based on scRNA-seq have produced much data in a range of fields, leading to exciting new information about the structure and interactions of plant cells.

The initial and paramount phase in performing scRNA-seq has been successfully isolating viable, single cells from the target location. The next step is to lyse isolated individual cells to collect as many RNA molecules as feasible. To specifically analyze polyadenylated mRNA molecules without collecting ribosomal RNAs, poly-T primers are often used. Then, reverse transcriptase produces complementary DNA from poly-T-primed mRNA (cDNA). PCR, nucleotide barcoding, or, in certain circumstances, reverse transcription is followed by more in vitro transcription to further amplify the small amounts of cDNA. Then, using the same techniques for creating libraries, platforms, and genomic-alignment tools as bulk samples, every cell's amplified and labeled cDNA is mixed and sequenced by NGS (Haque et al., 2017).

scRNA-seq has several challenges. The two primary issues are low cell capture efficiency and a lack of RNA material originating from individual cells. Single-cell RNA sequencing (scRNA-seq) is a common method for identifying cell types and trajectories in physiologically heterogeneous data despite its limits in identifying and quantifying poorly expressed genes.

Comparative Transcriptomics

Comparative transcriptome analysis compares how homologous genes in different species express themselves. Comparative transcriptome analysis is crucial for the functional annotation of genes in various plant species because most molecular mechanistic plant investigations have been carried out in model species, such as Arabidopsis and rice. Comparative transcriptomics has become increasingly prominent in genomic research, thanks to the advent of high-throughput technologies such as microarray and next-generation RNA sequencing that have produced a wealth of transcriptome data. Comparative transcriptome analysis seeks to find genes that are conserved across two or more species by comparing their levels of expression (Ruprecht et al., 2016). A comparative transcriptome study compares the expression patterns o

f homologous genes in several species together with their gene sequences. There have been reports of homologous genes being expressed at various embryonic stages, in various tissue types, or under various stress circumstances (Movahedi et al., 2011). This crucially supports the presence of functional variation in homologous genes among species, demonstrating the diversity of expression patterns (Netotea et al., 2014).

Establishing homologous links between proteins in the two species is the initial step in comparing transcriptomes between the two species. The second stage is finding expression data from tests performed on similar subjects or with similar tissue types. The expression patterns of the two datasets were compared in the third phase (Lee et al., 2019). Given that most molecular mechanistic plant investigations have been conducted in model species, principally Arabidopsis thaliana, comparative transcriptome analysis is crucial for plant research (Provart et al., 2016).

Comparative transcriptomics' primary use is to map the biological sample's transcriptional landscape. It considers the discovery of novel transcriptionally active loci and the projected transcription levels of these sites across the genome. More significantly, in addition to protein-coding genes, researchers have discovered transcribed loci in recent biological investigations that contain lncRNAs and other non-coding RNAs.

Comparative Transcriptomics Data Analysis

Five components comprise the generally used pipeline in comparative transcriptomics: identifying possible functional genes, revealing gene functions, and examining the regulatory links between these genes. Data preparation, analysis of the digital gene expression profile (DGEP), investigation of differentially expressed genes (DEGs), investigation of gene set enrichment (GSE), and investigation of the gene regulatory network (GRN), in that order, are the several steps. These five elements are interconnected throughout the pipeline, and the accompanying analyses mostly depend on bioinformatics data handling expertise. Plant comparative transcriptomics has investigated the functional processes and gene regulatory networks involved in anther formation and male sterility (Wan & Li, 2019).

The definition of what is similar is a significant issue when comparing expression patterns on a wide scale between species. Sequence comparison in comparative genomics makes it simple to identify orthologous genes, and various databases now provide accurate data (Sonnhammer et al., 2014). However, expression patterns are significantly more complicated gene-anatomical-temporal-pattern combinations. Thus, it is vital to provide criteria by which to compare different structures.

Comparative Transcriptomics for Functional Genomics and Gene Regulatory Mechanisms

By giving new functions to unidentified genes, functional genomics is a generic way to understand how an organism's genes interact. By comparing the known functions of related genes, it is possible to infer information about the hypothetical function of an unknown gene from its sequence structure. The examination of temporal and geographical patterns of gene expression is necessary for gene function analysis. Analysis of the various qualitative and quantitative alterations of messenger RNAs can provide the most conclusive information regarding changes in gene expression levels.

Comparative transcriptomics makes discovering functional orthologous genes responsible for plant development possible. A genome-wide comparative transcriptome study between Arabidopsis and maize was conducted to identify fundamental biological mechanisms and gene functions active during leaf development. Despite the intricacy of the orthology between such genetically distinct plants, 926 orthologous gene groups, encompassing 2829 Arabidopsis and 2974 maize genes, were discovered. This suggests that the underlying molecular networks have remained constant (Vercruysse et al., 2020).

Under specific comparison conditions, a gene is said to be differentially expressed (DEG) if there is a substantial difference in transcription levels between two sets of samples. In a specific comparison scenario, a gene is regarded as a DEG if the transcription levels between two sets of samples are significantly different. Notably, the DEG notion predominantly captures the expression changes of genes encoding proteins during the initial phases of expression information processing. However, the idea of DEG has been extended to noncoding transcripts, such as the differentially expressed (DE) miRNA and the DE lncRNA, due to the quick advancement of molecular biology and a better understanding of the functional component of the genome. Furthermore, transcriptional changes among treatment and control samples should be referred to as DE transcribed loci or DE loci if both coding and noncoding transcribed gene transcripts are considered in the comparative study of transcriptome data. The discovered DEG set is a functional gene set associated with the examined gene's function, phenotypic variation, stress resistance, or developmental process.

Furthermore, changed transcriptomes are caused by interactions between genes that control gene expression, and comparative transcriptomics studies may reveal novel gene regulatory networks. Function-related genes are frequently expressed together in a cell to create complexes or participate in a similar biological process. Therefore, it is possible to anticipate gene functions using a comparable gene expression pattern.

Gene expression atlases produce comparison data concurrently from many tissues and organs, which can provide knowledge of all regulatory systems required for vegetative and sexual development (Sharma et al., 2012). Even more detailed investigations into a particular tissue type, stage, or occurrence, such as yield, can either assist in clarifying the issue or be used directly for useful purposes, such as developing new kinds (Gandotra et al., 2013). In keeping with this, the transcriptomes of persimmon fruits treated with and without ethanol were analyzed to find the pathways and biosynthetic networks for proanthocyanidins, flavonols that give the fruit a bitter flavor and reduce its market price (Luo & Luo, 2014). From transcriptome data, pathway analysis and regulatory network construction might provide important hints about regulatory checkpoints (Yu et al., 2014). For example, the red color in apples is a desirable characteristic that increases economic worth. In a microarray study of the exocarps of apples, two regulatory genes encoding putative MYB and bHLH TFs were discovered. Red delicious fruits at different phases of fruit development. These genes are responsible for giving apples their red color (Vimolmangkang et al., 2014).

Applications

Although most cells in multicellular organisms share the same DNA and genes, not all genes are actively transcribed in every cell, suggesting that different types of cells exhibit distinct gene expression patterns. These variations play a role in differentiating between health and illness, creating a wide array of differences in physical, metabolic, and developmental traits seen among different cells and tissues.

Through the analysis of transcriptomes, researchers hope to understand where and when genes are activated or deactivated in various cell types and tissues. Counting the number of transcripts in a cell to estimate the degree of gene expression or activity is possible. Furthermore, a complete depiction of the genes that function at different growth phases may be created by considering the transcriptome. Numerous applications of transcriptome analysis have been made in plant-related research.

Identification of Gene Families

Gene families are groups of genes with similar sequences and structures that share a common ancestor. They carry out specialized tasks that vary between and within species. The same ancestor is duplicated and mutated to create a gene family. Additionally, it is stated that family members share the same domains. For instance, WRKY domain-containing WRKY genes are crucial for signal transduction related to plant defensive responses (Phukan et al., 2016). These domains typically possess conserved sequences that readily form stable three-dimensional structures, which dictate their specific functions. *Arabidopsis thaliana* tissue-specific transcriptomes were utilized to identify 66 gene families with a high frequency of "gradient genes," or genes that exhibit a notable expression gradient between tissues. Numerous members of the gradient gene families play various roles in secondary metabolism, cell wall construction and remodeling, hormone and peptide signaling, transcriptional control, and transport between cells (Kramer & Li, 2017).

SPL transcription factors play a crucial role in signal transduction, stress response, and developmental processes in plants due to their widespread distribution. The genome-wide identification, phylogenetic analysis, and expression study of the SPL gene family and its major involvement in salt stress in *Medicago sativa* L. were performed using real-time quantitative RT–PCR. Several model species, including *A. thaliana* and *G. max*, have been used to characterize the SPL gene family (He et al., 2022).

MYBs are one of the largest groups of transcription factors in plants. As a result, members of this family participate in a wide range of activities, such as development and specialized metabolism. Since the initial inquiry focused on *Arabidopsis thaliana* two decades ago, the MYB families of other plant taxa have been examined (Zhang et al., 2021). *Castanea crenata*'s annotated genome sequence and *Croton tiglium*'s transcriptome assembly were checked for MYBs (Pucker, 2022). The identified GmPYL-PP2C-SnRK2 gene families in soybean contain START-like, PP2Cc, and S TK domains. The expression patterns of these genes differ between types and depend on the tissue in which they are expressed (Zhang et al., 2020). Transcriptional investigation of the cotton plant *Gossypium hirsutum* L. identified the GhEXO gene family, and GhEXO7 At promotes plant growth and development through brassinosteroid signaling (Li et al., 2021). Three Brassica plant 2OGD superfamily genes were identified via genome-wide identification and expression profiling by qRT–PCR (Jiang et al., 2021).

RNA G-Quadruplex Analysis

Genetic cis-regulatory elements control gene expression. These components can be found in DNA and RNA in various structural configurations. Among these structural forms is the G-quadruplex (G4). G-rich sequences, constructed based on sets of four hydrogen-bonded guanine bases, result in G-quadruplexes (G4), which are nonhelical secondary structures of higher order for DNA and RNA. G4 production has been linked to genomic instability, hereditary disorders, and the advancement of cancer based on its recognized roles in chromatin remodeling and gene regulation. The genomic distribution of G4 motifs in and around genes has stimulated research into the roles of these motifs as cis-acting regulatory factors.

In vitro, RNA G-quadruplexes, complex RNA structures, may be formed from guanine-rich sequences. RNA G-quadruplexes are thought to occur in vivo and are connected with significant biological importance due to their great stability. The RNA structure's in vivo folding is closely related to its function and heavily reliant on the cellular environment. The intricacy of live cells may lead RNA folding in vivo to differ significantly from that in vitro. Although many complex RNA structures have been extensively investigated in vitro, little is known about folding in vivo.

Since they are sessile eukaryotic creatures, plants have independently evolved their regulatory systems. For instance, the most prevalent plant ion, cellular K+, is essential for plant growth and stress response (Wang et al., 2012). It is believed that plants tend to have more adaptable RG4 in vivo, given the significance of K+ in impacting RG4 folding and their capacity to maintain K+ equilibrium inside cells. Additionally, recent research on certain G-rich sequences with in vitro folding ability and other studies have revealed that plants may prefer RG4s. Whether RG4s exist in vivo in plants has not been answered and will likely never be.

The in vivo folding condition of G-rich areas in plants was investigated. In the Arabidopsis transcriptome, a profile of the topography of the G-rich areas with the capacity to fold into RG4s in vitro was first created. Additionally, it exposed the distinct RNA structural characteristics of these areas. It was discovered via chemical structural probing that Arabidopsis and rice had robustly folded G-rich areas with in vitro folding potential. Additionally, the proof of RG4 synthesis is sufficient to control gene expression, affecting plant development. These studies collectively identified RG4s as crucial plant growth regulators, providing the first direct evidence of worldwide RG4 synthesis in live eukaryotic cells.

Analysis in Organogenesis

For the molecular characterization of gene regulatory networks and biosynthetic pathways involved in plant development, transcriptome analysis offers an effective method to find genes or groups of genes that encode enzymes or transcription factors involved in various morphological pathways (Ebenezer et al., 2019; Pal et al., 2018). The transcriptomes of a super hybrid strain of rice (*Oryza sativa*) were compared to the transcriptomes of its two parents in a large study from the Beijing Genomics Institute, China (Bao et al., 2005). Several genes showed differential expression in the hybrid's three target tissues. The bulk of the upregulated genes enhanced carbon and nitrogen utilization, photosynthesis in foliage, absorption of nitrogen in roots, and expeditious development in both root systems and panicle growth.

According to a genome-scale transcriptome study of Populus shoot organogenesis, plant cell organogenesis's main regulatory processes occur at the beginning of dedifferentiation. The regulatory circuits exhibit the combined impacts of transcriptional regulation and hormone signaling and the resulting alterations in the light environment brought on by dedifferentiation (Bao et al., 2009).

High-throughput comparative transcriptome studies have been effectively used in recent years to pinpoint potential genes involved in the variation and growth of various tissues in various plant species concerning their temporal and spatial expression patterns. For instance, 79 tissue samples were used to create a gene expression atlas to study the development of maize roots (Stelpflug et al., 2016), maize embryo, endosperm, and seed RNA sequencing identified new transcripts that may be important in the development of seeds and embryos (Chen et al., 2014; Yi et al., 2019). According to a comparative analysis of the transcriptome of strawberry achenes and receptacles across four ripening stages, ethylene may have a significant role in strawberry receptacle ripening (Sánchez-Sevilla et al., 2017). A total of 1077 potential genes involved in *Miscanthus lutarioriparius* rhizome formation were discovered by transcriptome analysis of five organs (Hu et al., 2017). An RNA-seq study was performed to gain insight into the transcriptome throughout organogenesis (Puvvala et al., 2019). A total of 2075 transcription factors (TFs) presumably involved in floral development were identified by transcriptome analysis of the two phases of the *Zingiber zerumbet* flower (Zhao et al., 2020).

A transcriptome dataset specific to bermuda grass (*Cynodon dactylon* L) was generated to gain a deeper comprehension of the characteristics of its three stem types.

Analysis of Epigenetic Regulation

Epigenetic modifications induce alterations in gene regulation. Small RNA expression, chromatin modification, and some persistent protein structural modifications (prions) are all part of the field of epigenetics (i.e., histone tail modifications, such as acetylation, methylation, ubiquitination, phosphorylation, and DNA methylation). Methylation-sensitive restriction enzymes, antibodies against 5-methylcytosine, and bisulfite transformation are common techniques used to research epigenetics. These have been combined with microarray-based techniques to enable the study of DNA methylation over the whole genome (Buck & Lieb, 2004). NGS technologies have emerged as key tools when a substantial amount of sequence read coverage is needed for ChIPseq. NGS provided a breakthrough for epigenetics research. This paper thoroughly investigates epigenetic modifications specific to organs in maize and their impact on mRNA and sRNA utilizing Illumina technology (Wang et al., 2009). It was shown through an investigation of methylated areas in the genetic makeup of tomatoes using methods that have been merged with the conversion of bisulfite and subsequent sequencing using the Illumina platform that hormone therapy and epigenetic regulation work together to govern tomato fruit ripening (Zhong et al., 2013).

MicroRNAs (miRNAs), short endogenous RNAs (25 nt), have become important posttranscriptional regulators of gene expression in eukaryotes. They seem to be the main controllers of growth and different stress reactions. Most miRNAs have strong evolutionary conservation and complementarity with their target mRNAs. These locations are more commonly found in coding regions in plants than in the 5′ untranslated region. MiRNAs may be recognized and verified using a variety of methods. In the case of rice, quick and economical methods for in silico prediction that depend on conserved sequences and secondary structures are quite successful (Bonnet et al., 2004). Building miRNA libraries and then cloning and sequencing them is another approach. The modest expression of these molecules, which changes based on tissue type and period, limits this technique. Illumina technology has been used for radish (*Raphanus sativus* L.) roots to find miRNAs that could respond to cadmium exposure and the genes they target (Xu et al., 2013). Transcriptome analysis was utilized to find circular RNAs using Pacific Biosciences' SMRT technology. Due to their dual functions as competing endogenous RNAs and positive and negative regulators of their parent genes, these molecules play a substantial role in the operation of miRNAs and transcriptional regulation (Lu et al., 2015). According to recent findings, long noncoding RNAs (lncRNAs) serve as a crucial regulatory mechanism. No protein product encoded by these RNA molecules longer than 200 bp exists.

Extraction of Molecular Indicators

Numerous molecular markers exist, but the primary markers used most often are single-nucleotide polymorphisms (SNPs) and simple sequence repeats (SSRs). Simple sequence repeat (SSR) markers are di-trinucleotide repeats, with a core length of one to six base pairs. The variation in tandem repeat counts was thought to cause these markers' high polymorphism. SSRs have been extensively employed in genome map creation, research on diversity in genetics, and mapping of genes. NGS was initially only used to mine molecular markers in the rice and Arabidopsis model species. Even for species that lack genomes used as point references, such as red clover (Ištvánek et al., 2017), common bean (Cortés et al., 2011), and durum wheat (Trebbi et al., 2011); molecular markers have gradually been found over time. However, the rate of errors in sequencing methods due to inadequate reference genomes, repetitive sequences, or sequencing errors continues to limit the mining of genetic markers in many other commercially significant species. As a result, coverage-based consensus calling—a novel method for finding SNPs in chickpeas (*Cicer arietinum* L.)—was created by Azam et al. (2012) (CbCC). Maq, BowTie, Novoalign, and SOAP2 are its four freely downloadable local alignment tools. NGS provides several methods that can concurrently carry out genotyping and genome-wide SNP identification in a single step. The most popular genotyping techniques use restriction enzymes to record condensed genome representation (Miller et al., 2007).

SSR markers were used by Kapoor et al. (2020) to examine the genetic diversity of asparagus cultivars growing in various regions of northwestern India. There were 122 amplified alleles, with an average of 5 alleles per marker and allele counts ranging from three to eight. There are just two separate groupings of *Argylia adscendens* accessions. This demonstrates this species' extensive genetic basis compared to other species, according to a genetic diversity study of asparagus variations. These findings will greatly benefit the preservation of Asparagus species and hybrid breeding. Using Illumina paired-end sequencing technology, Bhandari et al. (2020) created a novel microsatellite marker for *Salvadora oleoides*.

Plant Breeding

Transcriptomic information may identify a genotype that performs better than its neighboring, related, or parental genotypes and reveal the phylogenic and evolutionary relationships between genotypes. These can help us gain a better grasp of the plant world and point us in the direction of plants with higher potential yields or unique agronomic characteristics. A method that combines genomics, transcriptomics, and bioinformatics tools reveals regulatory sequences and enables the identification of new genes. On the other hand, they make enormous sets of genetic markers, such as SNPs, available and link QTLs to gene identification and chromosomal location. The candidate genes are identified by gene expression, another product of genomics and transcriptomics, as the genes engaged in certain regulatory pathway processes associated with a trait related to agronomy. All these outcomes and findings could be beneficial for researchers to uncover genotypes that are more suited for selecting plants. The advancement of next-generation sequencing (NGS) technology, which can generate a vast quantity of genomic data, has enabled the mass sequencing of genomes and transcriptomes. The contribution of NGS technologies greatly enhances our understanding of how genes behave and control biological processes and molecular activities in a living creature. It includes how gene-based information resides in a cell. It is brought on by their quick creation and analysis and comparatively inexpensive cost. Studying gene expression has emerged as a vital technique for plant breeding and discovering essential genes imparting defense mechanisms against biotic and abiotic challenges, made possible by NGS technologies and RNA-seq. RNA-seq was used in a study on the wheat disease *Puccinia striiformis* f.sp. tritici Erikss to identify genes producing effector proteins that may be relevant for developing wheat cultivars immune to the particular pathogen (Garnica et al., 2013). Using 454 sequencing, analysis of the transcriptome of cucumber plants (*Cucumis sativus* L.) unveiled numerous genes associated with the stages of development (Ando et al., 2012). Methodologies such as microarray and Roche technologies were utilized by Nigam et al. (2014) to determine the genetic factors and by-products linked to cotton fiber quality.

Differential Gene Expression (DE) Analysis

One interesting method for finding plants with a preferred attribute for agronomy is differential expression (DE) analysis, which uses a DE profile spanning the whole-gene/transcript set expressed differently in the comparative conditions. Pairwise comparison of plants using differentially expressed genes and transcripts (DEGs) is a well-known and widely used method in transcriptomics (i.e., two opposing characters—vulnerable versus tolerant, etc.). Important traits in the crops studied, such as stress tolerance (Fracasso et al., 2017; Jazayeri et al., 2015), pathogen resistance, pest resistance (Gong et al., 2015; Li et al., 2016), biosynthesis of metabolites, such as alkaloids and photosynthesis efficiency, and gene expression profiles, have been analyzed by whole-transcriptome sequencing and RNA-Seq (Ding et al., 2015). In the next stages, plants with the same expression pattern were tested for characteristics using a list of candidate genes retrieved from the DEGs. Since it has been demonstrated and confirmed that a set of genes encompasses functional and regulatory elements active within a specific environment, the benefit of this strategy is that it uses the whole set of genes rather than just one gene.

Studies based on sequencing are consistently outnumbered by those using DNA microarrays. This contrast indicates that DNA microarrays continue to be employed to measure changes in gene expression, even when the publication of research project outcomes is delayed (Goralski et al., 2016).

Spatial Gene Expression Pattern Analysis

Morphogenesis, cell differentiation, and cell interaction are the three main processes responsible for the growth of plants and other multicellular eukaryotes. The fates of cell clusters, spatial organization, and collective development are coordinated during these processes by differential gene expression that is carefully controlled. Hence, a thorough analysis of gene expression patterns is crucial for comprehending the complex regulatory mechanisms that govern the development of plants. Regardless of their significance, scientists currently know little about how cells differentiate in whole organisms due to the technological challenges involved in separating cells from tissues for transcriptional profiling at the individual level. Recently, several methods known as "spatially resolved transcriptomics" have undergone development to investigate gene expression patterns in tissue represented in two dimensions (2D). These techniques use barcoded oligo-dT microarrays, targeted sequencing, or fluorescence microscopy. Only barcoded oligo-dT microarrays and single-molecule fluorescence in situ hybridization (smFISH) have been used on plant tissues thus far (Giacomello & Lundeberg, 2018).

Abiotic Stress

Physical or chemical factors such as saline stress and harsh climate conditions such as temperature and drought are the main causes of abiotic stress. The study of how abiotic stress alters gene expression in medicinal plants has found use in transcriptomics. This helps to discover the genes influencing medicinal plant growth, development, and resistance to environmental stresses. Transcriptomics has the potential to provide thorough genome-wide expression patterns under stressful conditions, spanning the whole range of transcriptional activity, helping to elucidate the complex regulatory network driving plant stress response and adaptation.

Salinity

Stresses caused by abiotic factors pose a severe danger to plant development and yield. Salt stress is an abiotic stressor often considered a severe danger to crop productivity. The negative impacts of salt stress on crops include ion toxicity, nutritional shortages, and oxidative and osmotic stressors. Integrated physiological, cellular, molecular, and metabolic reactions enable plants to tolerate salt. Salt tolerance is regulated by genes coding for transcription factors, ion transporters, protein kinases, and osmolytes. Multiple genes regulate the quantitative characteristics of salinity tolerance. Therefore, to enhance the salt tolerance network through genetic engineering, it is vital to identify the fundamental elements that underlie it.

Comparing RNA sequencing to earlier techniques for studying the transcriptome provides a substantially more accurate measurement of transcript quantities and isoforms than earlier transcriptome approaches. Recent studies used RNA-sequencing technology to examine the transcriptome profile of bread wheat shoots growing in salty environments. Shoot expression profiling was used to compare the salinity-tolerant mutant of *Triticum aestivum* with its vulnerable wild type. The results showed that the mutant further upregulated several genes related to salt tolerance. Additionally, they proved the significance of the oxidation–reduction (redox) balance for salt tolerance and identified a unique salt stress response pathway known as "butanoate metabolism" (Xiong et al., 2017). In reaction to salt pressure, Mahajan et al. (2017) employed RNA sequencing to create a transcriptome profile of flag leaves in the salt-tolerant cultivar Kharcha. They demonstrated that various biological functions were associated with the genes that reacted to salt stress (Mahajan et al., 2017).

Drought

Drought badly impacts numerous significant crops in terms of growth and output. The hormonal balance in plants, which is crucial for stress tolerance, is known to be disturbed by drought (Ullah et al., 2018). A characteristic of a plant's reaction to drought is an increased buildup of abscisic acid (ABA), which regulates stomatal closure to reduce transpiration in arid conditions. Transcriptome sequencing technology has emerged as a crucial tool for examining the molecular underpinnings of drought tolerance in plants and for identifying the genes responsible for it. RNA sequencing (RNA-Seq) may already offer extensive information on DEGs, transcript architecture, novel transcripts and isomers, alternative splicing, and allele-specific expression, among other topics (Wang et al., 2009). It has been demonstrated that comparing the transcriptomes of cultivars with considerably varying stress tolerances is an efficient method for examining the molecular stress responses in a particular crop.

Tang et al. (2013) used Roche's 454-GS FLX System to conduct RNA-seq research on olive, which thrives in arid or dry settings. Similar genes for drought tolerance were discovered when Illumina technology was employed to sequence the transcriptome of red clover. Consequently, three metabolites (pinito, proline, and malate), the quantities of which rose in leaves experiencing drought stress, were detected (Yates et al., 2014). Cuckoo (Cai et al., 2019), Yerba mate (Acevedo et al., 2019), cotton (Hasan et al., 2019), and other agricultural plants, including lentils (Hou et al., 2019), buckwheat (Morgil et al., 2019), and millet (Zhang et al. 2019a, 2019b), have all been effectively studied using RNA-Seq. Due to these studies, we have a wealth of useful knowledge on plant transcriptional resistance to drought stressors.

Temperature Stress

One of the crucial components in the environment that influences plant expansion and progress is temperature. Plant development is greatly influenced by daily and seasonal variations in surface temperature, which enables plants to grow and reproduce at opportune periods (Jenkitkonchai et al., 2021). The severe changes brought on by global warming have caused all living forms, especially sessile creatures such as plants, to adapt to rapid changes in temperature or risk extinction. This is in addition to the substantial negative impact on crop species in terms of decreased yields (Zhao et al., 2017).

Low Temperature

Low temperature is a primary ecological condition restricting plant production, maturation, and geographic distribution. Crop yield may be somewhat reduced due to the influence of cold temperature stress, which includes chilling stress (at 10 °C) and freezing stress (at 0 °C). Plants' reactions to cold temperatures depend on both their developmental stage and the length of their exposure. Low-temperature exposure alters plants' physiological and molecular makeup, resulting in metabolic diseases. Transcription factors, protein kinases, small noncoding RNAs, and enzymes in biosynthetic processes were among the genes whose expression changed as a reaction to exposure to cold temperatures, as found by transcriptome analysis. These genes manage target and downstream gene expression, protect cells against low-temperature harm and regulate their expression (Yang et al., 2018; Shen et al., 2014).

The full-length transcripts produced by the NGS method fully characterize the *Medicago falcata* transcriptome and may help discover additional genes associated with stress at low temperatures that are unique to *M. falcata* (Cui et al., 2019). Req-seq has been used to analyze transcription in the study "Expression and Function of Key Genes in Pepper Under Low-Temperature Stress" and "Cold Acclimatization" (Tang et al., 2022).

High Temperature

Abiotic stress, known as heat stress, has been extensively investigated in plants. Multiple factors can affect how a plant develops. As a result, there is a reduction in the overall amount of time needed for light perception and storage of chemical buildup in the plant.

Crop and model species abiotic stress functional genomics investigations reveal a developing image depicting shared and distinct regulatory transcriptional mechanisms (Sreenivasulu et al., 2007). Transcriptome analyses of crop growth have explored the consequences of prolonged exposure to elevated temperatures (Yamakawa et al., 2007). However, Qin et al. (2008) used oligonucleotides or microarrays to monitor the effect of short-term heat stress on wheat seedlings. Consequently, there is limited understanding regarding the initial stages of cereal grain reactions to heat stress.

Transcription factors (TFs), pentatricopeptide repeat-containing proteins, ubiquitin-protein ligases, Hsps, and other genes were all significantly altered by heat stress in Chieh-Qua (Wang et al., 2019). A thorough examination of the genes expressed in Korean fir (*Abies koreana*) during heat stress was performed using transcriptome analysis. A transcriptome study of the maize (*Zea mays* L.) inbred line B73 identified putative TFs that responded to high-temperature stress at the seedling stage (Qian et al., 2019).

Biotic Stress

Response to Pathogens

A plant's initial line of defense against pathogen invasion is determined by its capacity to recognize and react to disease-associated molecular patterns (PAMPs). Hosts may identify various possible pathogens for each PAMP because they often contain structural components that have been conserved throughout evolution (or epitopes within these compounds) (Zipfel et al., 2006).

Response to the Insects

Every plant is vulnerable to attacks from outside creatures, such as diseases and insects. Plants need various defense mechanisms to successfully defend themselves against invasion or attack by herbivorous insects. Many plant–insect interactions have been linked to the signaling chemicals ethylene (ET), jasmonic acid (JA), and salicylic acid (SA). After being attacked by herbivores, *A. thaliana* produces different amounts and timings of SA, JA, and ET that are considerably different. Additionally, genes associated with stress are overexpressed. It has been discovered by analyzing the transcriptional profiles that there may be a significant overlap between infections and insect-induced alterations of various assault strategies (De Vos et al., 2005). Hub genes that control cotton resistance against whitefly invasion include WRKY40 and transport proteins. The MPK-WRKY-JA and ET pathways were reportedly suppressed due to virus-induced gene silencing (VIGS), which increased vulnerability to whitefly infestation (Li et al., 2016). Another study found that the expression of the phenylpropanoid pathway-related gene and the accumulation of flavonoids were both boosted when the transcription factor AtMYB12 from *A. thaliana* was expressed in tobacco. AtMYB12 controls numerous pathways, but flavonol production is one of them. Increased accumulation of rutin aided the tobacco transgenic lines in developing immunity against the insects *Spodoptera litura* and *Helicoverpa armigera*. Through the implementation of artificial microRNAs, the production of flavonoids was suppressed, leading to a diminished capacity of the AtMYB12 tobacco plant to combat insects (Misra et al., 2010).

Symbiotic Relations

Rhizobia and legumes can coexist in symbiotic relationships at and below the species level, and these relationships are quite specialized. Legume-rhizobial symbiosis requires a variety of signal exchanges and regulation strategies between host plants and microbial symbionts for its commencement and development. NodD, a regulator of the LysR type that initiates transcription of the structural nodulation genes in rhizobia, is expressed as a result of plant-derived flavonoids. In several model species, global responses of plant cells to bacterial infection have been studied using next-generation sequencing techniques. To be more precise, RNA-seq transcriptional profiling enables the identification of every gene that exhibits differential expression during symbiosis. Citrus is one of the most significant fruit crops in the world, and its perennial woody plants heavily rely on arbuscular mycorrhizal symbiosis (AMS) to obtain the vital nutrients they need from the soil. When the arbuscular mycorrhizal (AM) fungus *Glomus versiforme* mycorrhized the roots of *Poncirus trifoliata*, RNA sequencing was employed to detect genes with distinct expression patterns. Comparative transcriptome studies with four herbaceous model plants assessed their conservation. The arbuscular fungus Rhizoglomus irregularly colonized the roots of sunflower plants, and the RNA-seq method was employed to assess the variance in the expression of genes during the early and late phases of mycorrhizal colonization.

Alfalfa expresses considerably diverse groups of genes in reaction to disease by several strains of rhizobia at below-species levels, according to plant transcriptome analyses (i.e., biotype or strain). Medsa002106, genes encoding nodules, NCR peptides, and proteins in the NBS-LRR family are among the candidates for the genes mediating the particular interactions. RNA-Seq was used to determine the transcriptional patterns of mycorrhizal *Sesbania cannabina* shoots and roots under salt stress. It was discovered that mycorrhizal *S. cannabina* coexpressed modules, networks, and hub genes in response to salt stress using weighted gene coexpression network analysis and pairwise comparisons. Transcriptome profiling revealed variance in gene expression concerning secreted proteases and distinctly specific gene repertoires linked to the Lactarius-pinus symbiotic relationship (Tang et al., 2022). *Zizania latifolia* is a perennial aquatic vegetable that forms inflated gall formations associated with the fungus *Ustilago esculenta* (part of the Basidiomycota family Ustilaginaceae). According to transcriptome analysis, *Zizania latifolia* gall formation by *Ustilago esculenta* is induced by a symbiotic mechanism (Li et al., 2021). Nine plant-associated bacteria from each of the main plant phyla were monocolonized on Arabidopsis leaves, and the cotranscriptomes of the plants and bacteria were analyzed. Plant-commensal couples' cotranscriptome signatures were similar and different.

In Disease Resistance Analysis

The complex regulatory network controls *pumpkin powdery mildew*. Gene expression profiles were examined from pumpkins affected by powdery mildew by analyzing the helix loop transcription factor and the mildew locus, heat stress transcription factor. In types resistant to powdery mildew, several genes were shown to be highly elevated or downregulated (Guo et al., 2018). Additionally, genes linked to wheat-resistance Thinopyrums to powdery mildew have been discovered. Analysis revealed that 12 genes showed high expression. The majority of the proteins that these single genes encoded were kinase enzymes, protease enzymes, synthase enzymes, and proteins involved in signal transduction, which may be crucial in the fight against powdery mildew.

Additionally, threefold transcriptome analyses of foliage from the wheat line vaccinated against pathogens such as powdery mildew and stripe rust were carried out. These disease-resistant wheat cell lines have shown various defense mechanisms to resist diseases. According to the comparative transcriptomics analysis, there were different gene expression patterns in reaction to different infections by different pathogens, such as powdery mildew and stripe rust. The pathogen and the wheat line share some microbial genetic material that can be exploited to defend against infection (Hou et al., 2015).

Transcriptomics Analysis in Economic Plants

Although some early data on the gene expression of such plants were accessible from microarray systems, NGS has proven to be a valuable technique for studying the transcriptomes of nonmodel plants with major economic importance. The development and stress of nonmodel plants have been the subject of useful research in several areas, including soybean (Le et al., 2012), cotton (Jiao et al., 2013), "Suli" pear (Liu et al., 2012), grape berry (Sweetman et al., 2012), Chinese cabbage (Wang et al., 2012), and *Jatropha curcas* (Jiang et al.). It will be possible to track the development of specific organs or developmental stages, such as the identification of the floral sex of cucumbers (Wu et al., 2010), estimation of the hormonal state of apple trees (Krost et al., 2012), and comparison of the cotton transcriptomes of two domesticated species (Lacape et al., 2012). The DEGs in the *Artemisia annua* developmental stages displaying varying amounts of the antimalarial phytomolecule artemisinin may be involved in its production (Nair et al., 2013). In addition to the comparative study of tissues and genotypes, transcriptome profiling and gene functionality controlling developmental processes may be utilized for directed research to find causative genes implicated in certain mechanisms and networks, such as secondary metastatic disease. This was demonstrated by the transcriptome of the Chinese medicinal plant Dendrobium officinale, which has many alkaloids, and revealed the genes associated with alkaloid production (Guo et al., 2013). Metabolic engineering may be influenced by information on the biosynthesis of critical chemicals in commercially significant plants. The promise of NGS is maximized by turning transcriptomics toward commercially significant plants that lack genomic information. This knowledge's true "fruits" will manifest when used for plant breeding to produce better kinds. Induction of flowering in sugarcane under controlled photoperiodic conditions: transcriptomic analysis of changes in gene expression (Manechini et al., 2021). A transcriptome investigation of potato (*Solanum tuberosum* L.) roots, shoots, and leaves revealed several consequences of nitrogen buildup and metabolism (Guo et al., 2022). Transcriptomics data are used to study genes involved in the plant's response to cucumber green mottle mosaic virus infection (Slavokhotova et al., 2021).

Advances in Plant Transcriptomics

Through the availability of genome and transcriptome sequences, recent developments in transcriptomics technology shed light on the obscure intergenic areas between protein-coding genes that had previously been labeled as transcriptional "noise," "junk DNA," or experimental artifacts. Genome assemblies include information on genes, transposable elements, and the overall structure of the genome. In-depth bioinformatic research, such as genome-wide analysis of gene families, is made possible and supported by the accessibility of a reference genome sequence for a species. The biggest problems after creating a genome sequence are gene prediction and creating functional annotations. Even if these techniques have significantly improved over the past several years, it is still advantageous to incorporate outside cues such as RNA-Seq reads. Resequencing can evaluate accession diversity if a species' high-quality genome and annotation are available. This aids in identifying small structural variations, such as insertions and deletions, and larger ones, such as inversions and transpositions. It takes sophisticated bioinformatic methods, many of which have been created recently, to identify these variations. The genetic mapping of characteristics can make use of sequence variations. Several mapping-by-sequencing techniques were created to identify the underlying genes for pertinent features in crops. These genomic strategies are supplemented by various transcriptome techniques, with RNA-Seq technology dominating the field. The associated cDNA molecules are sequenced to determine the abundance of the transcripts. RNA-Seq readings can be used for gene expression analysis or transcriptome assembly, for example, to compare the abundance of certain transcripts in various tissues, circumstances, or genotypes.

Single-cell RNA sequencing (scRNA-seq) has been quickly and enthusiastically adopted, proving that it is much more than another method for performing transcriptome research. To suggest that the development of scRNA-seq is changing the specifics of whole-transcriptome pictures from a tissue to a cell is not an exaggeration. It is now feasible to mine heterogeneity between different tissue types and within individual cells as never before, thanks to this game-changing technology. This makes it possible to identify uncommon and novel cell types more quickly, characterize several cell types and states simultaneously, and comprehend their functions in biological processes more thoroughly. However, the full potential of scRNA-seq applications is still waiting to be discovered. This is especially true for the study of plants, where single-cell transcriptome profiling is still in its infancy and faces many intriguing difficulties.

Spatial transcriptomics is a fast-growing area. Although single-cell and spatial transcriptomics investigations are increasingly common in the animal world, they have recently begun in the plant area. To obtain spatial single-cell resolution in plant tissues and expand the field of spatial transcriptomics to a wide range of species, from crops to forest trees, a new mixture of scRNA-seq or snRNA-seq whenever cell dissociation is not achievable is a potential approach. Due to the separation of the geographical placements of the many cell types prevalent in tissue, our understanding of cellular colocalization and cell–cell contact has significantly increased. Due to climate change and the increasing requirements of an expanding population in terms of both energy and food resources, the plant scientific community may now provide novel answers to pressing problems. This new information illuminated the intricate biological processes involved in plant development and reactions to abiotic and biotic influences outside the plant (Giacomello, 2021).

References

Acevedo, R. M., Avico, E. H., González, S., Salvador, A. R., Rivarola, M., Paniego, N., Nunes-Nesi, A., Ruiz, O. A., & Sansberro, P. A. (2019). Transcript and metabolic adjustments triggered by drought in Ilex paraguariensis leaves. *Planta, 250*(2), 445–462.

Adams, M. D., Kelley, J. M., Gocayne, J. D., Dubnick, M., Polymeropoulos, M. H., Xiao, H., Merril, C. R., Wu, A., Olde, B., Moreno, R. F., & Venter, J. C. (1991). Complementary DNA sequencing: Expressed sequence tags and human genome project. *Science, 252*(5013), 1651–1656.

Agnieszka Kisiel, & Ski, J. P. (2005). Expressed sequence tags and their application for plant. *Research, 27*(2), 157–161. https://doi.org/10.1007/s11738-005-0019-1

Ando, K., Carr, K. M., & Grumet, R. (2012). Transcriptome analyses of early cucumber fruit growth identifies distinct gene modules associated with phases of development. *BMC Genomics, 13*, 518.

Azam, S., Thakur, V., Ruperao, P., Shah, T., Balaji, J., Amindala, B., Farmer, A. D., Studholme, D. J., May, G. D., Edwards, D., Jones, J. D. G., & Varshney, R. K. (2012). Coverage-based consensus calling (CbCC) of short sequence reads and comparison of CbCC results to identify SNPs in chickpea (Cicer arietinum; Fabaceae), a crop species without a reference genome. *American Journal of Botany, 99*, 186–192.

Bao, J., Lee, S., Chen, C., Zhang, X., Zhang, Y., Liu, S., Clark, T., Wang, J., Cao, M., Yang, H., Wang, S. M., & Yu, J. (2005). Serial analysis of gene expression study of a hybrid rice strain (LYP9) and its parental cultivars. *Plant Physiology., 138*, 1216–1231.

Bao, Y., Dharmawardhana, P., Mockler, T. C., & Strauss, S. H. (2009). Genome scale transcriptome analysis of shoot organogenesis in Populus. *BMC Plant Biology, 9*(1), 1–15.

Bhandari, M. S., Meena, R. K., Shamoon, A., Saroj, S., & Pandey, S. (2020). First de novo genome specific development, characterization and validation of simple sequence repeat (SSR) markers in Genus *Salvadora*. *Molecular Biology Reports, 47*, 6997–7008. https://doi.org/10.1007/s11033-020-05758-z

Bonnet, E., Wuyts, J., Rouzé, P., & Van de Peer, Y. (2004). Detection of 91 potential conserved plant microRNAs in Arabidopsis thaliana and Oryza sativa identifies important target genes. *Proceedings of the National Academy of Sciences of the United States of America, 101*, 11511–11516.

Botchkareva, N. V. (2017). The molecular revolution in cutaneous biology: Noncoding RNAs: New molecular players in dermatology and cutaneous biology. *Journal of Investigative Dermatology, 137*(5), e105–e111.

Buck, M. J., & Lieb, J. D. (2004). ChIP-chip: Considerations for the design, analysis, and application of genome-wide chromatin immunoprecipitation experiments. *Genomics, 84*, 349–360.

Bustin, S. A. (2002). Quantification of mRNA using real-time reverse transcription PCR (RT-PCR): Trends and problems. *Journal of Molecular Endocrinology, 29*, 23–39.

Cai, Y. F., Wang, J. H., Zhang, L., Song, J., Peng, L. C., & Zhang, S. B. (2019). Physiological and transcriptomic analysis highlight key metabolic pathways in relation to drought tolerance in Rhododendron delavayi. *Physiology and Molecular Biology of Plants, 25*(4), 991–1008.

Chen, M., & Manley, J. L. (2009). Mechanisms of alternative splicing regulation: Insights from molecular and genomics approaches. *Nature Reviews Molecular Cell Biology, 10*(11), 741–754.

Chen, J., Zeng, B., Zhang, M., Xie, S., Wang, G., Hauck, A., et al. (2014). Dynamic transcriptome landscape of maize embryo and endosperm development. *Plant Physiology., 166*, 252–264. https://doi.org/10.1104/pp.114.240689

Chomczynski, P., & Sacchi, N. (2006). The single-step method of RNA isolation by acid guanidinium thiocyanate–phenol–chloroform extraction: Twenty-something years on. *Nature Protocols, 1*(2), 581–585.

Cortés, A. J., Chavarro, M. C., & Blair, M. W. (2011). SNP marker diversity in common bean (*Phaseolus vulgaris* L.). *Theoretical and Applied Genetics, 123*, 827–845.

Cui, G., Chai, H., Yin, H., Yang, M., Hu, G., Guo, M., Guo, M., Yi, R., & Zhang, P. (2019). Full-length transcriptome sequencing reveals the low-temperature-tolerance mechanism of Medicago falcata roots. *BMC Plant Biology, 19*(1), 1–16.

Davuluri, R. V., Suzuki, Y., Sugano, S., Plass, C., & Huang, T. H. M. (2008). The functional consequences of alternative promoter use in mammalian genomes. *Trends in Genetics, 24*(4), 167–177.

De Vos, M., Van Oosten, V. R., Van Poecke, R. M., Van Pelt, J. A., Pozo, M. J., Mueller, M. J., Buchala, A. J., Métraux, J.-P., Van Loon, L. C., Dicke, M., & Pieterse, C. M. (2005). Signal signature and transcriptome changes of Arabidopsis during pathogen and insect attack. *Molecular Plant-Microbe Interactions, 18*(9), 923–937.

Ding, Z., Weissmann, S., Wang, M., Du, B., Huang, L., Wang, L., Tu, X., Zhong, S., Myers, C., Brutnell, T. P., et al. (2015). Identification of photosynthesis-associated c4 candidate genes through comparative leaf gradient transcriptome in multiple lineages of c3 and c4 species. *PLoS One, 10*(10), e0140629.

Dries, R., Chen, J., Del Rossi, N., Khan, M. M., Sistig, A., & Yuan, G. C. (2021). Advances in spatial transcriptomic data analysis. *Genome Research, 31*(10), 1706–1718.

Ebenezer, T. E., Zoltner, M., Burrell, A., Nanakova, A., Vanclová, A. M. G. N., Prasad, B., et al. (2019). Transcriptome, proteome and draft genome of *Euglena gracilis*. *BMC Biology., 17*, 11. https://doi.org/10.1186/s12915-019-0626-8

Eddy, S. R. (2001). Non–coding RNA genes and the modern RNA world. *Nature Reviews Genetics, 2*(12), 919–929.

Femino, A. M., Fay, F. S., Fogarty, K., & Singer, R. H. (1998). Visualization of single RNA transcripts in situ. *Science, 280*(5363), 585–590.

Fracasso, A., Magnanini, E., Marocco, A., & Amaducci, S. (2017). Real-time determination of photosynthesis,transpiration, water-use efficiency and gene expression of two sorghum bicolor (moench) genotypes subjected to drydown. *Frontiers in Plant Science, 8*, 264303.

Gandotra, N., Coughlan, S. J., & Nelson, T. (2013). The Arabidopsis leaf provascular cell transcriptome is enriched in genes with roles in vein patterning. *The Plant Journal, 74*(1), 48–58.

Garnica, D. P., Upadhyaya, N. M., Dodds, P. N., & Rathjen, J. P. (2013). Strategies for wheat stripe rust pathogenicity identified by transcriptome sequencing. *PLoS One, 8*, e67150.

Giacomello, S. (2021). A new era for plant science: Spatial single-cell transcriptomics. *Current Opinion in Plant Biology, 60*, 102041.

Giacomello, S., & Lundeberg, J. (2018). Preparation of plant tissue to enable Spatial Transcriptomics profiling using barcoded microarrays. *Nature Protocols, 13*(11), 2425–2446. https://doi.org/10.1038/s41596-018-0046-1

Ginzinger, D. G. (2002). Gene quantification using real-time quantitative PCR: An emerging technology hits the mainstream. *Experimental Hematology, 30*, 503–512.

Gong, S., Hao, Z., Meng, J., Liu, D., Wei, M., & Tao, J. (2015). Digital gene expression analysis to screen disease resistance-relevant genes from leaves of herbaceous peony (paeonia lactiflora pall.) infected by botrytis cinerea. *PLoS One, 10*(7), e0133305.

Goralski, M., Sobieszczanska, P., Obrepalska-Steplowska, A., Swiercz, A., Zmienko, A., & Figlerowicz, M. (2016). A gene expression microarray for Nicotiana benthamiana based on de novo transcriptome sequence assembly. *Plant Methods, 12*(1), 1–10.

Guo, X., Li, Y., Li, C., Luo, H., Wang, L., Qian, J., Luo, X., Xiang, L., Song, J., Sun, C., Haibin, X., Yao, H., & Chen, S. (2013). Analysis of the Dendrobium officinale transcriptome reveals putative alkaloid biosynthetic genes and genetic markers. *Gene, 527*(1), 131–138.

Guo, W. L., Chen, B. H., Chen, X. J., Guo, Y. Y., Yang, H. L., Li, X. Z., & Wang, G. Y. (2018). Transcriptome profiling of pumpkin (Cucurbita moschata Duch.) leaves infected with powdery mildew. *PLoS One, 13*(1), e0190175.

Guo, H., Pu, X., Jia, H., Zhou, Y., Ye, G., Yang, Y., Na, T., & Wang, J. (2022). Transcriptome analysis reveals multiple effects of nitrogen accumulation and metabolism in the roots, shoots, and leaves of potato (Solanum tuberosum L.). *BMC Plant Biology, 22*(1), 1–12.

Guttman, M., Amit, I., Garber, M., French, C., Lin, M. F., Feldser, D., Huarte, M., Zuk, O., Carey, B. W., Cassady, J. P., Cabili, M. N., Jaenisch, R., Mikkelsen, T. S., Jacks, T., Hacohen, N., Bernstein, B. E., Kellis, M., et al. (2009). Chromatin signature reveals over a thousand highly conserved large non-coding RNAs in mammals. *Nature, 458*(7235), 223–227.

Haque, A., Engel, J., Teichmann, S. A., & Lönnberg, T. (2017). A practical guide to single-cell RNA-sequencing for biomedical research and clinical applications. *Genome Medicine, 9*(1), 1–12.

Hasan, M. M. U., Ma, F., Islam, F., Sajid, M., Prodhan, Z. H., Li, F., Shen, H., Chen, Y., & Wang, X. (2019). Comparative transcriptomic analysis of biological process and key pathway in three cotton (Gossypium spp.) species under drought stress. *International Journal of Molecular Sciences, 20*(9), 2076.

He, F., Long, R., Wei, C., Zhang, Y., Li, M., Kang, J., Yang, Q., Wang, Z., & Chen, L. (2022). Genome-wide identification, phylogeny and expression analysis of the SPL gene family and its important role in salt stress in Medicago sativa L. *BMC Plant Biology, 22*(1), 1–13.

Hou, L., Zhang, X., Li, X., Jia, J., Yang, H., Zhan, H., Qiao, L., Guo, H., & Chang, Z. (2015). Mapping of powdery mildew resistance gene pmCH89 in a putative wheat-Thinopyrum intermedium introgression line. *International Journal of Molecular Sciences, 16*(8), 17231–17244.

Hou, Z., Yin, J., Lu, Y., Song, J., Wang, S., Wei, S., Liu, Z., Zhang, Y., & Fang, Z. (2019). Transcriptomic analysis reveals the temporal and spatial changes in physiological process and gene expression in common buckwheat (Fagopyrum esculentum Moench) grown under drought stress. *Agronomy, 9*(10), 569.

Hu, R., Yu, C., Wang, X., Jia, C., Pei, S., He, K., et al. (2017). De novo transcriptome analysis of *Miscanthus Lutarioriparius* identifies candidate genes in rhizome development. *Frontiers in Plant Science, 8*, 492. https://doi.org/10.3389/fpls.2017.00492

Ištvánek, J., Dluhošová, J., Dluhoš, P., Pátková, L., Nedělník, J., & Řepková, J. (2017). Gene classification and mining of molecular markers useful in red clover (Trifolium pratense) breeding. *Frontiers in Plant Science, 8*, 367.

Jacquier, A. (2009). The complex eukaryotic transcriptome: Unexpected pervasive transcription and novel small RNAs. *Nature Reviews Genetics, 10*(12), 833–844.

Jazayeri, S. M., Melgarejo Munoz, L. M., & Romero, H. M. (2015). Rna-seq: A glance at technologies and methodologies. *Acta Biológica Colombiana, 20*(2), 23–35.

Jenkitkonchai, J., Marriott, P., Yang, W., Sriden, N., Jung, J. H., Wigge, P. A., & Charoensawan, V. (2021). Exploring PIF4's contribution to early flowering in plants under daily variable temperature and its tissue-specific flowering gene network. *Plant Direct, 5*(7), e339.

Jiang, D., Li, G., Chen, G., Lei, J., Cao, B., & Chen, C. (2021). Genome-wide identification and expression profiling of 2OGD superfamily genes from three Brassica plants. *Genes, 12*(9), 1399.

Jiao, X., Zhao, X., Zhou, X. R., Green, A. G., Fan, Y., Wang, L., Singh, S. P., & Liu, Q. (2013). Comparative transcriptomic analysis of developing cotton cotyledons and embryo axis. *PloS One, 8*(8), e71756.

Kampa, D., Cheng, J., Kapranov, P., Yamanaka, M., Brubaker, S., Cawley, S., Drenkow, J., Piccolboni, A., Bekiranov, S., Helt, G., Tammana, H., & Gingeras, T. R. (2004). Novel RNAs identified from an in-depth analysis of the transcriptome of human chromosomes 21 and 22. *Genome Research, 14*(3), 331–342.

Kapoor, M., Mawal, P., Sharma, V., & Gupta, R. C. (2020). Analysis of genetic diversity and population structure in *Asparagus* species using SSR markers. *Journal of Genetic Engineering and Biotechnology, 18*, 50. https://doi.org/10.1186/s43141-020-00065-3

Klein, D. (2002). Quantification using real-time PCR technology: Applications and limitations. *Trends in Molecular Medicine, 8*, 257–260.

Kramer, E. M., & Li, W. (2017). A transcriptomics and comparative genomics analysis Reveals gene families with a role in body plan complexity. *Frontiers in Plant Science, 8*, 869.

Krost, C., Petersen, R., & Schmidt, E. R. (2012). The transcriptomes of columnar and standard type apple trees (Malus x domestica)—a comparative study. *Gene, 498*(2), 223–230.

Lacape, J. M., Claverie, M., Vidal, R. O., Carazzolle, M. F., Guimaraes Pereira, G. A., Ruiz, M., Pré, M., Llewellyn, D., Al-Ghazi, Y., Jacobs, J., Dereeper, A., Huguet, S., Giband, M., & Lanaud, C. (2012). Deep sequencing reveals differences in the transcriptional landscapes of fibers from two cultivated species of cotton. *PLoS One, 7*(11), e48855.

Le, D. T., Nishiyama, R., Watanabe, Y., Tanaka, M., Seki, M., Ham, L. H., Yamaguchi-Shinozaki, K., Shinozaki, K., & Tran, L. S. P. (2012). Differential gene expression in soybean leaf tissues at late developmental stages under drought stress revealed by genome-wide transcriptome analysis. *PloS One, 7*(11), e49522.

Lee, J., Heath, L. S., Grene, R., & Li, S. (2019). Comparing time series transcriptome data between plants using a network module finding algorithm. *Plant Methods, 15*(1), 1–16.

Li, J., Zhu, L., Hull, J. J., Liang, S., Daniell, H., Jin, S., & Zhang, X. (2016). Transcriptome analysis reveals a comprehensive insect resistance response mechanism in cotton to infestation by the phloem feeding insect Bemisia tabaci (whitefly). *Plant Biotechnology Journal, 14*(10), 1956–1975.

Li, S., Liu, Z., Chen, G., Qanmber, G., Lu, L., Zhang, J., Ma, S., Yang, Z., & Li, F. (2021). Identification and analysis of GhEXO gene family indicated that GhEXO7_At promotes plant growth and development through brassinosteroid signaling in cotton (Gossypium hirsutum L.). *Frontiers in Plant Science*, 1840.

Lindberg, J., & Lundeberg, J. (2010). The plasticity of the mammalian transcriptome. *Genomics, 95*(1), 1–6.

Liu, G., Li, W., Zheng, P., Xu, T., Chen, L., Liu, D., Hussain, S., & Teng, Y. (2012). Transcriptomic analysis of 'Suli'pear (Pyrus pyrifolia white pear group) buds during the dormancy by RNA-Seq. *BMC Genomics, 13*(1), 1–18.

Livak, K. J., & Schmittgen, T. D. (2001). Analysis of relative gene expression data using real-time quantitative PCR and the 2− ΔΔCT method. *Methods, 25*(4), 402–408.

Lu, T., Cui, L., Zhou, Y., Zhu, Q., Fan, D., Gong, H., Zhao, Q., Zhou, C., Zhao, Y., Lu, D., Luo, J., Wang, Y., Tian, Q., Feng, Q., Huang, T., & Han, B. (2015). Transcriptome-wide investigation of circular RNAs in rice. *RNA, 21*, 2076–2087.

Luo, C. Z. Q., & Luo, Z. (2014). Genome-wide transcriptome analysis of Chinese pollination-constant nonastringent persimmon fruit treated with ethanol. *BMC Genomics, 15*, 112.

Mahajan, M. M., Goyal, E., Singh, A. K., Gaikwad, K., & Kanika, K. (2017). Transcriptome dynamics provide insights into long-term salinity stress tolerance in Triticum aestivum cv. Kharchia Local. *Plant Physiology and Biochemistry, 121*, 128–139.

Manechini, J. R. V., Santos, P. H. D. S., Romanel, E., Brito, M. D. S., Scarpari, M. S., Jackson, S., & LR & Vicentini, R. (2021). Transcriptomic analysis of changes in gene expression during flowering induction in sugarcane under controlled photoperiodic conditions. *Frontiers in Plant Science, 808*.

Mantione, K. J., Kream, R. M., Kuzelova, H., Ptacek, R., Raboch, J., Samuel, J. M., & Stefano, G. B. (2014). Comparing bioinformatic gene expression profiling methods: Microarray and RNA-Seq. *Medical Science Monitor Basic Research, 20*, 138.

McGettigan, P. A. (2013). Transcriptomics in the RNA-seq era. *Current Opinion in Chemical Biology, 17*(1), 4–11.

Mikkilineni, V., Mitra, R. D., Merritt, J., DiTonno, J. R., Church, G. M., Ogunnaike, B., & Edwards, J. S. (2004). Digital quantitative measurements of gene expression. *Biotechnology and Bioengineering, 86*, 117–124.

Miller, M. R., Dunham, J. P., Amores, A., Cresko, W. A., & Johnson, E. A. (2007). Rapid and cost-effective polymorphism identification and genotyping using restriction site associated DNA (RAD) markers. *Genome Research, 17*, 240–248.

Misra, P., Pandey, A., Tiwari, M., Chandrashekar, K., Sidhu, O. P., Asif, M. H., Chakrabarty, D., Singh, P. K., Trivedi, P. K., Nath, P., & Tuli, R. (2010). Modulation of transcriptome and metabolome of tobacco by Arabidopsis transcription factor, AtMYB12, leads to insect resistance. *Plant Physiology, 152*(4), 2258–2268.

Mitra, R. D., Shendure, J., Olejnik, J., Edyta Krzymanska, O., & Church, G. M. (2003). Fluorescent in situ sequencing on polymerase colonies. *Analytical Biochemistry, 320*, 55–65.

Morgil, H., Tardu, M., Cevahir, G., & Kavakli, İ. H. (2019). Comparative RNA-seq analysis of the drought-sensitive lentil (Lens culinaris) root and leaf under short-and long-term water deficits. *Functional & Integrative Genomics, 19*(5), 715–727.

Moses, L., & Pachter, L. (2022). Museum of spatial transcriptomics. *Nature Methods, 19*(5), 534–546.

Movahedi, S., Van de Peer, Y., & Vandepoele, K. (2011). Comparative network analysis reveals that tissue specificity and gene function are important factors influencing the mode of expression evolution in Arabidopsis and rice. *Plant Physiology, 156*(3), 1316–1330.

Nagalakshmi, U., Wang, Z., Waern, K., Shou, C., Raha, D., Gerstein, M., & Snyder, M. (2008). The transcriptional landscape of the yeast genome defined by RNA sequencing. *Science, 320*(5881), 1344–1349.

Nair, P., Misra, A., Singh, A., Shukla, A. K., Gupta, M. M., Gupta, A. K., Gupta, V., Khanuja, S. P., & Shasany, A. K. (2013). Differentially expressed genes during contrasting growth stages of Artemisia annua for artemisinin content. *PLoS One, 8*(4), e60375.

Netotea, S., Sundell, D., Street, N. R., & Hvidsten, T. R. (2014). ComPlEx: Conservation and divergence of co-expression networks in A. thaliana, Populus and O. sativa. *BMC Genomics, 15*(1), 1–17.

Nigam, D., Kavita, P., Tripathi, R. K., Ranjan, A., Goel, R., Asif, M., Shukla, A., Singh, G., Rana, D., & Sawant, S. V. (2014). Transcriptome dynamics during fibre development in contrasting genotypes of Gossypium hirsutum L. *Plant Biotechnology Journal, 12*, 204–218.

Okoniewski, M. J., & Miller, C. J. (2006). Hybridization interactions between probesets in short oligo microarrays lead to spurious correlations. *BMC Bioinformatics, 7*(1), 1–14.

Pal, T., Padhan, J. K., Kumar, P., Sood, H., & Chauhan, R. S. (2018). Comparative transcriptomics uncovers differences in photoautotrophic versus photoheterotrophic modes of nutrition in relation to secondary metabolites biosynthesis in *Swertia chirayita. Molecular Biology Reports, 45*, 77–98. https://doi.org/10.1007/s11033-017-4135-y

Parkinson, J. (2009). *Methods in molecular biology. Expressed sequence tags (ESTs) volume 533 || expressed sequence tags: An overview*. pp. 1–12. https://doi.org/10.1007/978-1-60327-136-3_1

Phukan, U. J., Jeena, G. S., & Shukla, R. K. (2016). WRKY transcription factors: Molecular regulation and stress responses in plants. *Frontiers in Plant Science, 7*, 760. https://doi.org/10.3389/fpls.2016.00760

Provart, N. J., Alonso, J., Assmann, S. M., Bergmann, D., Brady, S. M., Brkljacic, J., Browse, J., Chapple, C., Colot, V., Cutler, S., Dangl, J., Ehrhardt, D., Friesner, J. D., Frommer, W. B., Grotewold, E., Meyerowitz, E., Nemhauser, J., et al. (2016). 50 years of Arabidopsis research: Highlights and future directions. *New Phytologist, 209*(3), 921–944.

Pucker, B. (2022). Automatic identification and annotation of MYB gene family members in plants. *BMC Genomics, 23*(1), 1–11.

Pushkarev, D., Neff, N. F., & Quake, S. R. (2009). Single-molecule sequencing of an individual human genome. *Nature Biotechnology, 27*(9), 847–850.

Puvvala, S. S., Muddanuru, T., Thangella, P. A., Kumar, O. A., Chakravartty, N., Vettath, V. K., Mohan Katta, A. V. S. K., Lekkala, S. P., Kuriakose, B., Gupta, S., Mulpuri, S., & Lachagari,

V. B. R. (2019). Deciphering the transcriptomic insight during organogenesis in Castor (Ricinus communis L.), Jatropha (Jatropha curcas L.) and Sunflower (Helianthus annuus L.). *3 Biotech, 9*(11), 1–17.
Qian, Y., Ren, Q., Zhang, J., & Chen, L. (2019). Transcriptomic analysis of the maize (Zea mays L.) inbred line B73 response to heat stress at the seedling stage. *Gene, 692*, 68–78.
Qin, D., et al. (2008). Heat stress-responsive transcriptome analysis in heat susceptible and tolerant wheat (Triticum aestivum L.) by using Wheat Genome Array. *BMC Genomics., 9*, 432.
Ruprecht, C., Mendrinna, A., Tohge, T., Sampathkumar, A., Klie, S., Fernie, A. R., Nikoloski, Z., Persson, S., & Mutwil, M. (2016). FamNet: A framework to identify multiplied modules driving pathway expansion in plants. *Plant Physiology, 170*(3), 1878–1894.
Sánchez-Sevilla, J. F., Vallarino, J. G., Osorio, S., Bombarely, A., Posé, D., Merchante, C., et al. (2017). Gene expression atlas of fruit ripening and transcriptome assembly from RNA-seq data in octoploid strawberry (*Fragaria×ananassa*). *Scientific Reports, 7*, 13737.
Schena, M., Shalon, D., Davis, R. W., & Brown, P. O. (1995). Quantitative monitoring of gene expression patterns with a complementary DNA microarray. *Science, 270*, 467–470.
Sharma, R., Agarwal, P., Ray, S., Deveshwar, P., Sharma, P., Sharma, N., Nijhawan, A., Jain, M., Singh, A. K., Singh, V. P., Khurana, J. P., Tyagi, A. K., & Kapoor, S. (2012). Expression dynamics of metabolic and regulatory components across stages of panicle and seed development in indica rice. *Functional & Integrative Genomics, 12*(2), 229–248.
Shen, C., Li, D., He, R., Fang, Z., Xia, Y., Gao, J., Shen, H., & Cao, M. (2014). Comparative transcriptome analysis of RNA-seq data for cold-tolerant and cold-sensitive rice genotypes under cold stress. *Journal of Plant Biology, 57*(6), 337–348.
Slavokhotova, A., Korostyleva, T., Shelenkov, A., Pukhalskiy, V., Korottseva, I., Slezina, M., Istomina, E., & Odintsova, T. (2021). Transcriptomic analysis of genes involved in plant defense response to the cucumber green mottle mosaic virus infection. *Life, 11*(10), 1064.
Sonnhammer, E. L., Gabaldón, T., Sousa da Silva, A. W., Martin, M., Robinson-Rechavi, M., Boeckmann, B., Thomas, P. D., Dessimoz, C., & Quest for Orthologs Consortium. (2014). Big data and other challenges in the quest for orthologs. *Bioinformatics, 30*(21), 2993–2998.
Sreenivasulu, N., Sopory, S. K., & Kavi Kishor, P. B. (2007). Deciphering the regulatory mechanisms of abiotic stress tolerance in plants by genomic approaches. *Gene., 388*, 1–13.
Stelpflug, S. C., Sekhon, R. S., Vaillancourt, B., Hirsch, C. N., Buell, C. R., de Leon, N., et al. (2016). An expanded maize gene expression atlas based on RNA sequencing and its use to explore root development. *The Plant Genome., 9*, 1–16.
Sweetman, C., WonG, D. C. J., Ford, C. M., & Drew, D. P. (2012). Transcriptome analysis at four developmental stages of grape berry (Vitis vinifera cv. Shiraz) provides insights into regulated and coordinated gene expression. *BMC Genomics, 13*, 691.
Taft, R. J., Pheasant, M., & Mattick, J. S. (2007). The relationship between non-protein-coding DNA and eukaryotic complexity. *Bioessays, 29*(3), 288–299.
Tang, F., Barbacioru, C., Wang, Y., Nordman, E., Lee, C., Xu, N., .Wang X, Bodeau J, Tuch BB, Siddiqui A, Lao K, & Surani, M. A. (2009). mRNA-Seq whole-transcriptome analysis of a single cell. Nature Methods, 6(5), 377-382.
Tang, S., Liang, H., Yan, D., Zhao, Y., Han, X., Carlson, J. E., Xia, X., & Yin, W. (2013). Populus euphratica: The transcriptomic response to drought stress. *Plant Molecular Biology, 83*, 539–557.
Tang, B., Xie, L., Yang, H., Li, X., Chen, Y., Zou, X., Liu, F., & Dai, X. (2022). Analysis of the expression and function of key genes in pepper under low-temperature stress. *Frontiers in Plant Science, 1378.*
Trebbi, D., Maccaferri, M., de Heer, P., Sorensen, A., Giuliani, S., Salvi, S., Sanguineti, M. C., Massi, A., van der Vossen, E. A. G., & Tuberosa, R. (2011). High-throughput SNP discovery and genotyping in durum wheat (*Triticum durum* Desf.). *Theoretical and Applied Genetics, 123*, 555–569.

Ullah, A., Manghwar, H., Shaban, M., Khan, A., Akbar, A., Ali, U., et al. (2018). Phytohormones enhanced drought tolerance in plants: A coping strategy. *Environmental Science and Pollution Research, 25*, 33103–33118. https://doi.org/10.1007/s11356-018-3364-5

Van Dijk, E. L., Jaszczyszyn, Y., Naquin, D., & Thermes, C. (2018). The third revolution in sequencing technology. *Trends in Genetics, 34*(9), 666–681.

Velculescu, V. E., Zhang, L., Vogelstein, B., & Kinzler, K. W. (1995). Serial analysis of gene expression. *Science, 270*(5235), 484–487.

Vercruysse, J., Van Bel, M., Osuna-Cruz, C. M., Kulkarni, S. R., Storme, V., Nelissen, H., Gonzalez, N., Inzé, D., & Vandepoele, K. (2020). Comparative transcriptomics enables the identification of functional orthologous genes involved in early leaf growth. *Plant Biotechnology Journal, 18*(2), 553–567.

Vimolmangkang, S., Zheng, D., Han, Y., Khan, M. A., Soria-Guerra, R. E., & Korban, S. S. (2014). Transcriptome analysis of the exocarp of apple fruit identifies light-induced genes involved in red color pigmentation. *Gene, 534*(1), 78–87.

Wan, X., & Li, Z. (2019). *Plant comparative transcriptomics reveals functional mechanisms and gene regulatory networks involved in anther development and male sterility. In Transcriptome analysis*. IntechOpen.

Wang, Z., Gerstein, M., & Snyder, M. (2009). RNA-Seq: A revolutionary tool for transcriptomics. *Nature Reviews Genetics, 10*(1), 57–63.

Wang, F., Li, L., Li, H., Liu, L., Zhang, Y., Gao, J., & Wang, X. (2012). Transcriptome analysis of rosette and folding leaves in Chinese cabbage using high-throughput RNA sequencing. *Genomics, 99*(5), 299–307.

Wang, M., Jiang, B., Liu, W., Lin, Y. E., Liang, Z., He, X., & Peng, Q. (2019). Transcriptome analyses provide novel Insights into heat stress responses in Chieh-Qua (Benincasa hispida Cogn. var. Chieh-Qua How). *International Journal of Molecular Sciences, 20*(4), 883.

Wu, T., Qin, Z., Zhou, X., Feng, Z., & Du, Y. (2010). Transcriptome profile analysis of floral sex determination in cucumber. *Journal of Plant Physiology, 167*(11), 905–913.

Xiong, H., Guo, H., Xie, Y., Zhao, L., Gu, J., Zhao, S., Li, J., & Liu, L. (2017). RNAseq analysis reveals pathways and candidate genes associated with salinity tolerance in a spaceflight-induced wheat mutant. *Scientific Reports, 7*(1), 1–13.

Xu, L., Wang, Y., Zhai, L., Xu, Y., Wang, L., Zhu, X., Gong, Y., Yu, R., Limera, C., & Liu, L. (2013). Genome-wide identification and characterization of cadmium-responsive microRNAs and their target genes in radish (*Raphanus sativus* L.) roots. *Journal of Experimental Botany, 64*, 4271–4287.

Yamakawa, H., Hirose, T., Kuroda, M., & Yamaguchi, T. (2007). Comprehensive expression profiling of rice grain filling-related genes under high temperature using DNA microarray. *Plant Physiology., 144*, 258–277.

Yang, Y., Yu, Q., Yang, Y., Su, Y., Ahmad, W., Guo, J., Gao, S., Xu, L., & Que, Y. (2018). Identification of cold-related miRNAs in sugarcane by small RNA sequencing and functional analysis of a cold inducible ScmiR393 to cold stress. *Environmental and Experimental Botany, 155*, 464–476.

Yates, S. A., Swain, M. T., Hegarty, M. J., Chernukin, I., Lowe, M., Allison, G. G., Ruttink, T., Abberton, M. T., Jenkins, G., & Skøt, L. (2014). De novo assembly of red clover transcriptome based on RNA-Seq data provides insight into drought response, gene discovery and marker identification. *BMC Genomics, 15*, 453.

Yi, F., Gu, W., Chen, J., Song, N., Gao, X., Zhang, X., et al. (2019). High temporal-resolution transcriptome landscape of early maize seed development. *Plant Cell, 31*, 974–992. https://doi.org/10.1105/tpc.18.00961

Yu, Y., Guo, G., Lv, D., Hu, Y., Li, J., Li, X., & Yan, Y. (2014). Transcriptome analysis during seed germination of elite Chinese bread wheat cultivar Jimai 20. *BMC Plant Biology, 14*(1), 1–19.

Zhang, T. Q., Xu, Z. G., Shang, G. D., & Wang, J. W. (2019a). A single-cell RNA sequencing profiles the developmental landscape of Arabidopsis root. *Molecular Plant, 12*(5), 648–660.

Zhang, Y., Gao, X., Li, J., Gong, X., Yang, P., Gao, J., Wang, P., & Feng, B. (2019b). Comparative analysis of proso millet (Panicum miliaceum L.) leaf transcriptomes for insight into drought tolerance mechanisms. *BMC Plant Biology, 19*(1), 1–17.

Zhang, Z., Ali, S., Zhang, T., Wang, W., & Xie, L. (2020). Identification, evolutionary and expression analysis of PYL-PP2C-SnRK2s gene families in soybean. *Plants, 9*(10), 1356.

Zhang, X., He, Y., Li, L., Liu, H., & Hong, G. (2021). Involvement of the R2R3-MYB transcription factor MYB21 and its homologs in regulating flavonol accumulation in Arabidopsis stamen. *Journal of Experimental Botany, 72*(12), 4319–4332.

Zhao, W., He, X., Hoadley, K. A., Parker, J. S., Hayes, D. N., & Perou, C. M. (2014). Comparison of RNA-Seq by poly (A) capture, ribosomal RNA depletion, and DNA microarray for expression profiling. *BMC Genomics, 15*(1), 1–11.

Zhao, C., Liu, B., Piao, S., Wang, X., Lobell, D. B., Huang, Y., Huang, M., Yao, Y., Bassu, S., Ciais, P., Durand, J. L., & Asseng, S. (2017). Temperature increase reduces global yields of major crops in four independent estimates. *Proceedings of the National Academy of Sciences, 114*(35), 9326–9331.

Zhao, T., Specht, C. D., Dong, Z., Ye, Y., Liu, H., & Liao, J. (2020). Transcriptome analyses provide insights into development of the *Zingiber zerumbet* flower, revealing potential genes related to floral organ formation and patterning. *Plant Growth Regulation, 90*, 331–345. https://doi.org/10.1007/s10725-020-00575-7

Zhong, S., Fei, Z., Chen, Y. R., Vrebalov, J., Mcquinn, R., Gapper, N., & Giovannoni, J. (2013). Single-base resolution methylomes of tomato fruit development reveal epigenome modifications associated with ripening. *Nature Biotechnology, 31*(2), 154–159.

Zhu, J., Shendure, J., Mitra, R. D., & Church, G. M. (2003). Single molecule profiling of alternative pre-mRNA splicing. *Science, 301*, 836–838.

Zipfel, C., Kunze, G., Chinchilla, D., et al. (2006). Perception of the bacterial PAMP EF-Tu by the receptor EFR restricts Agrobacterium-mediated transformation. *Cell, 125*, 749–760.

Chapter 5
Plant Proteomic Approaches, Expression of Simple Peptides to Complex Proteins

Amjad Abbas, Muhammad Jabran, Adil Zahoor, Abdalmenem I. M. Hawamda, Amer Habib, Shahnaz Jamil, Muhammad Huzaifa Tanveer, and Muhammad Amjad Ali

Abstract Plants are diverse, and their interactions with living organisms make them more versatile. The plants respond to biotic and abiotic interactions through protein molecules. So, the expression of genes and formation of proteins are very important, and further regulation of genes is also vital in plant–microbe interactions. This book chapter addresses the multifaceted realm of plant proteomics, providing a comprehensive overview of the techniques, applications, and recent advancements in the study of plant proteins. It focuses on studying and expressing small peptides, like antimicrobial peptides, to complex proteins in heterologous expression systems and the related troubleshooting. By exploring the intricate network of plant proteomics, this chapter sheds light on the fundamentals of proteomics, outlining the various techniques employed in plant proteomic studies. From gel-based methods like two-dimensional gel electrophoresis (2-DE) to advanced mass spectrometry (MS)-based approaches, the readers gain insights into the diverse strategies employed to unravel the plant proteome. The power of these techniques lies in their ability to analyze complex protein mixtures, identify post-translational modifications, and quantify protein abundances, thereby enabling a holistic understanding of the plant proteome.

Keywords Proteomics · Proteome · Peptides · Protein engineering

---

A. Abbas (✉) · A. Habib · S. Jamil · M. H. Tanveer · M. A. Ali
Department of Plant Pathology, University of Agriculture, Faisalabad, Pakistan
e-mail: amjad.abbas@uaf.edu.pk; amer.habib@uaf.edu.pk; amjad.ali@uaf.edu.pk

M. Jabran
State Key Laboratory for Biology of Plant Diseases, Institute of Plant Protection, CAAS, Beijing, China

A. Zahoor
Department of Biotechnology, Chonnam National University, Yeosu, Republic of Korea

A. I. M. Hawamda
Department of Agricultural Biotechnology, Faculty of Agricultural Science and Technology, Palestine Technical University-Kadoorie (PTUK), Tulkarm, Palestine

S. Ijaz et al. (eds.), *Trends in Plant Biotechnology*,
https://doi.org/10.1007/978-981-97-0814-7_5

Introduction

Proteomics involves the investigation of proteins from their structure to interactions within cells. After completing the Human Genome Project, the next phase was to focus on and comprehend the complete proteins within the human body. The information in the DNA is centralized and can be understood only with the structure of the protein. This emphasis on the proteome study rather than focusing only on genomics is motivated by the fact that it covers the dynamics of the cell.

Proteomics guides in the right diagnosis of animal and plant diseases. Detection of protein markers helps to understand the nature of the disorder, possible treatment options, and more targeted therapy. It is a vast area, and still, detecting proteins with lower expression and quantification is challenging and invites research aiming at understanding diseases. The development in bioinformatics and computational biology coupled with proteomics technologies could be employed for uncovering the molecular basis of animal and plant diseases. Such comprehensive approaches could help understand the diseases' pathways and signaling cascades.

The proteomics study involves the comparative study of the entire proteome of an organism under certain biological conditions and could be displayed with SDS-PAGE. But such presentation masks many proteins lying on the same size in the gel and the others having faint expression. The study of two-dimensional gels separating the proteins based on size and charge is a more comprehensive approach. A particular protein could be further fished out or could be highlighted by using specific antibodies as well.

The modern research on proteomics is based on multiple cutting-edge approaches to understanding the function and structure of proteins. Mass spectroscopy allows for the identification and quantification of complex proteins. It determines the mass of the peptide as an intact molecule or first peptide is digested with an enzyme, and the fragments are analyzed for the arrangements of amino acids, and ultimately, the entire mass and sequence of the peptide is determined.

Tandem mass spectrometry, also known as MS/MS, uses the fragmentation approach to determine the mass of a protein or peptide. This technique further helps to identify posttranslational modifications of the proteins as well.

Western blotting is a widely used laboratory technique in proteomics and molecular biology for detecting and quantifying specific proteins in a complex mixture of proteins. It allows researchers to separate proteins based on their molecular weight and then identify or measure the abundance of a target protein of interest within that mixture. Here is a step-by-step overview of the Western blotting process in proteomics:

Protein Extraction: First, lysis buffers extract proteins from cells or tissues. These buffers disrupt cell membranes and nuclear envelopes, releasing the cellular proteins into solution. Cell lysis could be accomplished with the help of other sources like enzymes or sonication.

Protein Separation: The extracted proteins are typically separated by size using sodium dodecyl sulfate-polyacrylamide gel electrophoresis (SDS-PAGE). SDS is a detergent that denatures proteins and imparts a uniform negative charge to them, allowing separation based on size when subjected to an electric field in a gel matrix. Smaller proteins migrate faster through the gel, while larger ones move more slowly.

Protein Transfer: After separation, the proteins are transferred from the gel to a solid membrane, usually made of nitrocellulose or polyvinylidene difluoride (PVDF). This step is known as protein blotting. The transfer can be done by electroblotting, where an electric current is applied to move proteins from the gel to the membrane.

Blocking: To prevent nonspecific binding of antibodies and reduce background signals, the membrane is incubated with a blocking solution (e.g., milk, bovine serum albumin) that covers unoccupied binding sites on the membrane. It determines the specific binding of your protein to the antibody.

Antibody Incubation: Primary antibodies specific to the target protein are incubated with the membrane. These antibodies will bind specifically to the target protein if it is on the membrane.

Washing Step: After incubation with primary antibodies, the membrane is washed to remove unbound antibodies.

Secondary Antibody Incubation: A secondary antibody conjugated to an enzyme or a fluorescent marker is added to the membrane. The secondary antibody recognizes and binds to the primary antibody.

Detection: If an enzyme-linked secondary antibody is used, a substrate that reacts with the enzyme to produce a detectable signal (e.g., chemiluminescence or color change) is added. If using a fluorescent secondary antibody, the membrane is scanned using a fluorescence scanner.

Image Analysis: The signal produced on the membrane is captured and quantified using imaging systems or software. It allows researchers to determine the relative abundance of the target protein in their sample. Western blotting is a versatile technique widely used in proteomics research for various applications, including protein expression analysis, posttranslational modification analysis, and protein–protein interaction studies. It provides valuable information about specific proteins' presence, size, and relative abundance in complex biological samples.

MALDI-TOF-MS

MALDI-TOF-MS is a highly sensitive and accurate technology for rapidly analyzing complex molecules like proteins and other organic substances. It is widely used for protein identification in proteomics research, de novo peptide sequencing, and drug discovery. Overall, MALDI-TOF-MS is a versatile and powerful technique for analyzing a wide range of biomolecules, making it a valuable tool in various scientific and clinical applications.

Here is an overview of the key components and the process involved in MALDI-TOF-MS:

Sample Preparation

Samples are typically prepared by mixing the analyte of interest (e.g., peptides, proteins, nucleic acids) with a matrix compound. The matrix absorbs laser energy and facilitates desorption and ionization of the analyte when the laser is applied.

Desorption and Ionization

A pulsed laser is directed onto the sample, causing the matrix and analyte to evaporate, forming ions. The process is known as desorption and ionization. The laser energy absorbed by the matrix helps transfer charge to the analyte molecules, generating positively charged ions (cations).

Time-of-Flight Analysis

Once ionized, the ions are accelerated into the time-of-flight (TOF) mass analyzer. The TOF analyzer measures the time it takes for ions to travel a fixed distance from the ion source to the detector. Heavier ions (higher mass-to-charge ratios) travel more slowly, while lighter ions move faster.

Mass Measurement

The time it takes for ions to reach the detector is recorded, and this data is used to calculate the ions' mass-to-charge ratio (m/z). The resulting mass spectrum displays intensity (abundance) on the y-axis and m/z on the x-axis.

Data Analysis

The mass spectrum obtained from the analysis provides valuable information about the mass distribution of ions in the sample. Researchers can identify and quantify the molecules in the sample by comparing the experimental mass spectrum to a database of known mass spectra or through de novo analysis. Common approaches used during proteomics studies are given in Fig. 5.1.

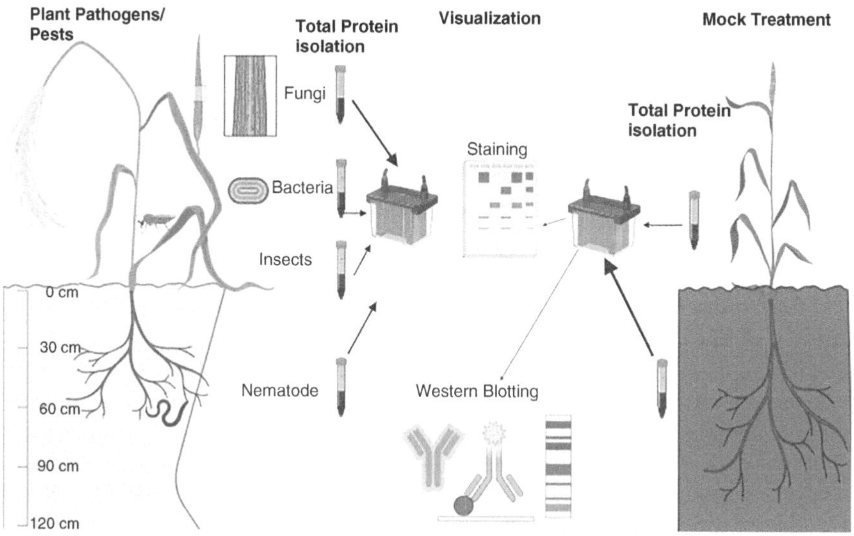

Fig. 5.1 Common approach of proteomics studies in the plants. The infected and mock-treated plants are proceeded for the total protein isolation. The upregulated and downregulated proteins are determined through expression analysis of proteins in control and treated plants on the SDS-PAGE. The expression of a specific protein or peptide can also be determined by western blotting

Plant and Pathogen Interactions

Being stationary by nature, plants are constantly subjected to various microorganisms, encompassing bacteria, fungi, viruses, and nematodes. These microorganisms feed on the plants and find refuge within the field crops. In addition to these pathogens, plants are also susceptible to attacks by various insects and animal pests.

Plants and animals employ distinct sets of immunity mechanisms to defend against pathogens. Vertebrate animals possess an innate immune system complemented by an adaptive immune system that generates specific antibodies targeting microbial pathogens. In contrast, plants, being immobile, confront pathogens without the option of fleeing. The plant's innate immune system comprises constitutive and inducible defense mechanisms (Greenberg, 1997; van Loon et al., 2006). Constitutive resistance mechanisms encompass physical barriers, such as the plant cell wall's sturdy structures and pre-existing low molecular compounds, peptides, and proteins.

On the other hand, the inducible defense system directed against microorganisms is more specialized, relying on detecting specific patterns known as *M*icroorganism/*P*athogen-*A*ssociated *M*olecular *P*atterns (MAMPs/PAMPs) specific to these organisms. So, the base of the induced resistance in plants relies on the recognition of pathogens through its tools to recognize and respond to these PAMPs. When microorganisms interact with plants, these MAMPs, such as peptidoglycans, lipopolysaccharides, and flagellin, are sensed by the plant's pattern recognition receptors (PRRs). This interaction results in the first line of defense of the plant, which is not overcome by all sorts of pathogens. These PRRs have a similar leucine-rich repeat (LRR) as animal PRRs (Jones and Dangl, 2006). One of the important transmembrane flagellin receptors in Arabidopsis, FLAGELLIN-SENSITIVE 2 (FLS2), is one example (Gomez-Gomez & Boller, 2000). It is present in the cell membrane and realizes the bacterial flagellin epitope, which is thus a bacterial MAMP. This recognition is important for the induction of defense response in plants.

The interaction of bacterial MAMPs and PRRs from plants results in different plant responses, such as the production of phytoalexins, PR-proteins, and antimicrobial peptides (AMPs), and involves the signaling compound salicylic acid (Gaffney et al., 1993). The signaling compounds trigger the overexpression of inducible markers like PR-proteins. AMPs are also present in a basal amount in plants and could be artificially induced by the external application of signaling hormones or pathogen infections. The AMPs are widespread in plants and animals, and several groups of these peptides have been described from plants, such as thionins, plant defensins, lipid transfer proteins, snakins, and hevein-like peptides. These plant AMPs are generally basic and contain several cysteine residues that form disulfide bridges (Sels et al., 2008). The recognition of PAMPs and plant receptors activates immunity in plants.

The immunity developed under the involvement of these receptors in plants is called pattern-triggered immunity (PTI). At this point, the pathogens attempt to overcome PTI by delivering effector molecules into the plant cell, suppressing the pathogen-triggered immunity. In response to these effectors, plants have evolved resistance (R) proteins, including intracellular immune receptors, which recognize pathogen effector molecules and induce a resistance response called effector-triggered immunity (ETI). It results in a hypersensitive response, a form of programmed cell death that communicates the alert signals to healthy tissues, which develop prior resistance against the threat, and is called systemic acquired resistance (SAR) (Gaffney et al., 1993; Durrant and Dong, 2004).

A possible scheme for the expression of proteins or peptides is shown in Fig. 5.2. Naturally, these entities are present in very low and undetectable amounts, so we need some expression system to understand the role of these peptides in organisms.

Antimicrobial Peptides (AMPs) Are an Important Plant Response

AMPs are small biomolecules ranging in size up to 10 kDa. AMPs are toxic and antimicrobial to a wide range of pathogens. They are mostly basic, cationic, secreted, and contain a relatively stable structure, often due to disulfide bridges (Lay & Anderson, 2005). However, many AMPs that do not contain cysteine are known (Pelegrini et al., 2008). They have antimicrobial properties against a broad range of microbes. They are present in many plant species (Bohlmann, 1994; Florack &

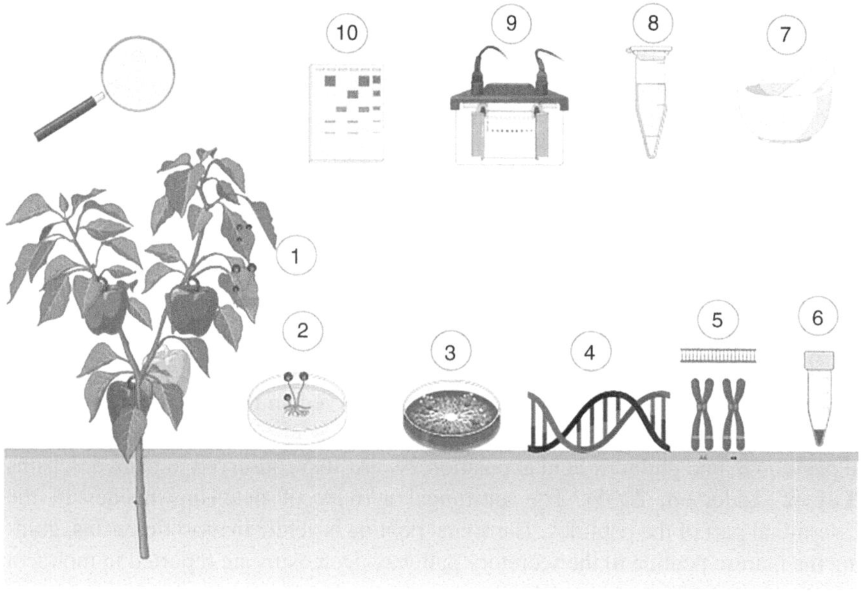

Fig. 5.2 Fungal pathogen is observed on the plant body. It is isolated under sterile lab conditions and is followed for cellular contents isolation. DNA is used to amplify selected genes and their subsequent expressions. The protein of interest is expressed and isolated in the same or heterologous expression system. The final step is visualization through SDS-PAGE and validation of expression

Stiekema, 1994; Kader, 1996; Lay & Anderson, 2005; Stec, 2006). Like other defense compounds, AMPs can be produced constitutively or induced by plant infection.

Classification of AMPs

Plant AMPs are proteinase inhibitors, plant defensins, thionins, lipid transfer proteins, cyclotides, knottins, heveins, and snakins. Details of important classes emphasizing thionins are discussed in the following chapters.

Plant Defensins

Plant defensins are expressed in stems, roots, leaves, and reproductive organs such as seed and floral parts, but they are most common in seeds, and their expression can be constitutive or inducible by pathogens (Lay & Anderson, 2005). With a size of mostly 5 kDa, they are basic and cysteine-rich. They are mostly produced as preproteins, but in some cases, C-terminal prodomains are also present (Lay et al., 2003). Along with antimicrobial activity, they can also have α-amylase activity (Zhang et al., 1997), insecticidal activity (Chen et al., 2004), protein synthesis inhibitor (Thevissen et al., 2000), and ion-channel blocking activity (Kushmerick et al., 1998). Their potential as antimicrobial agents is especially directed against filamentous fungi (Sels et al., 2008), but some are also active against bacteria (Segura et al., 1998). Initially, due to size and cysteine residues resemblance with thionins, they were assigned as γ-thionins (Colilla et al., 1990; Mendez et al., 1990), but later on, it became clear that their 3D structure shares similar characteristics with insects and mammalian defensins (Terras et al., 1995).

Plant defensins normally have four disulfide bonds, but flower defensins are exceptional, with five. Usually, the cysteine residues are conserved, and they form disulfide bridges. Glycine at position 34 is a conserved amino acid in the defensins. Some other conserved structural amino acids, such as glycine at position 13, serine at position 8, and glutamic acid at position 29, are also conserved in plant defensins (Lay & Anderson, 2005). The antifungal activity of defensins resides in the C-terminal part of the peptides. The signal peptide is acidic in most defensins, guiding the mature peptide to the secretory pathway. Defensins are reported in monocot and dicot plants. They are widely distributed among different plant families such as Chenopodiaceae, Amaranthaceae, Poaceae, Fabaceae, Rosaceae, Cucurbitaceae, Brassicaceae, and Ranunculaceae (Carvalho & Gomes, 2007; Rogozhin et al., 2009).

Lipid Transfer Proteins

Due to their ability to transfer lipids between donor and receptor, these proteins are called lipid transfer proteins (LTPs), which can be divided into two families. The LTP1 family contains peptides with a molecular weight of 10 kDa with 90–95 residues. These are highly basic, with an isoelectric point between 9 and 10. The tertiary structure is stabilized by four disulfide bonds (Kader, 1996).

The LPT2 family comprises peptides with a molecular mass near 7 kDa. The other features of LPT2 include a highly basic isoelectric point, conserved cysteine residues, and the ability to transfer lipids (Castro et al., 2003; Douliez et al., 2001). One important common feature among these two families is the presence of a signal peptide, which varies in size from 21 to 27 for the LTP1 and 27 to 35 for the LTP2 family (Arondel et al., 1991; Kalla et al., 1994). The disulfide patterns of LTP1 and LTP2 are different from each other.

From in vitro assays, it is clear that some nonspecific lipid transfer proteins (LTP) possess antimicrobial activities (da Rocha Pitta & Galdino, 2010) along with their involvement in the formation of cutin, beta-oxidation, cell signaling, development, and growth (Carvalho & Gomes, 2007). LTPs have been described in *Oryza sativa*, *Vigna radiata* var. *radiata*, *Sorghum vulgare*, *Brassica napus*, *Daucus carota*, *Triticum aestivum*, *Arabidopsis thaliana*, and many other plant species. They are expressed in different parts of the plant (Kader, 1996; Sels et al., 2008).

Thionins

Thionins have been detected in several plant families, especially *Poaceae*, *Viscaceae*, and *Brassicaceae*. They are produced as preproproteins, but only the mature thionin has been isolated from plants. Balls and colleagues discovered purothionin from wheat flour in 1942. It can inhibit yeast growth in flour (Balls et al., 1942). Now, it is clear that these peptides are found redundantly in plant species. Over one hundred sequences of thionins have been described from plant species. Thionins are commonly expressed in leaves (Bohlmann et al., 1988), stems, roots, seeds (Ohtani et al., 1977), and flowers (Stec, 2006).

Thionins contain an even number of cysteine residues, which are chemically linked in the complex structure of these peptides to make a disulfide bond between two cysteine residues. They contain six or eight cysteine residues. Thionins were grouped into five classes based on the number of cysteine residues and the pattern of disulfide bond formation (Garcia-Olmedo et al., 1989). Later, they were grouped into four classes by placing hordothionins and leaf thionins in the same class (Bohlmann et al., 1994).

Class I

These peptides are common in wheat (Jones et al., 1982; Wada, 1982), barley (Bohlmann & Apel, 1987), oat (Bekes and Lasztity, 1981), and rye (Hernandezlucas et al., 1978). These thionins usually contain 45 amino acids, are highly basic with a +10 isoelectric point, and contain eight cysteine amino acids. They are mostly deposited in cereal seeds, for instance, Purothionin α1 (Ohtani et al., 1977), in leaves of cereals, for instance, barley leaf-thionin DB4 (Bohlmann & Apel, 1987) but are also found in the seeds of *Pyrularia pubera* (Vernon et al., 1985).

Class II

These thionins have a total of 46–47 amino acids and contain six cysteine residues. These are less basic than class I thionins with an isoelectric point of +7. Some examples are viscotoxin A3 (Schrader and Apel, 1991), crambin (Teeter et al., 1981), and THI2.1 (Epple et al., 1995) from *Arabidopsis thaliana*.

Class III

These thionins have 45–46 amino acids comprising six cysteine residues, and their isoelectric point is +7. They lack the cysteines 2 and 8, such as the neutral seed-specific thionin in wheat (Castagnaro et al., 1992).

Class IV

These peptides are without cysteine number 2. New cysteines are between numbers 4 and 5. Thionin 4–1 from *Tulipa gesneriana* (De Boer, 1994).

Arabidopsis Thionins

In the model plant *Arabidopsis thaliana*, two thionin genes were originally discovered using PCR of genomic DNA with degenerate primers. When the genome sequence became available, it turned out that Arabidopsis contained two more thionin genes. According to GeneChip data (Zimmermann et al., 2004), these four genes are expressed in different tissues and organs of the plant. THI2.1 and THI2.2 have all indications of antimicrobial activity (basic pI, lysine at N-terminus, tyrosine at position 13), and overexpression in Arabidopsis has been shown to lead resistance to *Fusarium oxysporum* with hyper branching of fungal membranes in vivo (Epple et al., 1997b). It was reported that when Arabidopsis thionin gene *Thi2.1* was cloned in tomato crop, it enhanced resistance of the crop against phytopathogens, i.e., *Fusarium oxysporum* f. sp. *lycopersici* and *Ralstonia solanacearum* (Chan et al., 2005). THI2.1 was expressed in the BVE-E6E7 bovine endothelial cell line, and the recombinant fusion protein was purified and tested against different bacterial and fungal pathogens. These in vitro studies strongly confirmed the antimicrobial activity of Arabidopsis thionin (THI2.1) and could be used to treat contagious diseases (Loeza-Angeles et al., 2008). In the case of THI2.3 and THI2.4, it was not clear if they have antimicrobial activity. THI2.3 is seed-specific and highly acidic (and the acidic domain is basic) and has phenylalanine instead of tyrosine at position 13, indicating it is not toxic. THI2.4 is only slightly basic and has no lysine at the N-terminus.

To predict the 3D structure of thionin proteins, the sequence of the mature peptides was submitted to Phyre^2 online software (Kelley and Sternberg, 2009). It resulted in molecular structures of all Arabidopsis thionins resembling crambin with more than 90% precision. According to these structures, the Arabidopsis thionin peptides consist of two alpha helices and two antiparallel beta sheaths. The amino acid sequence of prothionins was also submitted to 3D structure prediction, but it resulted in incomplete or scrambled structures.

Processing of Thionin Preproproteins

Thionins have been isolated as 5 kDa peptides from a variety of plant species but are produced as preproproteins, consisting of a signal peptide, thionin motif, and a C-terminal domain called "acidic domain" (Ponz et al., 1986). All these precursors have a conserved cleavage site for signal sequence. This signal sequence leads the proprotein, consisting of the thionin and the acidic domain, into the secretory pathway. Arabidopsis thionin proproteins contain six cysteine amino acids in the thionin motif and six in the acidic domain. As the name says, the acidic domain is usually acidic, but there are some exceptions, for instance, THI2.3 from Arabidopsis, but in these cases, the isoelectric point of the thionin is in the acidic range. There is no information available about the function of the acidic domain. Overexpression of only the thionin domain for α-hordothionin in tobacco plants resulted in lower expression of the mature peptide (Florack et al., 1994). It indicates a possible involvement of the acidic domain in guiding the proprotein into the secretory pathway (Liu and Ganz, 1995). The assistance of the pro region in folding bovine pancreatic trypsin inhibitor (BPTI) was studied in detail. During in vitro analysis, it became clear that the formation rate of this peptide and the yield of its properly folded form were in positive correlation (Weissman and Kim, 1992).

From this, we can assume a possible role for the acidic domain in folding the mature thionin peptide. Another possible function of the acidic domain might be to neutralize the toxicity or to provide a shield to the cell against its AMPs because, in the case of barley, a toxic effect of leaf thionins has been shown against tobacco protoplasts and barley protoplasts (Lee et al., 1995; Reimann-Philipp et al., 1989).

By comparison with other thionins that have been isolated from plants (for instance, (Bohlmann & Apel, 1987; Epple et al., 1995), it is possible to predict the processing of the Arabidopsis thionin precursors. All prothionins are basic, except for proTHI2.4, while all thionins are basic, except for THI2.3. Physical properties for thionins and thionin proproteins.

In Vitro Antimicrobial Activity of Thionins

The first thionin peptide was isolated by Balls et al. (1942) from wheat flour and was named purothionin. It was found to be toxic in in vitro studies. Later variants of this compound with minor structural differences were discovered (Bohlmann, 1994; Stec, 2006). Afterward, related peptides were discovered in the seeds of other cereals and in the leaves of several mistletoe species named viscotoxins. Thionins from the endosperm of cereals have been intensively studied because they were suspected to inhibit baker yeast, which is used for bread production. Thionins are toxic to pathogenic fungi, bacteria, and animal cell lines. Because of their action on the membranes of pathogens, they are lethal to many pathogens. This results in the leakage of low molecular weight compounds and ions from different cells following treatment with thionins. Purothionin from wheat flour showed antimicrobial activity (Wada, 1982), and hordothionin from barley and purothionin showed antimicrobial activity against yeast (Okada and Yoshizum, 1970). Hordothionin and cell wall extracted thionins of barley leaves showed antifungal activity against *T. paradoxa* and *D. teres* (Bohlmann et al., 1988) and *Pyrularia* thionin from nuts of *Pyrularia pubera,* purothionin and hordothionin from wheat and barley flour, respectively, were reported to be toxic to phytopathogenic bacteria (Cammue et al., 1992; Evett et al., 1986; Florack et al., 1993), along with many other reports available in the literature (Bohlmann, 1994; Bohlmann & Apel, 1991; Florack & Stiekema, 1994; Pelegrini & Franco, 2005). Viscotoxins have also been intensively studied because mistletoe extracts are used for cancer treatment in alternative medicine. They were shown to be cytotoxic to a variety of cell cultures and to kill laboratory animals if injected (Bohlmann, 1994; Bohlmann & Apel, 1991).

Expression Systems for AMP Production

In the innate immune systems, AMPs are any organism's first line of defense. AMPs are intensively studied to replace antibiotics because the pathogens are getting resistant to them. These peptides show variation in sequence, structure, toxicity, physical state, and chemical properties. AMPs can be isolated from natural sources to study their activity in vitro; however, this requires mg amounts of the peptide, and in many cases, this would be impractical. AMPs can be produced synthetically or as recombinant proteins in different expression systems. Normally, the cost of producing synthetic peptides is high. However, producing AMPs in microbial expression systems encounters two major problems. First, AMPs are often toxic to the microbial expression host. Second, plant peptides contain several disulfide bridges, which are normally not formed within the cytoplasm. Posttranslational modification is usually not a problem because AMPs from plants have mostly no posttranslational modifications.

Plants as an Expression System

In contrast to prokaryotic microbial expression systems, plants have a good potential for posttranslational modification and have emerged as interesting hosts for the expression of foreign proteins (Barbosa Viana et al., 2012). Recombinant protein expression can be done either by stable transformation or by transient expression with agroinfiltration. Expression in transgenic plants can use constitutive or inducible promoters, while constitutive promoters are normally used for transient expression. Choosing a suitable promoter is important for high yields and the production of peptides in plants and other proteins (Hernandez-Garcia et al., 2010). In constitutive expression, viral-based promoters are common due to their high expression level; however, this can vary in different tissues (Govindarajulu et al., 2008). High-level production of foreign proteins can be harmful to the plant, and in those cases, using inducible or tissue-specific promoters could be helpful to overcome the problem of cell toxicity (Egelkrout et al., 2012). The main drawbacks of using plants as protein expression systems for AMPs (and other proteins) are the isolation and purification of AMPs. Plant metabolites, for instance, can interfere with the purification (Fitzgerald, 2003). The total amount after purification varied depending on the host plant and the type of protein to be expressed. Reporter genes like GFP have been reported to accumulate to 40% of TSP in *N. benthamiana* using transient expression (Marillonnet et al., 2005; Sheludko et al., 2007), while for interferon α2B, hGH, and alpha-1-antitrypsin expression levels up to 10%, 10%, and 5.8%, respectively, of TSP have been reported (Gils et al., 2005; Plesha et al., 2007; Sheludko et al., 2007). Retrocyclin-101 (RC101) and Protegrin-1 (PG1) AMPs expressed as a fusion with GFP in the chloroplast accumulated up to 32% and 17%, respectively, of the TSP (Lee et al., 2011).

Yeast

In heterologous studies, *P. pastoris* has an advantage over *E. coli* in posttranslational changes (Cregg et al., 2009). Similarly, compared to animal cells, this expression system is preferred because of its higher growth rate and low cost of production. For prokaryotic protein expression, the choice of a fusion partner is important, but in yeast, purification is easier with high harvests (de Bruin et al., 2005). Among fungi, yeast (*Saccharomyces cerevisiae*) was first used as a recombinant protein expression system because it has advantages comparable to *E. coli*, such that it needs a short time for growth and also has the additional benefits of a eukaryotic system with the features of improved folding (Brondyk, 2009). Yeast was the first fungus used for recombinant protein expression (Strausberg and Strausberg, 2001), but in the last decade, extensive research on *P. pastoris* made it a preferred microbe for heterologous protein production. The ability of *P. pastoris* to use methanol as a source of organic carbon and energy is important to exploit this organism for recombinant protein production. Its genome is similar to *Saccharomyces cerevisiae*, and stable lines can be produced by foreign DNA insertion using homologous recombination methods. In *P. pastoris*, the expression of cysteine-rich peptides, which involve the formation of disulfide bonds, is well-established and is achievable in intra- and extracellular spaces (Cereghino et al., 2002; Cregg et al., 2009).

The alcohol oxidase I promoter (PAOXI) is regulated by methanol. In this system, most recombinant heterologous proteins are synthesized using the potential of alcohol oxidase I promoter (PAOXI). This promoter is repressed when cells are grown in a glucose medium, but very high induction occurs when the fungus uses methanol as an organic carbon source (Tschopp et al., 1987). Isolation of recombinant proteins from the extracellular medium is always preferable because of a low level of unwanted proteins. This microbial system is efficient in secretion into the medium because it does not secrete many proteins. It is considered a good host for the heterologous production of AMPs. A comparative study of a defensin peptide produced in *P. pastoris* and *E. coli* showed antimicrobial activity against *Fusarium graminearum*. However, it was more toxic when expressed in *P. pastoris* (Kant et al., 2009). It has been reported that up to 750 mg/l heterologous protein have been obtained by using this expression system (Parachin et al., 2012), but still, very few strains are available, and knowledge about induced/constitutive promoters for AMP expression is limited in this organism.

E. coli as an Expression System for Peptides and Proteins

The Gram-negative bacterium *E. coli* is preferably used for heterologous protein expression because of fast growth, background knowledge about genetics, and low cost of production. The availability of many expression vectors and protein expression strains deficient for proteases is another reason for *E. coli* to be a major microbial expression system for pharmaceutical and industrial purposes (Schmidt, 2004). Recombinant protein can be produced in a simple way both in the cytoplasm and in the periplasmic space. However, producing toxic proteins containing cysteine residues, such as AMPs, is still challenging. Despite some drawbacks, according to a recent report, 97% of all heterologously produced AMPs are expressed in *E. coli* or yeast cells (Li & Chen, 2008).

AMP Expression in *E. coli*

To express AMPs directly in *E. coli* is difficult because of the toxic nature of AMPs, which can be lethal to the host. It is, therefore, common practice to use carrier proteins to express AMPs as fusion proteins (Piers et al., 1993). These fusion partners can increase the solubility and protect the host organism against the toxic effects of AMPs. Normally, such fusion is supported by an affinity tag and a specific cleavage site to release the mature AMP after digestion. These fusion proteins are less toxic to the host, and AMPs regain their activity after the protease cleavage separation. The expression of enterocin A and enterocin B as a fusion with cellulose binding protein and s tag in the periplasm of *E. coli* directed by the native signal peptide resulted in recombinant protein, which could be digested with Factor Xa protease. After protease digestion, the mature enterocin peptide showed antimicrobial activity (Klocke et al., 2005). Commonly used fusion partners include maltose-binding protein, thioredoxin, glutathione-S-transferase, and NusA. Many other proteins are also used to enhance the solubility of recombinant proteins and to avoid protein storage in inclusion bodies (Parachin et al., 2012). Maltose-binding protein is native to *E. coli* and emerged as the best fusion partner to increase the solubility of six target peptides compared to thioredoxin and NusA (Kapust & Waugh, 1999). In a similar study, heterologous expression of viscotoxin was tested with 13 different fusion partners. Mass spectroscopy analysis showed the high purity of viscotoxin and the formation of disulfide bonds in the cytoplasm (Bogomolovas et al., 2009). In this study, thioredoxin emerged as the best fusion partner. With features of high solubility and chaperon-like behavior, thioredoxin was the fusion partner in 20% of all recombinantly expressed AMPs in *E. coli* (Li, 2009). The solubility of AMPs and thioredoxin fusions can be further increased by choosing low-temperature induction and suitable expression strains (Shlyapnikov et al., 2008).

E. coli Strains for Protein Expression

For the expression of fusion proteins in *E. coli*, promoters with high expression levels are preferred along with supportive host strains. On the other hand, the involvement of rare codons of the foreign proteins, mispairing of cysteine residues, and thermosensitivity of proteins during isolation and purification steps can result in misfolding of proteins (Berkmen, 2012). To avoid such problems, different strategies were adapted to minimize these structural and yield losses. The involvement of two different translation initiation factors from the hyperthermophilic sulfate-reducing archaeon *Archaeoglobus fulgidus* increased the expression of recombinant fusion protein in C41(DE3) and C43(DE3) strains over BL21(E3).

The two formerly discussed expression strains were raised by combining rare codons in the plasmid (Sorensen et al., 2003). In the case of AMPs, stability is not a problem. The presence of disulfide bridges stabilizes the structure and can help in the isolation and purification process of such peptides in heterologous form or from the source. However, if the formation of disulfide bonds is incorrect, it can result in misfolding of the protein. Until recently, the *E. coli* cytoplasm was not considered suitable to produce proteins with disulfide bridges due to its reduced activity (Locker & Griffiths, 1999) compared to secretion into the periplasmic space. However, the amount of protein that can be produced in the cytoplasm of *E. coli* is much higher than the amount of secretion into the periplasm.

The inability of the cytoplasm to support disulfide bond formation is mainly due to the active involvement of the pathways of thioredoxin reductase and glutaredoxin. It was found that the mutant strain for these reductases enhanced the expression of disulfide-containing protein in the cytoplasm of *E. coli* (Bessette et al., 1999). Recently, the SHuffle strain C3030 (New England Biolabs) was introduced in which the lethal phenotype of lacking the *gor* and *trxB* reductases is suppressed by a mutation in the peroxidase *ahpC* (Faulkner et al., 2008), thus providing an oxidizing environment in the cytoplasm. Additionally, expression of the periplasmic disulfide bond isomerase (DsbC) in the cytoplasm of these strains is supposed to enhance proper disulfide bond formation (Maskos et al., 2003; Tait & Straus, 2011).

Periplasmic Expression

The periplasm of *E. coli* contains a single chaperon (Skp) to capture unfolded proteins with other modulators. The important feature of the periplasm is the oxidizing environment. In prokaryotes like *E. coli*, periplasm is the only place where disulfide-containing proteins can be synthesized due to the presence of oxidoreductases known as Dsb proteins (Kadokura et al., 2003). The properties of this protein family have been extensively characterized; DsbA has a molecular weight of 21 kDa and is a strong oxidase; it oxidizes the substrate when it arrives in the periplasm. Reduced DsbA again oxidizes by transferring electrons to DsbB, an inner membrane protein. The reduced DsbB then donates its electron to ubiquinone and is again ready to reduce DsbA. In the process of disulfide bond formation, DsbC is used as a proofreading mechanism (Kadokura et al., 2004). The reduced state of DsbC is maintained by the inner membrane protein DsbD using the energy of NADH oxidation, which is in the cytoplasm of bacteria.

Protein expression as a fusion protein in the periplasm provides a simple way of purifying the protein of interest with the help of affinity tags. The chances of contaminating proteins are also low. The proteins are harvested by an osmotic shock procedure (Jonasson et al., 2002). The use of different signal peptides at the N-terminus of the protein leads to the fusion of proteins into the periplasm. The signal peptide contains three parts: hydrophobic h region, positively charged n region, and c region, which is neutral. The c-c region provides the cleavage site for the signal peptidase (Yoon et al., 2010). During the secretory pathway of proteins from the cytoplasm to the periplasm, the signal sequence is cleaved by signal peptidase to yield a mature protein product. The signal peptide is inserted at the N terminus to secrete a fusion protein into the periplasm. Frequently used signal peptides for recombinant protein expression in the periplasm of *E. coli* are derived from LamB, OmpA, PhoA, MalE, PelB, LTB, and OmpT (Blight et al., 1994; Choi & Lee, 2004). The disadvantage of protein expression in the periplasm is the low yield, which might be acceptable for research purposes but is unacceptable for commercial production of proteins or peptides.

References

Arondel, V., Tchang, F., Baillet, B., Vignols, F., Grellet, F., Delseny, M., Kader, J. C., & Puigdomenech, P. (1991). Multiple mRNA coding for phospholipid-transfer protein from Zea mays arise from alternative splicing. *Gene, 99*, 133–136.

Balls, A. K., Hale, W. S., & Harris, T. H. (1942). A crystalline protein obtained from a lipoprotein of wheat flour. *Cereal Chemistry, 19*, 279–281.

Barbosa Viana, A. A., Pelegrini, P. B., & Grossi-de-Sá, M. F. (2012). Plant biofarming: Novel insights for peptide expression in heterologous systems. *Peptide Science, 98*, 416–427.

Bekes, F., & Lasztity, R. (1981). Isolation and determination of amino-acid-sequence of avenothionin, a new purothionin analog from oat. *Cereal Chemistry, 58*, 360–361.

Berkmen, M. (2012). Production of disulfide-bonded proteins in Escherichia coli. *Protein Expression and Purification, 82*, 240–251.

Bessette, P. H., Aslund, F., Beckwith, J., & Georgiou, G. (1999). Efficient folding of proteins with multiple disulfide bonds in the Escherichia coli cytoplasm. *Proceedings of the National Academy of Sciences, 96*, 13703–13708.

Blight, M. A., Chervaux, C., & Holland, I. B. (1994). Protein secretion pathway in Escherichia coli. *Current Opinion in Biotechnology, 5*, 468–474.

Bogomolovas, J., Simon, B., Sattler, M., & Stier, G. (2009). Screening of fusion partners for high yield expression and purification of bioactive viscotoxins. *Protein Expression and Purification, 64*, 16–23.

Bohlmann, H. (1994). The role of thionins in plant-protection. *Critical Reviews in Plant Sciences, 13*, 1–16.

Bohlmann, H., & Apel, K. (1987). Isolation and characterization of cDNAs coding for leaf-specific thionins closely related to the endosperm-specific hordothionin of barley (Hordeum-Vulgare-L). *Molecular & General Genetics, 207*, 446–454.

Bohlmann, H., & Apel, K. (1991). Thionins. *Annual Review of Plant Physiology, 42*, 227–240.

Bohlmann, H., Clausen, S., Behnke, S., Giese, H., Hiller, C., Reimannphilipp, U., Schrader, G., Barkholt, V., & Apel, K. (1988). Leaf-specific thionins of barley—A novel class of cell-wall proteins toxic to plant-pathogenic fungi and possibly involved in the defense-mechanism of plants. *The EMBO Journal, 7*, 1559–1565.

Bohlmann, H., Apel, K., & Garcia-Olmedo, F. (1994). Thionins. *Plant Molecular Biology Reporter, 12*, S75.

Brondyk, W. H. (2009). Selecting an appropriate method for expressing a recombinant protein. *Methods in Enzymology, 463*, 131–147.

Cammue, B. P. A., Debolle, M. F. C., Terras, F. R. G., Proost, P., Vandamme, J., Rees, S. B., Vanderleyden, J., & Broekaert, W. F. (1992). Isolation and characterization of a novel class of plant antimicrobial peptides from Mirabilis-Jalapa L seeds. *The Journal of Biological Chemistry, 267*, 2228–2233.

Carvalho, A. D., & Gomes, V. M. (2007). Role of plant lipid transfer proteins in plant cell physiology—A concise review. *Peptides, 28*, 1144–1153.

Castagnaro, A., Marana, C., Carbonero, P., & Garciaolmedo, F. (1992). Extreme divergence of a novel wheat thionin generated by a mutational burst specifically affecting the mature protein domain of the precursor. *Journal of Molecular Biology, 224*, 1003–1009.

Castro, M. S., Gerhardt, I. R., Orru, S., Pucci, P., & Bloch, C. (2003). Purification and characterization of a small (7.3 kDa) putative lipid transfer protein from maize seeds. *Journal of Chromatography B, 794*, 109–114.

Cereghino, G. P. L., Cereghino, J. L., Ilgen, C., & Cregg, J. M. (2002). Production of recombinant proteins in fermenter cultures of the yeast Pichia pastoris. *Current Opinion in Biotechnology, 13*, 329–332.

Chan, Y. L., Prasad, V., Sanjaya, C., & K.H., Liu, P.C., Chan, M.T., Chiu-Ping, C. (2005). Transgenic tomato plants expressing an Arabidopsis thionin (Thi2.1) driven by fruit-inactive promoter battle against phytopathogenic attack. *Planta, 221*, 386–393.

Chen, J. J., Chen, G. H., Hsu, H. C., Li, S. S., & Chen, C. S. (2004). Cloning and functional expression of a mungbean defensin VrD1 in Pichia pastoris. *Journal of Agricultural and Food Chemistry, 52*, 2256–2261.

Choi, J. H., & Lee, S. Y. (2004). Secretory and extracellular production of recombinant proteins using Escherichia coli. *Applied Microbiology and Biotechnology, 64*, 625–635.

Colilla, F. J., Rocher, A., & Mendez, E. (1990). Gamma-purothionins: Amino acid sequence of two polypeptides of a new family of thionins from wheat endosperm. *FEBS Letters, 270*, 191–194.

Cregg, J. M., Tolstorukov, I., Kusari, A., Sunga, J., Madden, K., & Chappell, T. (2009). Expression in the yeast Pichia pastoris. *Methods in Enzymology, 463*, 169–189.

da Rocha Pitta, M. G., & Galdino, S. L. (2010). Development of novel therapeutic drugs in humans from plant antimicrobial peptides. *Current Protein and Peptide Science, 11*, 236–247.

De Boer, D. A. D. (1994). *Tulipa gesneriana thionin cDNAs*. EMBL database.
de Bruin, E. C., Duitman, E. H., de Boer, A. L., Veenhuis, M., Bos, I. G., & Hack, C. E. (2005). Pharmaceutical proteins from methylotrophic yeasts. *Methods in Molecular Biology, 308*, 65–76.
Douliez, J. P., Pato, C., Rabesona, H., Molle, D., & Marion, D. (2001). Disulfide bond assignment, lipid transfer activity and secondary structure of a 7-kDa plant lipid transfer protein, LTP2. *European Journal of Biochemistry, 268*, 1400–1403.
Durrant, W. E., & Dong, X. (2004). Systemic acquired resistance. *Annual Review of Phytopathology, 42*, 185–209.
Egelkrout, E., Rajan, V., & Howard, J. A. (2012). Overproduction of recombinant proteins in plants. *Plant Science, 184*, 83–101.
Epple, P., Apel, K., & Bohlmann, H. (1995). An Arabidopsis-thaliana thionin gene is inducible via a signal-transduction pathway different from that for pathogenesis-related proteins. *Plant Physiology, 109*, 813–820.
Epple, P., Apel, K., & Bohlmann, H. (1997b). Overexpression of an endogenous thionin enhances resistance of Arabidopsis against Fusarium oxysporum. *Plant Cell, 9*, 509–520.
Evett, G. E., Donaldson, D. M., & Vernon, L. P. (1986). Biological properties of Pyrularia-thionin prepared from nuts of Pyrularia-pubera. *Toxicon, 24*, 622–625.
Faulkner, M. J., Veeravalli, K., Gon, S., Georgiou, G., & Beckwith, J. (2008). Functional plasticity of a peroxidase allows evolution of diverse disulfide-reducing pathways. *Proceedings of the National Academy of Sciences, 105*, 6735–6740.
Fitzgerald, D. A. (2003). Revving up the green express. *Scientist, 17*, 45–47.
Florack, D. E. A., & Stiekema, W. J. (1994). Thionins—Properties, possible biological roles and mechanisms of action. *Plant Molecular Biology, 26*, 25–37.
Florack, D. E. A., Visser, B., Devries, P. M., Vanvuurde, J. W. L., & Stiekema, W. J. (1993). Analysis of the toxicity of Purothionins and Hordothionins for plant-pathogenic bacteria. *Netherlands Journal of Plant Pathology, 99*, 259–268.
Florack, D. E. A., Dirkse, W. G., Visser, B., Heidekamp, F., & Stiekema, W. J. (1994). Expression of biologically-active hordothionins in tobacco. Effects of pre-sequences and pro-sequences at the amino and carboxyl termini of the hordothionin precursor on mature protein expression and sorting. *Plant Molecular Biology, 24*, 83–96.
Gaffney, T., Friedrich, L., Vernooij, B., Negrotto, D., Nye, G., Uknes, S., Ward, E., Kessmann, H., & Ryals, J. (1993). Requirement of salicylic-acid for the induction of systemic acquired-resistance. *Science, 261*, 754–756.
Garcia-Olmedo, F., Rodriguez-Palenzuela, P., Hernandez-Lucas, C., Ponz, F., Marana, C., Carmona, M. J., Lopez-Fando, J., Fernandez, J. A., & Carbonero, P. (1989). The thionins: A protein family that includes purothionins, viscotoxins and crambins. *Plant Molecular and Cell Biology, 6*, 31–60.
Gils, M., Kandzia, R., Marillonnet, S., Klimyuk, V., & Gleba, Y. (2005). High-yield production of authentic human growth hormone using a plant virus-based expression system. *Plant Biotechnology Journal, 3*, 613–620.
Gomez-Gomez, L., & Boller, T. (2000). FLS2: An LRR receptor-like kinase involved in the perception of the bacterial elicitor flagellin in Arabidopsis. *Molecular Cell, 5*, 1003–1011.
Govindarajulu, M., Elmore, J. M., Fester, T., & Taylor, C. G. (2008). Evaluation of constitutive viral promoters in transgenic soybean roots and nodules. *Molecular Plant-Microbe Interactions, 21*, 1027–1035.
Greenberg, J. T. (1997). Programmed cell death in plant-pathogen interactions. *Annual Review of Plant Physiology, 48*, 525–545.
Hernandezlucas, C., Carbonero, P., & Garciaolmedo, F. (1978). Identification and purification of a purothionin homolog from rye (Secale-cereale L). *Journal of Agricultural and Food Chemistry, 26*, 794–796.

Hernandez-Garcia, C. M., Bouchard, R. A., Rushton, P. J., Jones, M. L., Chen, X. F., Timko, M. P., & Finer, J. J. (2010). High level transgenic expression of soybean (Glycine max) GmERF and Gmubi gene promoters isolated by a novel promoter analysis pipeline. *BMC Plant Biology, 10.*

Jonasson, P., Liljeqvist, S., Nygren, P. A., & Stahl, S. (2002). Genetic design for facilitated production and recovery of recombinant proteins in Escherichia coli. *Biotechnology and Applied Biochemistry, 35*, 91–105.

Jones, J. D. G., & Dangl, J. L. (2006). The plant immune system. *Nature, 444*, 323–329.

Jones, B. L., Lookhart, G. L., Mak, A., & Cooper, D. B. (1982). Sequences of purothionins and their inheritance in diploid, tetraploid, and hexaploid wheats. *The Journal of Heredity, 73*, 143–144.

Kader, J. C. (1996). Lipid-transfer proteins in plants. *Annual Review of Plant Physiology, 47*, 627–654.

Kadokura, H., Katzen, F., & Beckwith, J. (2003). Protein disulfide bond formation in prokaryotes. *Annual Review of Biochemistry, 72*, 111–135.

Kadokura, H., Tian, H. P., Zander, T., Bardwell, J. C. A., & Beckwith, J. (2004). Snapshots of DsbA in action: Detection of proteins in the process of oxidative folding. *Science, 303*, 534–537.

Kalla, R., Shimamoto, K., Potter, R., Nielsen, P. S., Linnestad, C., & Olsen, O. A. (1994). The promoter of the barley aleurone-specific gene encoding a putative 7-Kda lipid transfer protein confers aleurone cell-specific expression in transgenic rice. *The Plant Journal, 6*, 849–860.

Kant, P., Liu, W. Z., & Pauls, K. P. (2009). PDC1, a corn defensin peptide expressed in Escherichia coli and Pichia pastoris inhibits growth of Fusarium graminearum. *Peptides, 30*, 1593–1599.

Kapust, R. B., & Waugh, D. S. (1999). Escherichia coli maltose-binding protein is uncommonly effective at promoting the solubility of polypeptides to which it is fused. *Protein Science, 8*, 1668–1674.

Kelley, L. A., & Sternberg, M. J. E. (2009). Protein structure prediction on the Web: A case study using the Phyre server. *Nature Protocols, 4*, 363–371.

Klocke, M., Mundt, K., Idler, F., Jung, S., & Backhausen, J. E. (2005). Heterologous expression of enterocin A, a bacteriocin from Enterococcus faecium, fused to a cellulose-binding domain in Escherichia coli results in a functional protein with inhibitory activity against listeria. *Applied Microbiology and Biotechnology, 67*, 532–538.

Kushmerick, C., Castro, M. D., Cruz, J. S., Bloch, C., & Beirao, P. S. L. (1998). Functional and structural features of gamma-zeathionins, a new class of sodium channel blockers. *FEBS Letters, 440*, 302–306.

Lay, F. T., & Anderson, M. A. (2005). Defensins—Components of the innate immune system in plants. *Current Protein and Peptide Science, 6*, 85–101.

Lay, F. T., Brugliera, F., & Anderson, M. A. (2003). Isolation and properties of floral defensins from ornamental tobacco and petunia. *Plant Physiology, 131*, 1283–1293.

Lee, D. W., Apel, K., & Binding, H. (1995). The formation of leaf thionins and their effect on the viability of isolated protoplasts of barley (Hordeum-Vulgare L). *Journal of Plant Physiology, 145*, 131–136.

Lee, S. B., Li, B. C., Jin, S. X., & Daniell, H. (2011). Expression and characterization of antimicrobial peptides Retrocyclin-101 and Protegrin-1 in chloroplasts to control viral and bacterial infections. *Plant Biotechnology Journal, 9*, 100–115.

Li, Y. F. (2009). Carrier proteins for fusion expression of antimicrobial peptides in Escherichia coli. *Biotechnology and Applied Biochemistry, 54*, 1–9.

Li, Y., & Chen, Z. (2008). RAPD: A database of recombinantly-produced antimicrobial peptides. *FEMS Microbiology Letters, 289*, 126–129.

Locker, J. K., & Griffiths, G. (1999). An unconventional role for cytoplasmic disulfide bonds in vaccinia virus proteins. *The Journal of Cell Biology, 144*, 267–279.

Loeza-Angeles, H., Sagrero-Cisneros, E., Lara-Zarate, L., Villagomez-Gomez, E., Lopez-Meza, J. E., & Ochoa-Zarzosa, A. (2008). Thionin Thi2.1 from Arabidopsis thaliana expressed in endothelial cells shows antibacterial, antifungal and cytotoxic activity. *Biotechnology Letters, 30*, 1713–1719.

Liu, L., & Ganz, T. (1995). The pro region of human neutrophil defensin contains a motif that is essential for normal subcellular sorting. *Blood, 85*, 1095–1103.

Marillonnet, S., Thoeringer, C., Kandzia, R., Klimyuk, V., & Gleba, Y. (2005). Systemic Agrobacterium tumefaciens-mediated transfection of viral replicons for efficient transient expression in plants. *Nature Biotechnology, 23*, 718–723.

Maskos, K., Huber-Wunderlich, M., & Glockshuber, R. (2003). DsbA and DsbC-catalyzed oxidative folding of proteins with complex disulfide bridge patterns in vitro and in vivo. *Journal of Molecular Biology, 325*, 495–513.

Mendez, E., Moreno, A., Colilla, F., Pelaez, F., Limas, G. G., Mendez, R., Soriano, F., Salinas, M., & de Haro, C. (1990). Primary structure and inhibition of protein synthesis in eukaryotic cell-free system of a novel thionin, gamma-hordothionin, from barley endosperm. *European Journal of Biochemistry/FEBS, 194*, 533–539.

Ohtani, S., Okada, T., Yoshizumi, H., & Kagamiyama, H. (1977). Complete primary structures of 2 subunits of purothionin-a, a lethal protein for brewers-yeast from wheat-flour. *The Journal of Biochemistry–Tokyo, 82*, 753–767.

Okada, T., & Yoshizum, H. (1970). A lethal toxic substance for brewing yeast in wheat and barley .2. Isolation and some properties of toxic principle. *Agricultural and Biological Chemistry Tokyo, 34*, 1089.

Parachin, N. S., Mulder, K. C., Viana, A. A., Dias, S. C., & Franco, O. L. (2012). Expression systems for heterologous production of antimicrobial peptides. *Peptides, 38*, 446–456.

Pelegrini, P. B., & Franco, O. L. (2005). Plant gamma-thionins: Novel insights on the mechanism of action of a multi-functional class of defense proteins. *The International Journal of Biochemistry & Cell Biology, 37*, 2239–2253.

Pelegrini, P. B., Murad, A. M., Silva, L. P., dos Santos, R. C. P., Costa, F. T., Tagliari, P. D., Bloch, C., Noronha, E. F., Miller, R. N. G., & Franco, O. L. (2008). Identification of a novel storage glycine-rich peptide from guava (Psidium guajava) seeds with activity against Gram-negative bacteria. *Peptides, 29*, 1271–1279.

Piers, K. L., Brown, M. H., & Hancock, R. E. W. (1993). Recombinant-DNA procedures for producing small antimicrobial cationic peptides in bacteria. *Gene, 134*, 7–13.

Plesha, M. A., Huang, T. K., Dandekar, A. M., Falk, B. W., & McDonald, K. A. (2007). High-level transient production of a heterologous protein in plants by optimizing induction of a chemically inducible viral amplicon expression system. *Biotechnology Progress, 23*, 1277–1285.

Ponz, F., Pazares, J., Hernandezlucas, C., Garciaolmedo, F., & Carbonero, P. (1986). Cloning and nucleotide-sequence of a Cdna-encoding the precursor of the barley toxin alpha-hordothionin. *European Journal of Biochemistry, 156*, 131–135.

Reimann-Philipp, U., Schrader, G., Martinoia, E., Barkholt, V., & Apel, K. (1989). Intracellular thionins of barley. A second group of leaf thionins closely related to but distinct from cell wall-bound thionins. *The Journal of Biological Chemistry, 264*, 8978–8984.

Rogozhin, E. A., Odintsova, T. I., Musolyamov, A. K., Smirnov, A. N., Babakov, A. V., Egorov, T. A., & Grishin, E. V. (2009). The purification and characterization of a novel lipid transfer protein from caryopsis of barnyard grass (Echinochloa crusgalli). *Applied Biochemistry and Microbiology, 45*, 363–368.

Schmidt, F. R. (2004). Recombinant expression systems in the pharmaceutical industry. *Applied Microbiology and Biotechnology, 65*, 363–372.

Schrader, G., & Apel, K. (1991). Isolation and characterization of Cdnas encoding viscotoxins of mistletoe (Viscum-album). *European Journal of Biochemistry, 198*, 549–553.

Segura, A., Moreno, M., Molina, A., & Garcia-Olmedo, F. (1998). Novel defensin subfamily from spinach (Spinacia oleracea). *FEBS Letters, 435*, 159–162.

Sels, J., Mathys, J., De Coninck, B. M. A., Cammue, B. P. A., & De Bolle, M. F. C. (2008). Plant pathogenesis-related (PR) proteins: A focus on PR peptides. *Plant Physiology and Biochemistry, 46*, 941–950.

Sheludko, Y. V., Sindarovska, Y. R., Gerasymenko, I. M., Bannikova, M. A., & Kuchuk, N. V. (2007). Comparison of several Nicotiana species as hosts for high-scale Agrobacterium-mediated transient expression. *Biotechnology and Bioengineering, 96*, 608–614.

Shlyapnikov, Y. M., Andreev, Y. A., Kozlov, S. A., Vassilevski, A. A., & Grishin, E. V. (2008). Bacterial production of latarcin 2a, a potent antimicrobial peptide from spider venom. *Protein Expression and Purification, 60*, 89–95.

Sorensen, H. P., Sperling-Petersen, H. U., & Mortensen, K. K. (2003). Production of recombinant thermostable proteins expressed in Escherichia coli: Completion of protein synthesis is the bottleneck. *Journal of Chromatography B, 786*, 207–214.

Stec, B. (2006). Plant thionins—The structural perspective. *Cellular and Molecular Life Sciences, 63*, 1370–1385.

Strausberg, R. L., & Strausberg, S. L. (2001). Overview of protein expression in Saccharomyces cerevisiae. *Current Protocols in Protein Science Chapter, 5*(Unit5), 6.

Tait, A. R., & Straus, S. K. (2011). Overexpression and purification of U24 from human herpesvirus type-6 in E. coli: Unconventional use of oxidizing environments with a maltose binding protein-hexahistine dual tag to enhance membrane protein yield. *Microbial Cell Factories, 10*, 51.

Teeter, M. M., Mazer, J. A., & L'Italien, J. J. (1981). Primary structure of the hydrophobic plant protein crambin. *Biochemistry, 20*, 5437–5443.

Terras, F. R. G., Eggermont, K., Kovaleva, V., Raikhel, N. V., Osborn, R. W., Kester, A., Rees, S. B., Torrekens, S., Vanleuven, F., Vanderleyden, J., Cammue, B. P. A., & Broekaert, W. F. (1995). Small cysteine-rich antifungal proteins from radish—Their role in host-defense. *Plant Cell, 7*, 573–588.

Thevissen, K., Osborn, R. W., Acland, D. P., & Broekaert, W. F. (2000). Specific binding sites for an antifungal plant defensin from dahlia (Dahlia merckii) on fungal cells are required for antifungal activity. *Molecular Plant-Microbe Interactions, 13*, 54–61.

Tschopp, J. F., Brust, P. F., Cregg, J. M., Stillman, C. A., & Gingeras, T. R. (1987). Expression of the lacZ gene from two methanol-regulated promoters in Pichia pastoris. *Nucleic Acids Research, 15*, 3859–3876.

van Loon, L. C., Rep, M., & Pieterse, C. M. J. (2006). Significance of inducible defense-related proteins in infected plants. *Annual Review of Phytopathology, 44*, 135–162.

Vernon, L. P., Evett, G. E., Zeikus, R. D., & Gray, W. R. (1985). A Toxic thionin from Pyrularia-pubera - purification, properties, and amino-acid sequence. *Archives of Biochemistry and Biophysics, 238*, 18–29.

Wada, K. (1982). Localization of purothionins and genome expression in wheat plants. *Plant & Cell Physiology, 23*, 1357–1361.

Weissman, J. S., & Kim, P. S. (1992). The pro region of Bpti facilitates folding. *Cell, 71*, 841–851.

Yoon, S. H., Kim, S. K., & Kim, J. F. (2010). Secretory production of recombinant proteins in Escherichia coli. *Recent Patents on Biotechnology, 4*, 23–29.

Zhang, N. Y., Jones, B. L., & Tao, H. P. (1997). Purification and characterization of a new class of insect alphaamylase inhibitors from barley. *Cereal Chemistry, 74*, 119–122.

Zimmermann, P., Hirsch-Hoffmann, M., Hennig, L., & Gruissem, W. (2004). Genevestigator. Arabidopsis microarray database and analysis toolbox. *Plant Physiology, 136*, 2621–2632.

Chapter 6
Plant Metabolomics

Elsayed Fathi Abd-Allah, Abeer Hashem, and Ghalia Nadeem

Abstract The demand for functional genomics has significantly increased in recent years, accelerating the development in other omics fields, such as metabolomics. Metabolomics is a rapidly developing field that is used for the identification and quantification of small metabolites that are present in biological systems, including the cells and tissues of plants. This chapter provides an overview of the rapidly developing field of metabolomics and its use in plant research. It outlines the main analytical methods frequently used in plant metabolomics, including mass spectrometry and nuclear magnetic resonance spectroscopy, and how they may be used to identify and quantify metabolites. It highlights the basic information regarding the genetic and biochemical mechanisms underlying plant development, growth, and response to stress with an emphasis on the contributions of metabolomics to real-world uses such as crop quality improvement, determining the safety of food, and plant breeding. This chapter also emphasizes this developing sector's existing status and potential to facilitate more research that might result in better crop production and quality.

Keywords Metabolites · LC · MS · Hydrodistillation · Metabolome · Chromatography

---

E. F. Abd-Allah (✉)
Plant Production Department, College of Food and Agricultural Sciences, King Saud University, Riyadh, Saudi Arabia
e-mail: eabdallah@ksu.edu.sa

A. Hashem
Botany and Microbiology Department, College of Science, King Saud University, Riyadh, Saudi Arabia

G. Nadeem
Centre of Agricultural Biochemistry and Biotechnology, University of Agriculture Faisalabad, Faisalabad, Pakistan

S. Ijaz et al. (eds.), *Trends in Plant Biotechnology*,
https://doi.org/10.1007/978-981-97-0814-7_6

Introduction

It is well known that the plant kingdom has a wide variety of metabolites, with an estimated 1,000,000 distinct compounds. As part of their defense systems against pathogens and environmental challenges, plants create small chemicals known as metabolites (Yonekura-Sakakibara et al., 2019; Rao & Ravishankar, 2002). Due to their immobility, plants are more susceptible to environmental alterations and assault from other organisms, which can harm their growth and development. Therefore, the ability of plants to survive depends heavily on complex adaptive responses that entail signal transduction, stress detection, and activation of several stress-related genes and metabolites. Plant metabolism is crucial to control various developmental processes that allow plants to withstand these environmental challenges. Understanding how plants modify their metabolism in response to changing conditions has been made possible by measuring primary metabolites such as sugars, amino acids, and organic acids. Additionally, the evaluation of particular secondary metabolites, such as phytohormones (cytokinins, gibberellins, auxins, abscisic acid, and salicylic acid), which are essential for signaling and communication between plants and their environment, has greatly advanced our knowledge of the development and defense mechanisms of plants (Jorge et al., 2016a, 2016b).

Studying the plant metabolome can be challenging because plants contain diverse metabolites, including carbohydrates, amino acids, sugars, flavonoids, and alkaloids. Another difficulty in studying the plant metabolome is the changing nature of plant metabolites. Numerous genetic and environmental variables, including nutrition availability, temperature and light intensity variations, and biotic or abiotic stress, can impact a plant's metabolome. Additionally, several variables, including the plant species, stage of development, tissue type, and environmental circumstances, affect the concentration of metabolites in plants (Schripsema, 2010; Borges et al., 2017). The biological activities associated with these metabolites are often related to their biochemical structures. Thus, it is crucial to investigate plant metabolites and their structural makeup to comprehend their biological functions and prospective therapeutic applications (Thoppil & Bishayee, 2011; Ullah et al., 2020).

The study of small molecules, or metabolites, found in cells, tissues, and organisms is called metabolomics. It has grown in relevance in the twenty-first century as a relatively young discipline (Roessner & Bowne, 2009; Augustijn et al., 2021). A subfield of metabolomics called "plant metabolomics" aims to understand the variety of plant metabolites. Plant metabolomics is a quantitative approach that can enhance plant biochemistry and metabolism knowledge. It provides exact measurements of the quantities of several known metabolites in plant samples, which can then be submitted for statistical analysis (Fiehn et al., 2000). The study and measurement of plant metabolites found in a sample and understanding of how their levels are altered in response to environmental stimuli, such as nutrient availability or stress, are performed using various analytical methods, such as mass spectrometry and NMR spectroscopy. To improve the separation and identification of plant metabolites, mass spectrometry is used with chromatography methods, including liquid chromatography (LC) and gas chromatography (GC) (Augustijn et al., 2021). These are used to detect a wide variety of metabolites, including secondary metabolites such as alkaloids, flavonoids, and terpenoids, as well as primary metabolites such as amino acids, sugars, and organic acids (Carrieri et al., 2009).

Plant metabolomics is a cutting-edge science expanding quickly and has applications in various scientific fields, such as agriculture, health, and industry. The primary goal is the identification and measurement of a wide variety of plant metabolites. Metabolomics may identify metabolites related to plant characteristics such as disease and insect resistance, nutritive value, or yield (Mashabela et al., 2022). It may be used to produce new plant varieties that can adapt to changing environmental circumstances or to devise strategies for enhancing plant growth, yield, and resilience to environmental stress (Hong et al., 2016). It is an effective method that can offer insightful information on the biochemical networks that underlie plant metabolism. It can assist researchers in identifying new targets for crop development, monitoring plant health and quality, and improving plant breeding (Hall et al., 2008; Sharma et al., 2018).

History

The cutting-edge scientific area of metabolomics thoroughly examines small metabolites found in biological samples. Building on advancements in other "omics" sciences, such as genomics and proteomics, the scientific field of metabolomics evolved in the late 1990s and early 2000s. Steven Oliver initially proposed the term "metabolomics" in the 1990s, and it was first published in an article in 1998 (Alseekh & Fernie, 2018). The main objective of the rapidly growing field of plant metabolomics is the study of all the metabolites found in plants, including primary and secondary metabolites. Bienaimé Caventou, a French chemist, originally identified the plant pigment chlorophyll in 1817, which marked the beginning of the study of plant metabolomics. To better understand plants' biological functions and chemical makeup, additional metabolites, such as phytohormones, alkaloids, and several others, were subsequently found and isolated (Katam et al., 2022).

Plant compounds have been extracted and purified since the nineteenth century (Fitzgerald et al., 2020). In 1903, Russian botanist Mikhail Tswett utilized chromatography as the first analytical method to separate plant pigments (Walters, 2019). Since then, various plant metabolites, including carbohydrates, amino acids, and alkaloids, have been identified. Plant metabolism research started in the 1980s and 1990 when scientists used these advanced analytical techniques (Dunn et al., 2005). Initially, studies focused on analyzing individual metabolic compounds, such as amino acids and carbohydrates, in specific plant tissues or under various environmental circumstances. The ability to detect, extract, and measure metabolites on a large scale was made feasible using modern analytical methods, such as nuclear magnetic resonance spectroscopy and mass spectrometry (Griffiths, 2008). Plant metabolomics, a broad field of study, helps analyze and investigate many biological processes, such as plant growth and development and stress responses. It has also identified several novel plant metabolites with potential uses in industry, pharmaceuticals, and agriculture (Sharma et al., 2021).

Types of Plant Metabolites

Numerous metabolites are produced by plants, which can be generally divided into primary and secondary metabolites. Primary metabolites are vital for plants to survive, thrive, and protect themselves from infections and other stresses (Zaynab et al., 2019). Modern chemistry and biology have helped shed light on the role of primary metabolites in essential biological activities such as cell division, respiration, and storage (Bourgaud et al., 2001). Plants produce these compounds through regular metabolic pathways and are present in almost all plants. Carbohydrates, amino acids, nucleic acids, organic acids, and lipids are some examples of primary metabolites (Chinou, 2008).

Secondary metabolites are substances that plants need to protect themselves against diseases and herbivores but are not directly necessary for their growth and development. Albrecht Kossel, a German scientist, coined the term "secondary metabolite" in 1891 (Thirumurugan et al., 2018). Plants produce these chemicals, but only in very minute amounts. In higher plants, these secondary compounds are synthesized from carbohydrates, amino acids, and lipids. These compounds are usually stored in specific cells, tissues, or organs of plants and are released in response to tissue damage or specific signals, i.e., temperature changes, drought, salinity, supply of water, CO_2, and minerals, and pathogen attacks. This feature enables plants to adapt to changing environmental conditions and overcome various biotic and abiotic stress conditions. These compounds are reported to have antifungal, antibiotic, and antiviral properties and, therefore, can provide resistance to pathogens in plants. Secondary metabolites also play a role in plants' distinct aromas, flavors, and colors. Recently, these compounds have been suggested as attractive candidates for use as fine chemicals, cosmetics, and nutraceuticals. In addition, these compounds have pharmacological properties and have been used in traditional medicine for centuries. Secondary metabolites in plants, such as phenolic compounds, nitrogen compounds, and carotenoids, have attracted attention in pharmacy for their antimicrobial, anti-inflammatory, and antioxidant properties (Hussein & El-Anssary, 2019; Akbari et al., 2022).

Secondary metabolites are structurally diverse compounds, many of which have been discovered. The basic properties of secondary metabolites in plants are their distinct carbon skeleton structures. The secondary compounds in plants are classified based on their biosynthetic pathways. Secondary metabolites in plants are alkaloids, flavonoids, phenolics, and terpenoids (Bourgaud et al., 2001). Alkaloids are organic substances that assist plants in protecting themselves against different herbivores and insects (Ali et al., 2019). Plants require flavonoids and terpenoids to combat oxidative stress (Graßmann, 2005; Kumar & Pandey, 2013).

Experimental Draft for Plant Metabolomics

For a good experimental design in plant metabolomics, it is critical to consider the preliminary stages of the experiment. This requires choosing the proper growing parameters, such as temperature, relative humidity, and photoperiod, and applying randomization techniques throughout the workflow. Additionally, it is advised to monitor any environmental changes that can occur throughout the experiment, such as little variation in temperature or light intensity in the greenhouse or growth chamber, which may significantly impact the plant's biochemical condition (Jorge et al., 2016a, 2016b).

Sample Preparation

Sample preparation is a critical technique in plant metabolomics that can significantly impact the accuracy and quality of the results. Each step of this process needs to be carefully planned, including the harvesting and quenching of plant tissue for the extraction of metabolites. The primary goal of sample preparation is to separate and purify the target metabolites from the complex biological makeup of plant tissues. Sample preparation is a time-consuming process and results in a high labor cost. It is essential to keep the process as fast as possible to handle a variety of samples and to avoid possible sample contamination. In addition, the process should be cost-effective and uphold the integrity of the samples. Sample preparation generally includes four steps, i.e., harvesting the plant material, drying or quenching, metabolite extraction, and detection of samples. Based on the specific analytical method and the sample characteristics, a few steps of the sample preparation may be omitted, i.e., drying or extraction. Despite the analytical tools being used, the drying and harvesting steps of the plant material are generally similar. However, the efficiency of the extraction process can vary depending on the choice of solvent (Razzaq et al., 2019).

Harvesting

Researchers must cultivate their plants under appropriate conditions in most experiments to ensure accurate and reliable results. Approximately 10–100 mg of plant materials are typically needed per biological sample (Patel et al., 2021). Environmental factors, i.e., light intensity and temperature, should be considered, as these can cause variation in the experiment's outcome. Improper handling during the collection of samples is probably the most common source of variation in the study of plant metabolomics. Harvesting plant materials is essential to minimize the rapid changes in the plant metabolome caused by oxidation or enzymatic degradation. The harvesting of plant materials should be performed quickly and efficiently to lessen any physical damage.

Moreover, the availability of sterile tools and equipment is also necessary to prevent contamination due to microorganisms that can cause changes in the metabolome. Several other things need to be considered while harvesting the plant materials for metabolomics studies, i.e., the age of the plant, ripening variations, seasonal changes, weather conditions, and harvesting time of the day (Halabalaki et al., 2014). The age of the plant tissue is necessary to consider, as the metabolic content of young plants differs from that of mature plants. The harvesting time of the plant samples is also important since the plant metabolite levels change throughout the day. For example, if plant samples are not harvested simultaneously or under similar environmental conditions, the amount of certain metabolites may vary significantly.

Additionally, the choice of plant part is another factor that should be considered during the harvesting process since the content of the metabolites differs throughout the plant. Different plant organs, i.e., roots, stems, leaves, fruits, and flowers, often contain different metabolites (Kim & Verpoorte, 2010). It is important to optimize the harvesting method to minimize changes in metabolite levels during sample processing.

Drying of Plant Material

Drying the plant material is a crucial stage in metabolomics studies. After harvesting, fresh plant materials typically have a high water content (Zubr, 1986). Water serves as the medium for the enzyme-mediated processes that may occur during the sample preparation process, resulting in the degradation of metabolites. Moreover, the varying water levels in the sample can influence metabolite quantification. It can hinder the proper functioning of instruments required to analyze the samples, particularly in the case of nuclear magnetic resonance (NMR) spectroscopy, where even small amounts of water can distort the spectral resolution. Water can also impact gas chromatography (GC) columns, leading to a loss of reproducibility (Rangarajan & Ghosh, 2011).

The basic purpose of the drying step is to eliminate excess water from plant specimens before further analysis since water can interfere with precise metabolite detection. Drying is the easiest and most inexpensive method to remove excess moisture from plant samples. In addition, drying makes it easier to handle and store the samples, which is important for large-scale metabolomics studies. Grinding plant samples into a fine powder before drying can enhance the drying process and extraction efficiency. The increased surface area of the finely ground plant material allows for better moisture and heat transfer during drying, leading to a quicker and more even drying process. The enhanced surface area also improves extraction efficiency by providing more contact points between the solvent and the plant matrix.

This facilitates the extraction of target metabolites, particularly those tightly bound to the plant matrix (Kim & Verpoorte, 2010).

The selection of the drying method should be made by considering crucial factors such as the drying rate and size of the sample, efficiency of the drying method, and preservation of important metabolites. The efficiency of the drying method was evaluated by measuring its capacity to provide a high yield in a short period. Furthermore, it is crucial to consider the possibility of losing volatile compounds during the process (Mushtaq et al., 2014; Hernandez-Bolio et al., 2019).

Common drying techniques include microwave heating, freeze-drying, oven drying, and hot air drying. Among these methods, freeze-drying or lyophilization is the commonly used method of drying plant material. It is used particularly for products that have a short storage life or are prone to heat. Lyophilization is the "gold standard" among drying methods and is essential in maintaining purity and boosting the shelf life of plant extracts. The first step of lyophilization is to freeze the plant samples at low temperatures (−80 °C to −20 °C) to solidify the liquid materials, such as moisture or solvent. This is followed by sublimation (where a solid directly changes into a gas without passing through the liquid phase) of the frozen liquid (Azwanida, 2015; ElNaker et al., 2021). Compared to other techniques that utilize evaporation to remove water from samples, it is a quick and mild approach. Hundreds of plant samples can typically be freeze-dried at once for 24–48 h. One advantage of lyophilization is that under vacuum conditions, the low oxygen concentration minimizes oxidative damage that can take place in metabolites. However, the lyophilization method has limitations that can impact the experimental outcomes. The lyophilization process can remove only unbound water molecules from the sample. It may be difficult to completely remove all the water using lyophilization since some water molecules have strong hydrogen bonding. Removing water molecules tightly bound to amino ($-NH_2$) and carboxyl ($-COOH$) groups of proteins and hydroxyl groups of most polysaccharides is even more challenging. The leftover water after freeze-drying will not result in the growth of microbes in the plant samples. The lyophilization method is more expensive than other drying methods because of the specialized equipment, i.e., a vacuum pump, a freeze dryer, and costly materials. In addition, the operational cost is another limitation of the lyophilization process. This time-consuming process requires skilled personnel to handle the equipment efficiently (Kim & Verpoorte, 2010; ElNaker et al., 2021).

Oven drying is another method commonly used to prepare plant samples for further analysis. In this method, the plant samples are heated at a particular temperature over a specific period to remove moisture from the plant samples. The utilization of oven drying for plant samples is significant due to the consistency of the resulting dried samples. The morphology of the dried products might vary depending on the drying methods used. However, studies show that plant nutrients remain largely unaffected when the oven air temperature does not exceed 80 °C. Oven drying, a cost-effective alternative to lyophilization, is usually performed using conventional ovens because they are inexpensive and accessible (Alara et al., 2019; ElNaker et al., 2021).

Microwave heating is another technique that has certain advantages over other drying methods. The procedure involves exposing the sample to electromagnetic radiation, which causes the water molecules to vibrate and produce heat, resulting in the drying of the sample. The main advantage of this technique is its drying speed since it can quickly dry the sample because of direct heat generation inside the sample. Additionally, it results in a more uniform drying process because the heat is evenly distributed across the sample (Liu et al., 2019). It is essential to consider any potential adverse effects before using microwave technology. High temperatures and insufficient power can cause structural damage to the dried materials during the microwave drying process. Without proper temperature control, the evaporation rate might increase to a maximum level, resulting in the rapid loss of moisture and potentially damaging the structure of the sample. Therefore, it is recommended to adjust the microwave conditions, such as power, temperature, and exposure time, to minimize the chances of metabolite degradation (Hu et al., 2021).

Extraction

The next step after collecting and drying samples is the extraction of metabolites from plant tissues. Metabolite extraction is a crucial stage in the process of metabolomics since its efficiency has a direct impact on the quality of the outcome. The primary objective of extraction is to obtain the highest possible number of metabolites, preferably all the metabolites in the sample. It also aims to eliminate any interfering substances, such as proteins and salts, that could complicate the analysis and, if needed, concentrate trace metabolites prior to analysis (Álvarez-Sánchez et al., 2010). Other important considerations include preventing the decomposition of crucial metabolites or forming of artifacts. Artifacts refer to compounds that do not originally exist in an intact metabolome but instead are produced during various stages of metabolomics experiments, i.e., sample preparation, harvesting, drying, extraction, isolation, and detection. These compounds can arise from the reaction of analytes with either solvents or impurities present in the solvents and extraction conditions (Jones & Kinghorn, 2005; Verpoorte et al., 2022).

The quantity of plant tissue used in metabolomics experiments can vary depending on the extraction method. For instance, primary metabolites generally require up to 100 mg of tissue, while phytohormones may need even more, up to 2 g. Various factors should be considered for extraction, i.e., choice of appropriate solvent, solvent properties, ratio of sample to solvent, optimizing extraction conditions such as temperature, and duration of the extraction. It is necessary to thoroughly extract metabolites from plant tissues while keeping in mind that the chemical stability of metabolites remains unaffected (Mushtaq et al., 2014).

Choice of Extraction Solvents

Selecting the appropriate solvent is a critical and significant step during the extraction procedure. It is impossible to extract all plant metabolites using a single solvent since there is a wide variety of plant metabolites with different polarities (Kim et al., 2010). It is important to carry out several extractions using various solvents to obtain a complete view of the metabolome. When choosing a solvent, numerous parameters must be considered, such as solubilizing power, dissolving rate, selectivity, chemical reactivity, toxicity, and pH (Schripsema, 2010). The choice of solvent will also depend on the specific compounds required to be extracted.

The ideal solvent should be affordable and have a low boiling point to prevent damage caused by high temperatures (Abubakar & Haque, 2020). Additionally, it should have a strong solubilizing capability and be nontoxic. While standard safety procedures may be used to control a solvent's toxicity (Maltese et al., 2009), solubilization power is a property that is inherent to all solvents. The rate at which the solvent dissolves the metabolites from the plant tissue is called the "dissolution rate." A slow dissolution rate indicates that the extraction procedure has to be prolonged or that more energy needs to be given to the system by raising the temperature using various techniques (such as microwave heating, conventional heating, or ultrasonic treatment). Dissolution can also be hindered by the chemical and physical interactions between the metabolites and the sample matrix, which can be viewed as an adsorbent. The dissolution rate is influenced by several factors, including chemical and physical forces and the diffusion rate of the sample particles into the solvent. Solvent selectivity is another significant factor in addition to solubilization power, which refers to the solvent's ability to dissolve particular compounds with almost similar polarities.

The tissue-to-solvent ratio is another factor to consider to ensure accurate and reproducible results. For each metabolite, there is a maximum quantity that can be dissolved in any given solvent. Both temperature and time function similarly in increasing the yield of metabolites. It is also important to consider the chemical reactivity of the solvents when employing heat or performing a longer extraction (Kim & Verpoorte, 2010). In some cases, the optimal solvent selection relies on the analytical technique. Nonpolar solvents, e.g., chloroform, either alone or combined with methanol, are preferred for GC–MS analysis.

However, polar solvents have been utilized for NMR- and LC-based analysis (Verpoorte et al., 2008). Some commonly used solvents for extraction are water, alcohol, chloroform, ether, and ionic liquids. Water is a highly polar solvent that is frequently used to extract polar molecules because of its capability to dissolve a variety of compounds. It is also inexpensive, nonflammable, and nontoxic, making it a solvent of choice for extraction techniques. However, there are some drawbacks as well. It promotes the growth of bacteria and molds and requires significant heat energy to extract. Alkaloids and terpenoids may be extracted from specific molecules using ether, a nonpolar solvent. Using ionic liquids as extraction solvents might be useful for metabolomic analysis. Ionic liquids are salts (made up of a cation and an anion), usually liquid at room temperature. Ionic liquids are heat-stable solvents suitable for microwave-assisted extraction and other extraction techniques requiring high temperatures. These liquids have the potential to replace traditional solvents in metabolomic analysis, as they are capable of dissolving compounds with medium polarity, including secondary and primary metabolites found in plants, and are more effective than water or nonpolar solvents. For instance, the solubility rates of certain flavonoids are up to 100 times higher in ionic liquids than in water (Kim et al., 2011; Abubakar & Haque, 2020).

It is recommended to use simple solvents with minimal chance of producing artifacts, i.e., water and alcohol. However, some alcohols may react with functional groups, such as carbinolamines, or result in the formation of esters. If methanol is used as an extraction solvent, it becomes challenging to differentiate between naturally occurring methoxy groups or those arising from the solvent. Ethanol does not pose this issue since ethoxy groups are uncommon (Verpoorte et al., 2008). Solvents such as acetonitrile and methanol, which are polar, can extract both hydrophobic and hydrophilic compounds. Therefore, a combination of polar (e.g., water and water–methanol) and nonpolar solvents such as chloroform is frequently employed to separate a polar phase containing hydrophilic metabolites from a nonpolar phase containing hydrophobic metabolites, i.e., lipids.

Extraction Methods

An efficient extraction method may produce results quickly or with minimal effort. The extraction method is considered a crucial step in achieving reliable chemical analysis, as it can significantly impact the quality and reliability of analytical results (Teo et al., 2013). The conventional methods for extracting plant materials using solvents rely mostly on selecting appropriate solvents and using heat or agitation to extract the desired compounds from the plant material. Agitation helps to improve the contact between the solvent and the plant material, facilitating the transfer of the compounds from the plant matrix to the solvent. However, these conventional techniques often require a longer duration of extraction and thus pose a high chance of thermal degradation for the desired compounds. The presence of numerous secondary metabolites in a single plant emphasizes the need to develop advanced and efficient extraction methods that can extract these compounds quickly and effectively. New extraction techniques have been developed that prioritize shorter durations for extraction, minimize solvent usage, and preserve thermolabile compounds (Mandal et al., 2007).

Conventional Methods

Conventional plant extraction methods involve using solvents to extract desired compounds from plant materials. Some novel conventional methods include maceration, hydrodistillation, and Soxhlet extraction (Azmir et al., 2013).

Maceration

Maceration has been utilized for a long time to make a homemade tonic that is both popular and cost-effective for obtaining bioactive compounds and essential oils. When performing small-scale extractions using maceration, the process typically involves several steps. Initially, plant materials are ground into fine particles. Second, a suitable solvent is added to the ground plant material in a closed vessel. The solvent is chosen based on its ability to dissolve the desired plant constituents. After sufficient time, the liquid is strained off, separating the solid plant material from the liquid extract. The solid residue is known as the marc. The marc is then pressed to recover any occluded liquid that remains. This pressed-out liquid is typically mixed with the initially strained liquid. Afterward, the resulting liquid is filtered to eliminate solid particles or impurities, producing a purified plant extract. The maceration process can be aided by occasional shaking to facilitate extraction either by increased diffusion or by removal of concentrated solution from the surface of the solid material, which allows for the new solvent to come in contact with the solvent to improve the extraction yield (Azmir et al., 2013).

The maceration process involves basic utensils and equipment and does not require a skilled operator. It is an energy-efficient method and is well-suited for a few substances with low solubility that only need prolonged exposure to the solvent. However, the maceration technique has certain disadvantages as well. The extraction process can be lengthy and may last for several weeks. The maceration process may prove ineffective if the solvent used for extraction cannot extract all the active compounds. In addition, a larger amount of solvent is necessary for the extraction process (Rasul, 2018).

Hydrodistillation

Hydrodistillation is a conventional approach used to extract bioactive compounds, i.e., secondary metabolites and essential oils, from plants. This extraction method does not require organic solvents. Hydrodistillation can be divided into three types: water and steam, direct, and water distillation (Vankar, 2004). The process involves packing plant materials into a still container, adding sufficient water, and boiling the mixture. Alternatively, steam is injected directly into the plant material. The heat from boiling water or steam helps release bioactive compounds from plant tissue. After that, the vapor mixture of water and oil is condensed by indirect cooling using water, which then flows to a separator where bioactive compounds and oil automatically separate from the water (Silva et al., 2005).

Hydrodistillation is an environmentally friendly and cost-effective extraction technique since no organic solvents are needed. The process may not always result in the complete extraction of all the desired compounds. Hydrodistillation may result in the loss of volatile components at high extraction temperatures, restricting its use for extracting heat-sensitive compounds (Azmir et al., 2013; Rasul, 2018).

Soxhlet Extraction

Soxhlet extraction, also named hot continuous extraction, is a conventional method to isolate target compounds from plant samples. In 1879, Franz Ritter von Soxhlet, a German chemist, introduced the Soxhlet extractor primarily to extract lipids. However, this apparatus is no longer restricted to this purpose and can be used for other extractions (Azmir et al., 2013). It has been regarded as the most important among all traditional methods for the last 126 years (Letellier et al., 1999). It has a dual function of extracting desired compounds and serving as a fixed standard for evaluating new extraction methods. Various factors, such as the solvent-to-sample ratio, agitation speed, and temperature, must be considered for the extraction process. The technique involves placing a finely powdered sample inside a porous thimble typically made of cellulose or high-quality filter paper. The porous thimble containing the sample is placed in the Soxhlet apparatus, which consists of a flask, a condenser, and a siphon arm. The solvent is heated in the flask, causing it to vaporize and then condense. Once the liquid reaches the siphon arm, it is drained back into the flask, and the process continues. Various parts of the plant, i.e., aerial parts, leaves, bark, stems, roots, seeds, flowers, and fruit, are used for Soxhlet extraction. However, it is necessary to choose healthy plants that are free of insects and weeds. An ideal solvent for Soxhlet extraction should be inert and easy to remove. However, a small amount of solvent is required for the extraction method (Azwanida, 2015; Gopalasatheeskumar, 2018).

The major drawback is the lengthy extraction period, ranging from 8 to 24 h or even more in some cases (Pastor et al., 1997), resulting in a significant consumption of time and heat energy. The method is more time-consuming and labor-intensive because of the prolonged extraction period. Moreover, a large amount of hazardous organic solvents is required for the process, and additional steps are needed, followed by evaporation to concentrate the extract. As a result, the process becomes more difficult and detrimental, which might harm the environment. Additionally, high-purity solvents are needed for the extraction, which might raise the cost of the process (Mandal et al., 2007; Azwanida, 2015).

Advanced Methods of Extraction

Advanced extraction methods include microwave-assisted extraction, accelerated solvent extraction, supercritical fluid extraction, and ultrasound-assisted extraction. Using these extraction techniques, it is possible to extract target compounds from the plant matrix quickly and effectively. These techniques may be effective at high pressures and temperatures, shortening the extraction duration (Wang & Weller, 2006).

Microwave-Assisted Extraction

Microwave-assisted extraction is an advanced technique for separating bioactive compounds from plant samples. Bioactive compounds, i.e., secondary metabolites, are produced by plants through enzymatic biosynthetic pathways. Microwave-assisted extraction (MAE) uses the energy of microwave radiation to help separate target molecules from the sample matrix and into the extraction solvent. Microwaves are a kind of electromagnetic radiation that have a frequency range of 300 MHz to 300 GHz (Routray & Orsat, 2012). When microwave radiation interacts with polar materials such as samples and solvents during microwave-assisted extraction, heat is produced close to the surface of the polar materials. The heat is then transferred through conduction. The dipole rotation of the sample molecules caused by microwave radiation energy disrupts the hydrogen bonding between the molecules in the sample. This disruption results in the transfer of dissolved ions and solvent diffusivity into the matrix, allowing for efficient extraction of the target compounds (Azwanida, 2015).

Various parameters should be taken into account while extracting plant secondary metabolites using microwave-assisted extraction, including the frequency and power of the microwave, duration of exposure to microwave radiation, particle size and moisture content of plant samples, as well as solvent type and volume, concentration, solid-to-liquid ratio, extraction temperature, pressure, and number of cycles. Among these factors, the solvent is considered the most significant parameter because it plays a crucial role in determining the solubility of the target elements and the absorption of microwave energy in the sample (Zhang et al., 2011). The efficiency of MAE is typically affected by the size of particles in plants. Fine powder can improve extraction by having a larger surface area, facilitating contact between the solvent and the plant matrix (Wang & Weller, 2006).

Generally, two types of commercially available apparatus are used for microwave-assisted extraction, i.e., a closed extraction vessel and a focused microwave oven. Closed extraction vessels frequently extract plant secondary metabolites at high temperatures and pressures, particularly volatile metabolites. Compared to closed extraction vessels, the focused microwave oven is a safer option since it may function in less severe conditions, including air pressure (Kaufmann & Christen, 2002).

However, it is reported to be a time-consuming process and may also reduce extraction efficiency (Wang & Weller, 2006).

Various compounds, especially secondary metabolites from plants, can be recovered using microwave-assisted extraction. One benefit of MAE is the quicker extraction time attained when utilizing microwaves. In microwave-assisted extraction, the heating process occurs in a closed system, resulting in accurate and targeted heating with less heat loss into the surrounding environment. In contrast to Soxhlet extraction, this specific heating method may significantly reduce the extraction time, which normally takes less than 30 min. The target compounds may be selectively extracted via MAE, leaving behind unwanted compounds. This can result in improved quality of the extracted compounds. Additionally, MAE allows for the simultaneous processing of many samples and much lower organic solvent consumption (Mandal et al., 2007; Zhang et al., 2011; Chaturvedi, 2018).

While it can be an effective method for some compounds, its use has certain limitations. In some target compounds, such as secondary metabolites, microwave irradiation can accelerate chemical reactions or changes, which could lower extraction yields. Additionally, the chemical structures of the target compounds may change under certain MAE operating conditions, resulting in a lower yield. When working with nonpolar solvents or target compounds or when the solvent viscosity is particularly high, the effectiveness of microwave-assisted extraction may be hampered. Additionally, this method may not be viable for extracting heat-sensitive compounds (Zhang et al., 2011).

Supercritical Fluid Extraction

Supercritical fluid extraction (SFE) is a popular separation technique that depends on the critical temperature and pressure of the supercritical fluid (Sairam et al., 2012). A substance enters a supercritical state when the pressure and temperature exceed its critical value. Charles Cagniard first identified supercritical fluids in 1822. Supercritical fluids have both liquid and gas characteristics, providing several advantages over liquid solvents. They can diffuse into solids, such as gases, and dissolve substances, such as liquids. Supercritical solvents possess distinct physicochemical properties, such as viscosity, diffusivity, density, and dielectric constant. The dissolving power of supercritical solvents depends on density, which may be altered by adjusting the temperature and pressure. Additionally, the high diffusivity and low viscosity of supercritical fluids make it easier for them to diffuse through solid materials and increase the extraction rate.

It is critical to consider several factors for developing an efficient supercritical fluid extraction. Some of these parameters include the choice of supercritical fluids, preparation of plant material, coextractors or modifiers, and extraction conditions. Various solvents, including butane, hexane, pentane, sulfur hexafluoride, and nitrous oxide, can be utilized for this extraction technique (Reverchon & De Marco, 2006). CO_2 is the preferred solvent used in supercritical fluid extraction techniques. It is a colorless, odorless, and inert solvent considered safe (nonexplosive, nonflammable). The primary benefit of SFE is that CO_2 completely evaporates, leaving no solvent residue behind (Ingle et al., 2017).

Its selection can be attributed to several factors, such as its affordability, low toxicity, easy recovery, high purity, and low critical pressure and temperature (304 K). Additionally, it is highly suitable for extracting nonpolar and heat-sensitive compounds. The nonpolar nature of CO_2 is a significant drawback that can be overcome by adding coextractors or modifiers. A modifier is added either with the solvent or with the sample. Argon, which is more inert and inexpensive than CO_2, may also be an alternative solvent (Gupta et al., 2012).

The size of plant particles plays a crucial role in achieving a successful SFE process. If the particles are large, the extraction process may take longer. The extraction procedure, however, may be accelerated by using fine powder, although it may be difficult to maintain the proper flow rate. Another important aspect to consider for SFE is the solvent flow rate. A high flow rate often increases the extraction capacity; however, a higher flow rate might lead to a reduction in some cases. SFE is more efficient than other extraction methods in terms of increased yields and shorter extraction periods and may be used in combination with analytical methods such as gas chromatography or supercritical fluid chromatography (Wang & Weller, 2006; Mushtaq et al., 2014; Da Silva et al., 2016).

Accelerated Solvent Extraction

Accelerated solvent extraction (ASE), often called pressurized fluid extraction, has emerged as a viable substitute for conventional solvent extraction for separating chemicals in various analytical and industrial processes. The first study on the pressurized liquid extraction method was published in the late 1990s (Azmir et al., 2013).

Accelerated solvent extraction (ASE), a solid–liquid extraction, functions at temperatures between 50 and 200 °C and pressures between 10 and 15 MPa. Compared to Soxhlet extraction, ASE has been more effective since it uses the same solvents but requires only small amounts, making it possible to extract specific compounds more quickly and precisely. Organic solvents are the solvent of choice used in accelerated solvent extraction. However, it is critical to carefully monitor accelerated solvent extraction at high temperatures, as this may cause heat-sensitive compounds to degrade. Target compounds may be selectively extracted by modifying the ASE temperature and pressure (Wang & Weller, 2006; Cho et al., 2007).

Ultrasound-Assisted Extraction

Ultrasound-assisted extraction (UAE) is a promising and well-known technique for extracting plant bioactive compounds such as flavonoids and phenolics. In ultrasonic-assisted extraction (UAE), high-frequency and high-intensity sound waves interact with materials. During the process, a frequency greater than 20 kHz is useful in extracting bioactive compounds from plants. The rapid movement of solid and liquid particles caused by the application of ultrasonic waves speeds up the diffusion of the solute from the solid phase into the solvent. High-intensity ultrasound waves in the liquid can cause the molecular structure to break down, as the intermolecular forces cannot hold it intact. This phenomenon, called cavitation, leads to the formation of tiny empty spaces or bubbles within the liquid surrounding the material. As these bubbles expand and contract, they initiate a series of chemical, physical, and mechanical reactions capable of disrupting biological membranes. This disruption proves beneficial in releasing targeted compounds, thereby facilitating the penetration of solvents into cellular components and enhancing the overall mass transfer process. Ultrasound-assisted extraction is environmentally friendly because it can extract the targeted compounds using less solvent in a minimum period. It is a promising technique since it is simple and affordable and does not require expensive tools. It is also appropriate for extracting heat-sensitive compounds (Dai & Mumper, 2010; Gupta et al., 2012; Yusoff et al., 2022).

Metabolite Profiling or Detection

Metabolic profiling has emerged as an efficient approach to studying plant biology. Metabolite profiling identifies and quantifies metabolites in a biological sample (Hall, 2006). Metabolite profiling aims to identify a variety of metabolites, including secondary metabolites such as phenolics, terpenoids, and alkaloids, as well as primary metabolites such as organic acids, amino acids, and carbohydrates (Hill & Roessner, 2013).

Identifying the diverse range of metabolites in plant samples is impossible using a single analytical instrument. Plant metabolomics studies use various methods, including GC–MS, CE-MS, LC–MS, IMS, MALDI, and NMR. These methods provide highly sensitive and specific quantitative and qualitative analyses of metabolites. Each of these analytical techniques, however, has its advantages and disadvantages. Metabolite profiling can be carried out more successfully by combining various analytical methods (Patel et al., 2021). A combination of various techniques may also help achieve maximum metabolite coverage (Razzaq et al., 2019).

Detection Methods

Gas Chromatography–Mass Spectrometry (GC–MS)

GC–MS is the most commonly used analytical method for identifying metabolites in plant metabolomics. Because of its distinct benefits, GC–MS is regarded as the "gold standard" in metabolomics and is used as a benchmark for more recent techniques (Fiehn, 2016). This technique has become popular due to the groundbreaking research done in the early 2000s, which emphasizes using GC–MS metabolite profiling for simultaneous and thorough quantitative detection of up to 150 metabolites in Solanum tuberosum (potato tubers) tissues, including amino acids, sugars, and organic acids. Afterward, further GC–MS techniques were quickly developed and utilized to identify metabolites in other plant tissues. Approximately 200 compounds were found in a single tomato leaf tissue and fruit extract, while 87 metabolites were found in different Lotus japonicus organs (Jorge et al., 2016a, 2016b). GC–MS is a very effective technique for identifying and quantifying small metabolites weighing approximately 500 Da. These compounds include amino acids, fatty acids, hydroxyl acids, sterols, sugars, alcohols, and amines. These compounds are usually rendered volatile using chemical derivatization to facilitate gas chromatography analysis (Fiehn, 2016). In mass spectrometry, derivatization involves the modification of a chemical compound by using a derivatization reagent to generate a chemically similar derivative (Orata, 2012). During mass spectrometry analysis, chemical modification of an analyte or molecule enhances its volatility, ionization effectiveness, and detection.

Several derivatization techniques can also be used, including acylation, alkylation, methoximation, silylation, and trimethylsilylation. A two-step derivatization procedure is required to extract and identify metabolites using GC–MS. The initial step is the conversion of all carbonyl (C=O) groups into oximes using methoxyamine hydrochloride. In the next step, a trimethylsilylation reaction is performed to improve the thermal stability and volatility of the metabolites. In this step, the hydrogen atoms of particular functional groups, such as –SH, –NH, –OH, and –COOH, are replaced with [$-Si(CH_3)3$] groups. The resultant TMS derivatives are less polar, more volatile, and thermally stable than the original metabolites, making it easier to separate and identify them using GC–MS (Patel et al., 2021).

GC–MS is a valuable tool for analyzing samples with complex matrices, such as plant extracts. Metabolomics enables effective metabolite separation and resolution and identifies and measures known and new metabolites despite their low concentrations (Dettmer et al., 2007). The two main ionization techniques utilized in GC–MS are chemical and electric. Most GC–MS techniques in metabolomics have relied on EI. This technique includes the interaction of molecules with activated electrons at a standard energy level of 70 eV, producing reproducible mass spectra that may be utilized to build mass spectral libraries for comparison and acquisition of global metabolomics data (Jorge et al., 2016a, 2016b).

A mass analyzer is a mass spectrometer (MS) component that separates ions based on their mass-to-charge (*m/z*) values. Various mass analyzers have been employed in metabolomics for various applications. The quadrupole (Q) mass analyzer was introduced in the 1950s by the Nobel Prize winner Wolfgang Paul. It is particularly suitable for GC–MS instrumentation due to its small size, durability, and affordability. However, the quadrupole analyzer has limitations related to poor mass resolution. In contrast, the time-of-flight (ToF) analyzer is a relatively simple mass analyzer first introduced in the mid-twentieth century. GC coupled with TOF-MS (mass spectrometry) is a favored method for metabolic profiling because it offers higher mass accuracy (El-Aneed et al., 2009; Patel et al., 2021).

GC–MS technology has a significant drawback, as it can only be utilized to analyze thermally stable and volatile metabolites or those metabolites that can be chemically altered to generate volatile derivatives. However, the analysis of heat-labile di- and triphosphates and larger oligosaccharides is impossible even with the derivatization method due to their lower volatility (Weckwerth, 2003).

Liquid Chromatography–Mass Spectrometry (LC–MS)

LC–MS is a powerful analytical technique used to study the plant metabolome. It can measure a wide variety of complex metabolites. This method eliminates the requirement for metabolite volatilization and is particularly suitable for plant metabolites with molecular weights greater than 500 kDa, polar metabolites, heat-sensitive functional groups, and those with higher vapor points. Choosing a suitable column for LC–MS analysis is a critical stage in the analytical process since it significantly influences the accuracy and quality of the results. The column type most appropriate for the analysis depends significantly on the polarity and physicochemical characteristics of the studied metabolites. A C8 or C18 reversed-phase (RP) chromatography column is used for most applications. In reversed-phase liquid chromatography, the mobile phase is typically a combination of water and organic solvents, such as methanol or acetonitrile. Reversed-phase chromatography is a widely used technique for extracting mid-polarity and nonpolar analytes. However, highly polar compounds may not be retained on reversed-phase (RP) stationary phases and may elute with the void volume instead. Hydrophilic interaction liquid chromatography (HILIC) is a suitable alternative for studying highly polar compounds in complex mixtures, such as plant extracts. Electron spray ionization (ESI) and atmospheric pressure ionization are the two most often used ionization methods for LC–MS. LC–MS is a highly efficient tool for profiling key secondary metabolites such as alkaloids, flavonoids, steroids, and terpenes. Compared to other analytical methods, sample preparation for LC–MS is rather easy.

Additionally, it may be employed with various separation modes and does not need complex derivatization steps. A drawback of this approach is the requirement for standard reference compounds. These are pure compounds with known chemical structures and characteristics that are utilized as a reference for metabolite identification and quantification. It might be challenging to accurately identify and quantify the metabolites in a sample if a standard reference compound is unavailable commercially (Dettmer et al., 2007; Patel et al., 2021).

Capillary Electrophoresis-Mass Spectrometry

CE-MS analyzes various charged and polar metabolites depending on their charge and mass ratio. CE is a highly efficient method of separating a wide range of chemical compounds and has a greater separation efficiency than LC (Ramautar et al., 2009). Electrospray ionization (ESI) is frequently utilized in CE-MS for the ionization process, and TOF-MS has been used as the main detector. The fast scan speed of TOF-MS is one of its benefits. It is appropriate for full scan analyses for metabolomics research since it can quickly assess many metabolites. Capillary electrophoresis is carried out inside a capillary tube made of fused silica. Both ends of the capillary tube are immersed in buffer solution, and high voltages ranging from 20 to 30 kV are applied across it (Jorge et al., 2016a, 2016b).

While CE-MS can detect many of the same metabolites as GC–MS, it does not require any derivatization process, resulting in significant savings of both time and consumables. CE offers certain advantages compared to other metabolomics tools, primarily due to its utilization of a low separation volume, making it particularly suitable for the analysis of the plant metabolome. However, CE is not a suitable platform for investigating complex plant metabolites because of its poor migration time and low reproducibility and sensitivity (Soga & Imaizumi, 2001; Patel et al., 2021).

MALDI-Imaging Mass Spectrometry

The matrix-assisted laser desorption or ionization-mass spectrometry imaging (MALDI-MSI) technique has enabled the direct visualization of molecules within plant tissues and surfaces. The MALDI-MSI technique was first published in 1994, and since 1997, it has been used to visualize different biomolecules (Sturtevant et al., 2016; Maia et al., 2022). Using MALDI-MSI, researchers have examined the spatial distribution of sugars, lipids, and other metabolites in various plant sections, including leaves, roots, and flowers. This method has a spatial resolution range of 1.4–100 mm. Identifying and quantifying the metabolites are only one aspect of studying plant tissues. Studying their distribution is also important since it might reveal which plant organ has the highest concentration of the desired secondary metabolites. The distribution of metabolites can also shed light on the mechanisms of both primary and secondary metabolism in plants (Susniak et al., 2020).

A crucial stage in MALDI-mass spectrometry imaging is sample preparation. It is essential to perform each sample preparation stage to achieve optimal sensitivity carefully. Additionally, the choice of matrix and its deposition method should be carefully considered, as they can substantially influence the spatial resolution of the molecules to be examined (Sturtevant et al., 2016). While performing MALDI-mass imaging, a thin matrix film of sinapinic acid, 2,5-dihydroxybenzoic acid, or α-cyano-4-hydroxycinnamic acid (CHCA) is applied to the tissue surface. The tissue surface coated with the matrix is then exposed to a laser beam to produce a mass spectrum. MALDI is the most commonly used ionization technique for imaging proteins and metabolites. It is frequently used with several mass analyzers, such as ToF, QToF (quadrupole time of flight), and ion-trap (Goodwin et al., 2008). Other ionization techniques have also been investigated, such as secondary ion mass spectrometry (SIMS), laser ablation electrospray ionization (LAESI), and desorption electrospray ionization (DESI) (Lee et al., 2012). The thickness of the tissue segment can significantly impact the quality of the data generated by MALDI-MSI experiments. Initially, MALDI-MSI users employed sections that ranged in thickness from 50 to 100 μm. The thickness has, however, been decreased to 10–20 μm in more recent applications (Dong et al., 2016).

A reference sample is not needed for the identification of metabolites. Each metabolite is distinguished based on its *m/z* value (Susniak et al., 2020). MALDI-MSI has been used to quantify surface molecules on Arabidopsis thaliana leaves, roots, and flowers, including epicuticular lipids and secondary metabolites (i.e., flavonoids). Additionally, MALDI-MSI has been used to map the localization of primary metabolites in wheat and potato seeds, including arginine, glucose-6-phosphate, and sucrose (Grassl et al., 2011).

Nuclear Magnetic Resonance (NMR) Spectrometry

NMR has a unique capability for the in-depth analysis of plant metabolism, making it an efficient tool for plant metabolite profiling (Bligny & Douce, 2001). NMR is an effective analytical method for examining the structure and purity of molecules in a sample. It has been widely used to assess secondary metabolites in plants on a quantitative and qualitative level. A sample containing metabolites is placed in a high magnetic field and then given a radio frequency pulse in NMR-based metabolite identification. This radiation is absorbed by atomic nuclei such as ^{1}H and ^{13}C, which release it at a distinctive radio frequency. The composition and structure of the metabolites present in the sample can be determined by measuring the frequency of this emission using NMR spectroscopy (Patel et al., 2021).

NMR spectroscopy offers greater flexibility in the choice of solvents; however, a mid-polarity solvent, such as aqueous MeOH, is commonly used to cover a wide range of extracted metabolites. The pH must be carefully examined when using aqueous solvents since artifacts may arise at either low or high pH values during extraction. Various solvents can be combined since no single solvent can extract all metabolites. One efficient technique is a two-phase solvent solution with a 2:1:1 v/v ratio of MeOH, water, and chloroform. This combination of solvents has been demonstrated to be highly effective in extracting both polar and nonpolar metabolites. NMR is sensitive to changes in pH, so buffered solutions are commonly used to maintain a stable pH. For NMR analysis of both primary and secondary metabolites, the most effective buffering solution is a mixture of methanol and aqueous phosphate buffer at a 1:1 volume ratio with a pH of 6.0 or ionic liquids such as 1-butyl-3-methylimidazolium chloride.

The popularity of NMR in metabolomics has significantly increased in recent years due to its ability to perform analyses without the need for sample derivatization or chromatographic separation. Additionally, the nondestructive nature of NMR, combined with excellent repeatability and easy sample preparation, makes it suitable for high-throughput and quick analysis in metabolomics. However, a significant limitation of the NMR technique is its low sensitivity. This is because NMR has a high detection limit compared to other analytical techniques and considerable overlap among NMR signals (Kim & Verpoorte, 2010; Patel et al., 2021).

Bioinformatics Tools for Processing, Annotation, and Analysis of Metabolomics Data

The large amount of data generated by metabolomic experiments presents a significant challenge for researchers, and thus, bioinformatics tools are required for data processing and analysis. Various software programs can be utilized to perform in silico analysis of large amounts of metabolite spectrum data produced by different analytical instruments. These tools allow researchers to extract meaningful information from the large and complex datasets generated by analytical techniques. These bioinformatic tools are XCMS, MeltDB, MetaboAnalyst, metaP-server, METLIN, MolFind, MZmine2, and MetAlign.

XCMS is an online bioinformatic platform where users can easily submit raw metabolomic data for processing and analysis. One key feature of the XCMS software package is automated data transfer, which allows users to easily transmit raw data from the LC–MS instrument to the platform for analysis. This significant feature saves time and effort by eliminating the need for manual data processing and transfer, which leads to more accurate analysis (Montenegro-Burke et al., 2017). MeltDB is an online tool used in plant metabolomics to process data, perform statistical analysis, and review results (Kessler et al., 2013). MetaboAnalyst is an important Web-based tool for metabolomic data analysis and interpretation. It can process metabolomic data from different metabolomic platforms, such as MS and NMR, and is appropriate for various metabolomic experiments, including quantitative, targeted, and untargeted experiments. This software program aims to provide greater accessibility due to its ease of use and online availability. Users must perform manual processing procedures to upload data because of the software's heavy reliance on high-level data inputs (Xia et al., 2009). The MetaP server is an effective Web-based tool for analyzing metabolomic data. This platform provides an automated and standardized method for analyzing quantitative metabolomic data (Kastenmüller et al., 2011).

Metabolite annotation involves comparing the MS/MS spectra of unknown metabolites with those in spectral libraries that are openly accessible. These spectral matching algorithms and libraries recognize unknown compounds using spectral similarities. In some instances, the structural information provided by the MS spectrum of certain compounds may help identify unknown metabolites (Graça et al., 2022). METLIN, an online database, has been utilized in several studies investigating plant metabolic profiling in response to stress. Metabolomics offers a practical and effective method that makes it possible to quickly analyze and annotate metabolites of interest (Smith et al., 2005). Another program called MolFind employs a pipeline-style method for annotating unknown metabolites (Menikarachchi et al., 2012).

Applications of Plant Metabolomics

Plant metabolomics is an emerging discipline with various applications in various fields. Some of the key applications of plant metabolomics are discussed below:

Improvement of the Nutritional Quality of Crops

Plant-based foods constitute the main source of human nutrition globally, regardless of economic status or geographic location. Moreover, in some cultures, even the whole diet comprises plant-based items. The best strategy for obtaining nutritional enhancement may be to modify staple crops. The discipline of metabolomics provides better insight into food components, their synthesis, and the genetic and environmental variables that affect food composition. It can improve our knowledge of complex metabolic regulation and how even small alterations can lead to several biochemical outcomes. Afterward, this information may be utilized to create more targeted crop improvement strategies, especially regarding nutritional value and biochemical composition. Analytical techniques such as GC–MS, LC–MS, and NMR are often used to better understand the changes in food composition regarding quality and nutrition (Hall et al., 2008).

Plant Breeding

Metabolomics has various uses in plant breeding, and one of its applications is to find the specific metabolic pathways important for developing desirable characteristics. This method enables researchers to identify the genes that control these pathways to breed or genetically modify crops selectively (Hall et al., 2008). Additionally, targeted metabolite profiling is useful for finding plant varieties that yield desired amounts of secondary metabolites (Bueno & Lopes, 2020).

Evaluation of the Safety of GM Crops

Metabolomics is being utilized to acquire the knowledge to implement targeted approaches and evaluate the extent and effectiveness of the introduced modifications. Metabolomics is being utilized to evaluate the possible unintended consequences of genetically modified (GM) products, specifically in the context of biosafety. Metabolomics has a significant role in gaining public acceptance for GM crops by offering essential data for managing associated risks (Rischer & Oksman-Caldentey, 2006; Hall et al., 2008).

Future Prospects

Plant metabolomics has significantly advanced in recent years due to the increased interest in using metabolomic technologies for various biological purposes. Metabolomics integration with other "omics" and the capabilities of currently available analytical platforms have provided a deeper understanding of the biochemical and genetic aspects of cellular function. Although there are some limitations, plant metabolomics is undoubtedly an essential tool that is revolutionizing the field of plant biology.

In the future, research in this field will encompass two main areas. The first involves enhancing the metabolomic platform to accurately identify and quantify a broad range of metabolites, particularly secondary metabolites, and precise data interpretation. The second area of interest is the thorough investigation of plant metabolic variations. This will improve our plant metabolism knowledge and help increase crop quality and yield (Hong et al., 2016).

References

Abubakar, A. R., & Haque, M. (2020). Preparation of medicinal plants: Basic extraction and fractionation procedures for experimental purposes. *Journal of Pharmacy & Bioallied Sciences, 12*, 1.

Akbari, B., Baghaei-Yazdi, N., Bahmaie, M., & Mahdavi Abhari, F. (2022). The role of plant-derived natural antioxidants in reduction of oxidative stress. *BioFactors, 48*, 611–633.

Alara, O. R., Abdurahman, N. H., & Olalere, O. A. (2019). Mathematical modelling and morphological properties of thin layer oven drying of Vernonia amygdalina leaves. *Journal of the Saudi Society of Agricultural Sciences, 18*, 309–315.

Ali, A. H., Abdelrahman, M., & El-Sayed, M. A. (2019). Alkaloid role in plant defense response to growth and stress. In *Bioactive molecules in plant defense: Signaling in growth and stress* (pp. 145–158).

Alseekh, S., & Fernie, A. R. (2018). Metabolomics 20 years on: What have we learned and what hurdles remain? *The Plant Journal, 94*, 933–942.

Álvarez-Sánchez, B., Priego-Capote, F., & de Castro, M. D. L. (2010). Metabolomics analysis II. Preparation of biological samples prior to detection. *TrAC, Trends in Analytical Chemistry, 29*, 120–127.

Augustijn, D., de Groot, H. J. M., & Alia, A. (2021). HR-MAS NMR applications in plant metabolomics. *Molecules, 26*, 931.

Azmir, J., et al. (2013). Techniques for extraction of bioactive compounds from plant materials: A review. *Journal of Food Engineering, 117*, 426–436.

Azwanida, N. N. (2015). A review on the extraction methods use in medicinal plants, principle, strength and limitation. *Medicinal and Aromatic Plants, 4*, 412–2167.

Bligny, R., & Douce, R. (2001). NMR and plant metabolism. *Current Opinion in Plant Biology, 4*, 191–196.

Borges, C. V., Minatel, I. O., Gomez-Gomez, H. A., & Lima, G. P. P. (2017). Medicinal plants: Influence of environmental factors on the content of secondary metabolites. In *Medicinal plants and environmental challenges* (pp. 259–277).

Bourgaud, F., Gravot, A., Milesi, S., & Gontier, E. (2001). Production of plant secondary metabolites: A historical perspective. *Plant Science, 161*, 839–851.

Bueno, P. C. P., & Lopes, N. P. (2020). Metabolomics to characterize adaptive and signaling responses in legume crops under abiotic stresses. *ACS Omega, 5*, 1752–1763.

Carrieri, D., et al. (2009). Identification and quantification of water-soluble metabolites by cryoprobe-assisted nuclear magnetic resonance spectroscopy applied to microbial fermentation. *Magnetic Resonance in Chemistry, 47*, S138–S146.

Chaturvedi, A. K. (2018). Extraction of nutraceuticals from plants by microwave assisted extraction. *Systematic Reviews in Pharmacy, 9*, 31–35.

Chinou, I. (2008). Primary and secondary metabolites and their biological activity. *Chromatographic Science Series, 99*, 59.

Cho, S.-K., Abd El-Aty, A. M., Choi, J.-H., Kim, M. R., & Shim, J. H. (2007). Optimized conditions for the extraction of secondary volatile metabolites in Angelica roots by accelerated solvent extraction. *Journal of Pharmaceutical and Biomedical Analysis, 44*, 1154–1158.

Da Silva, R. P. F. F., Rocha-Santos, T. A. P., & Duarte, A. C. (2016). Supercritical fluid extraction of bioactive compounds. *TrAC Trends in Analytical Chemistry, 76*, 40–51.

Dai, J., & Mumper, R. J. (2010). Plant phenolics: Extraction, analysis and their antioxidant and anticancer properties. *Molecules, 15*, 7313–7352.

Dettmer, K., Aronov, P. A., & Hammock, B. D. (2007). Mass spectrometry-based metabolomics. *Mass Spectrometry Reviews, 26*, 51–78.

Dong, Y., et al. (2016). Sample preparation for mass spectrometry imaging of plant tissues: A review. *Frontiers in Plant Science, 7*, 60.

Dunn, W. B., Bailey, N. J. C., & Johnson, H. E. (2005). Measuring the metabolome: Current analytical technologies. *Analyst, 130*, 606–625.

El-Aneed, A., Cohen, A., & Banoub, J. (2009). Mass spectrometry, review of the basics: Electrospray, MALDI, and commonly used mass analyzers. *Applied Spectroscopy Reviews, 44*, 210–230.

ElNaker, N. A., et al. (2021). A metabolomics approach to evaluate the effect of lyophilization versus oven drying on the chemical composition of plant extracts. *Scientific Reports, 11*, 22679.

Fiehn, O. (2016). Metabolomics by gas chromatography–mass spectrometry: Combined targeted and untargeted profiling. *Current Protocols in Molecular Biology, 114*, 30–34.

Fiehn, O., et al. (2000). Metabolite profiling for plant functional genomics. *Nature Biotechnology, 18*, 1157–1161.

Fitzgerald, M., Heinrich, M., & Booker, A. (2020). Medicinal plant analysis: A historical and regional discussion of emergent complex techniques. *Frontiers in Pharmacology, 10*, 1480.

Goodwin, R. J. A., Pennington, S. R., & Pitt, A. R. (2008). Protein and peptides in pictures: Imaging with MALDI mass spectrometry. *Proteomics, 8*, 3785–3800.

Gopalasatheeskumar, K. (2018). Significant role of soxhlet extraction process in phytochemical research. *Mintage Journal of Pharmaceutical and Medical Sciences, 7*, 43–47.

Graßmann, J. (2005). Terpenoids as plant antioxidants. *Vitamins and Hormones, 72*, 505–535.

Graça, G., et al. (2022). Automated annotation of untargeted all-ion fragmentation LC–MS metabolomics data with MetaboAnnotatoR. *Analytical Chemistry, 94*, 3446–3455.

Grassl, J., Taylor, N. L., & Millar, A. (2011). Matrix-assisted laser desorption/ionisation mass spectrometry imaging and its development for plant protein imaging. *Plant Methods, 7*, 1–11.

Griffiths, J. (2008). A brief history of mass spectrometry. *Analytical Chemistry, 80*, 5678–5683.

Gupta, A., Naraniwal, M., & Kothari, V. (2012). Modern extraction methods for preparation of bioactive plant extracts. *International Journal of Applied and Natural Sciences, 1*, 8–26.

Halabalaki, M., et al. (2014). Sample preparation issues in NMR-based plant metabolomics: Optimisation for Vitis wood samples. *Phytochemical Analysis, 25*, 350–356.

Hall, R. D. (2006). Plant metabolomics: From holistic hope, to hype, to hot topic. *The New Phytologist, 169*, 453–468.

Hall, R. D., Brouwer, I. D., & Fitzgerald, M. A. (2008). Plant metabolomics and its potential application for human nutrition. *Physiologia Plantarum, 132*, 162–175.

Hernández-Bolio, G. I., et al. (2019). The influence of drying temperatures on the metabolic profiles and antioxidant activity of Manilkara zapota leaves. *Metabolites, 9*, 217.

Hill, C. B., & Roessner, U. (2013). Metabolic profiling of plants by GC–MS. In *Handbook of plant metabolomics* (pp. 1–23).

Hong, J., Yang, L., Zhang, D., & Shi, J. (2016). Plant metabolomics: An indispensable system biology tool for plant science. *International Journal of Molecular Sciences, 17*, 767.

Hu, Q., et al. (2021). Microwave technology: A novel approach to the transformation of natural metabolites. *Chinese Medicine, 16*, 1–22.

Hussein, R. A., & El-Anssary, A. A. (2019). Plants secondary metabolites: The key drivers of the pharmacological actions of medicinal plants. *Herbal Medicine, 1*(3), 11–30.

Ingle, K. P., et al. (2017). Phytochemicals: Extraction methods, identification and detection of bioactive compounds from plant extracts. *Journal of Pharmacognosy and Phytochemistry, 6*, 32–36.

Jones, W. P., & Kinghorn, A. D. (2005). Extraction of plant secondary metabolites. *Natural Products Isolation, 20*(2), 323–351.

Jorge, T. F., Mata, A. T., & António, C. (2016a). Mass spectrometry as a quantitative tool in plant metabolomics. *Philosophical Transactions of the Royal Society A Mathematical, Physical and Engineering Science, 374*, 20150370.

Jorge, T. F., et al. (2016b). Mass spectrometry-based plant metabolomics: Metabolite responses to abiotic stress. *Mass Spectrometry Reviews, 35*, 620–649.

Kastenmüller, G., Römisch-Margl, W., Wägele, B., Altmaier, E., & Suhre, K. (2011). metaP-server: A web-based metabolomics data analysis tool. *Journal of Biomedicine & Biotechnology, 2011.*

Katam, R., Lin, C., Grant, K., Katam, C. S., & Chen, S. (2022). Advances in plant metabolomics and its applications in stress and single-cell biology. *International Journal of Molecular Sciences, 23*, 6985.

Kaufmann, B., & Christen, P. (2002). Recent extraction techniques for natural products: Microwave-assisted extraction and pressurized solvent extraction. *Phytochemical Analysis International Journal of Plant Chemical and Biochemical Techniques, 13*, 105–113.

Kessler, N., et al. (2013). MeltDB 2.0–Advances of the metabolomics software system. *Bioinformatics, 29*, 2452–2459.

Kim, H. K., Choi, Y. H., & Verpoorte, R. (2010). NMR-based metabolomic analysis of plants. *Nature Protocols, 5*, 536–549.

Kim, H. K., Choi, Y. H., & Verpoorte, R. (2011). NMR-based plant metabolomics: Where do we stand, where do we go? *Trends in Biotechnology, 29*, 267–275.

Kim, H. K., & Verpoorte, R. (2010). Sample preparation for plant metabolomics. *Phytochemical Analysis An International Journal of Plant Chemical and Biochemical Techniques, 21*, 4–13.

Kumar, S., & Pandey, A. K. (2013). Chemistry and biological activities of flavonoids: An overview. *Scientific World Journal, 2013*.

Lee, Y. J., Perdian, D. C., Song, Z., Yeung, E. S., & Nikolau, B. J. (2012). Use of mass spectrometry for imaging metabolites in plants. *The Plant Journal, 70*, 81–95.

Letellier, M., Budzinski, H., Charrier, L., Capes, S., & Dorthe, A. M. (1999). Optimization by factorial design of focused microwave assisted extraction of polycyclic aromatic hydrocarbons from marine sediment. *Fresenius' Journal of Analytical Chemistry, 364*, 228–237.

Liu, Y. Q., Strappe, P., Zhou, Z. K., & Blanchard, C. (2019). Impact on the nutritional attributes of rice bran following various stabilization procedures. *Critical Reviews in Food Science and Nutrition, 59*, 2458–2466.

Maia, M., et al. (2022). Grapevine leaf MALDI-MS imaging reveals the localisation of a putatively identified sucrose metabolite associated to Plasmopara viticola development. *Frontiers in Plant Science, 13*, 1012636.

Maltese, F., van der Kooy, F., & Verpoorte, R. (2009). Solvent derived artifacts in natural products chemistry. *Natural Product Communications, 4*(3), 1934578X0900400326.

Mandal, V., Mohan, Y., & Hemalatha, S. (2007). Microwave assisted extraction—An innovative and promising extraction tool for medicinal plant research. *Pharmacognosy Reviews, 1*, 7–18.

Mashabela, M. D., Masamba, P., & Kappo, A. P. (2022). Metabolomics and chemoinformatics in agricultural biotechnology research: Complementary probes in unravelling new metabolites for crop improvement. *Biology (Basel), 11*, 1156.

Menikarachchi, L. C., et al. (2012). MolFind: A software package enabling HPLC/MS-based identification of unknown chemical structures. *Analytical Chemistry, 84*, 9388–9394.

Montenegro-Burke, J. R., et al. (2017). Data streaming for metabolomics: Accelerating data processing and analysis from days to minutes. *Analytical Chemistry, 89*, 1254–1259.

Mushtaq, M. Y., Choi, Y. H., Verpoorte, R., & Wilson, E. G. (2014). Extraction for metabolomics: Access to the metabolome. *Phytochemical Analysis, 25*, 291–306.

Orata, F. (2012). Derivatization reactions and reagents for gas chromatography analysis. In *Advanced gas chromatography agricultural biomedical and industrial applications* (Vol. 91).

Pastor, A., Vázquez, E., Ciscar, R., & De la Guardia, M. (1997). Efficiency of the microwave-assisted extraction of hydrocarbons and pesticides from sediments. *Analytica Chimica Acta, 344*, 241–249.

Patel, M. K., et al. (2021). Plants metabolome study: Emerging tools and techniques. *Plants, 10*, 2409.

Ramautar, R., Somsen, G. W., & de Jong, G. J. (2009). CE-MS in metabolomics. *Electrophoresis, 30*, 276–291.

Rangarajan, R., & Ghosh, P. (2011). Role of water contamination within the GC column of a GasBench II peripheral on the reproducibility of 18O/16O ratios in water samples. *Isotopes in Environmental and Health Studies, 47*, 498–511.

Rao, S. R., & Ravishankar, G. A. (2002). Plant cell cultures: Chemical factories of secondary metabolites. *Biotechnology Advances, 20*, 101–153.

Rasul, M. G. (2018). Conventional extraction methods use in medicinal plants, their advantages and disadvantages. *International Journal of Basic Sciences and Applied Computing, 2*, 10–14.

Razzaq, A., Sadia, B., Raza, A., Khalid Hameed, M., & Saleem, F. (2019). Metabolomics: A way forward for crop improvement. *Metabolites, 9*, 303.

Reverchon, E., & De Marco, I. (2006). Supercritical fluid extraction and fractionation of natural matter. *Journal of Supercritical Fluids, 38*, 146–166.

Rischer, H., & Oksman-Caldentey, K.-M. (2006). Unintended effects in genetically modified crops: Revealed by metabolomics? *Trends in Biotechnology, 24*, 102–104.

Roessner, U., & Bowne, J. (2009). What is metabolomics all about? *BioTechniques, 46*, 363–365.

Routray, W., & Orsat, V. (2012). Microwave-assisted extraction of flavonoids: A review. *Food and Bioprocess Technology, 5*, 409–424.

Sairam, P., Ghosh, S., Jena, S., Rao, K. N. V., & Banji, D. (2012). Supercritical fluid extraction (SFE)—An overview. *Asian Journal of Research in Pharmaceutical Science, 2*, 112–120.

Schripsema, J. (2010). Application of NMR in plant metabolomics: Techniques, problems and prospects. *Phytochemical Analysis International Journal of Plant Chemical and Biochemical Techniques, 21*, 14–21.

Sharma, K., et al. (2018). *Plant metabolomics: An emerging technology for crop improvement. New visions plant science* (1st ed., pp. 65–79). IntechOpen.

Sharma, V., et al. (2021). Metabolomics intervention towards better understanding of plant traits. *Cells, 10*, 346.

Silva, L. V., Nelson, D. L., Drummond, M. F. B., Dufossé, L., & Glória, M. B. A. (2005). Comparison of hydrodistillation methods for the deodorization of turmeric. *Food Research International, 38*, 1087–1096.

Smith, C. A., et al. (2005). METLIN: A metabolite mass spectral database. *Therapeutic Drug Monitoring, 27*, 747–751.

Soga, T., & Imaizumi, M. (2001). Capillary electrophoresis method for the analysis of inorganic anions, organic acids, amino acids, nucleotides, carbohydrates and other anionic compounds. *Electrophoresis, 22*, 3418–3425.

Sturtevant, D., Lee, Y.-J., & Chapman, K. D. (2016). Matrix assisted laser desorption/ionization-mass spectrometry imaging (MALDI-MSI) for direct visualization of plant metabolites in situ. *Current Opinion in Biotechnology, 37*, 53–60.

Susniak, K., Krysa, M., Gieroba, B., Komaniecka, I., & Sroka-Bartnicka, A. (2020). Recent developments in MALDI MSI application in plant tissues analysis. *Acta Biochimica Polonica, 67*, 277–281.

Teo, C. C., Chong, W. P. K., & Ho, Y. S. (2013). Development and application of microwave-assisted extraction technique in biological sample preparation for small molecule analysis. *Metabolomics, 9*, 1109–1128.

Thirumurugan, D., Cholarajan, A., Raja, S. S., & Vijayakumar, R. (2018). An introductory chapter: Secondary metabolites. In *Secondary metabolites-sources and applications* (Vol. 1, pp. 3–21).

Thoppil, R. J., & Bishayee, A. (2011). Terpenoids as potential chemopreventive and therapeutic agents in liver cancer. *World Journal of Hepatology, 3*, 228.

Ullah, A., et al. (2020). Important flavonoids and their role as a therapeutic agent. *Molecules, 25*, 5243.

Vankar, P. S. (2004). Essential oils and fragrances from natural sources. *Resonance, 9*, 30–41.

Verpoorte, R., Choi, Y. H., Mustafa, N. R., & Kim, H. K. (2008). Metabolomics: Back to basics. *Phytochemistry Reviews, 7*, 525–537.

Verpoorte, R., Kim, H. K., & Choi, Y. H. (2022). Trivialities in metabolomics: Artifacts in extraction and analysis. *Frontiers in Molecular Biosciences, 9*, 972190.

Walters, C. C. (2019). Molecular characterization from the earliest days to the present. In *AAPG Hedberg conference, the evolution of petroleum systems analysis*.

Wang, L., & Weller, C. L. (2006). Recent advances in extraction of nutraceuticals from plants. *Trends in Food Science and Technology, 17*, 300–312.

Weckwerth, W. (2003). Metabolomics in systems biology. *Annual Review of Plant Biology, 54*, 669–689.

Xia, J., Psychogios, N., Young, N., & Wishart, D. S. (2009). MetaboAnalyst: A web server for metabolomic data analysis and interpretation. *Nucleic Acids Research, 37*, W652–W660.

Yonekura-Sakakibara, K., Higashi, Y., & Nakabayashi, R. (2019). The origin and evolution of plant flavonoid metabolism. *Frontiers in Plant Science, 10*, 943.

Yusoff, I. M., Taher, Z. M., Rahmat, Z., & Chua, L. S. (2022). A review of ultrasound-assisted extraction for plant bioactive compounds: Phenolics, flavonoids, thymols, saponins and proteins. *Food Research International, 157*, 111268.

Zaynab, M., et al. (2019). Role of primary metabolites in plant defense against pathogens. *Microbial Pathogenesis, 137*, 103728.

Zhang, H.-F., Yang, X.-H., & Wang, Y. (2011). Microwave assisted extraction of secondary metabolites from plants: Current status and future directions. *Trends in Food Science and Technology, 22*, 672–688.

Zubr, J. (1986). Methanogenic fermentation of fresh and ensiled plant materials. *Biomass, 11*, 159–171.

Chapter 7
Genome Editing

Barbaros Çetinel and Hafiza Arooj Razzaq

Abstract The genome editing prospect has altered molecular biology research and presented diverse options for detecting, imaging, manipulating, and annotating various organisms' specific genomic or transcriptomic sequences. These systems facilitate programmable site-directed mutation and are most commonly used to determine the precise functioning of a gene and the deletion of toxic genes. The key players in these systems are programmable engineered endonucleases. These endonucleases induce targeted breaks at one or both strands of DNA, filled by endogenous DNA repair mechanisms. These repairs are either of the two types: nonhomologous end joining or homology-directed repair. Several genome editing systems have been utilized for targeted mutagenesis, of which the most prominent are the ZFN, TALEN, and CRISPR–Cas systems. All these systems have pros and cons, but recent advancements in the CRISPR–Cas system make it a good choice for genome editing. This chapter discusses the genome editing systems with their limitations and their application in plant sciences.

Keywords Genome editing · Double-stranded break · Homologous recombination · Nonhomologous end joining

Introduction

An organism's genetic composition determines genotype, whereas observable characteristics determine phenotype. Genetic mutations are heritable changes in a genotype that are found naturally or may be artificially induced. The resultant organisms

B. Çetinel (✉)
Republic of Türkiye Ministry of Agriculture and Forestry, Directorate of Plant Protection Research Institute, Bornova-Izmir, Turkey

H. A. Razzaq
Centre of Agricultural Biochemistry and Biotechnology, University of Agriculture Faisalabad, Faisalabad, Pakistan

S. Ijaz et al. (eds.), *Trends in Plant Biotechnology*,
https://doi.org/10.1007/978-981-97-0814-7_7

are mutants (Holmes & Jobling, 1996). In comparison, genome editing is a group of techniques that offers possible alteration in an organism's DNA in terms of addition, deletion, or substitution. It enables targeted genome modification of several organisms, revolutionizing biological sciences of basic and applied research. In the last twentieth century, genome editing approaches were introduced to the DNA world when a rare genome cutter, I-SceI endonuclease, produced genome mutations and introduced homologous donor sequences at the targeted site (Rouet et al., 1994). However, in earlier approaches, cross-linking agents (psoralen or bleomycin), chemical agents (peptide nucleic acids or polyamides), self-splicing introns, and intron-encoded homing endonucleases were used to modify DNA or RNA (Jacquier & Dujon, 1985; Yang et al., 1996; Faruqi et al., 1998; Chevalier et al., 2002).

In recent years, sequence-specific nucleases have facilitated genome editing techniques more precisely. The most promising genome editing approaches include zinc finger nucleases, transcription activator-like nucleases, and the CRISPR–Cas system. The ease of these techniques brings scientific breakthroughs in synthetic biology; medical sciences, i.e., gene therapy, neuroscience, disease modeling, and drug discovery and development; and agricultural sciences. The main principle of these techniques is to produce cuts on both DNA strands at selected sites. Natural DNA repair mechanisms fix these breaks on continuous DNA strands, either by nonhomologous end joining (NHEJ) or site-directed mutagenesis or by homology-directed repair (HDR) or site-directed integration. As a result, gene mutation or gene replacement occurred. In multicellular organisms, NHEJ is the most common DNA repair mechanism in which the disrupted ends of DNA are directly ligated, requiring no template (Shim et al., 2017). However, it is an error-prone mechanism (Yanik et al., 2017), resulting in insertions, deletions, inversions, and translocations of one to several nucleotides (Hendel et al., 2015). In the HDR mechanism, a homologous sequence can replace the genetic information at DNA breaks (Christian et al., 2010) (Fig. 7.1).

Genome Editing Technologies

Zinc Finger Nucleases (ZFNs)

The zinc finger motifs were first identified in transcription factor IIIA of *Xenopus laevis* (Miller et al., 1985). Nine consecutive repeats of nearly 30 amino acids (AAs) with conserved pairs of two AA residues, histidine (His) and cystine (Cys), at specified positions were found. These conserved residues chelate the central metal ion, Zn^{+2}, and stabilize the independent fold-forming zinc finger (ZF). Notably, the fingerlike appearance in a sequence schematic suggested its name, the zinc finger. Functionally, these structures interact with DNA, RNA, and other proteins. These characteristics enhance their importance, particularly in gene regulation, where they bind to DNA sequentially to ON/OFF the gene. These ZF domains are highly abundant and versatile (Isalan, 2013). With the phage display method, several ZF domains with a distinct recognition of triplet base pairs have been identified (Wu

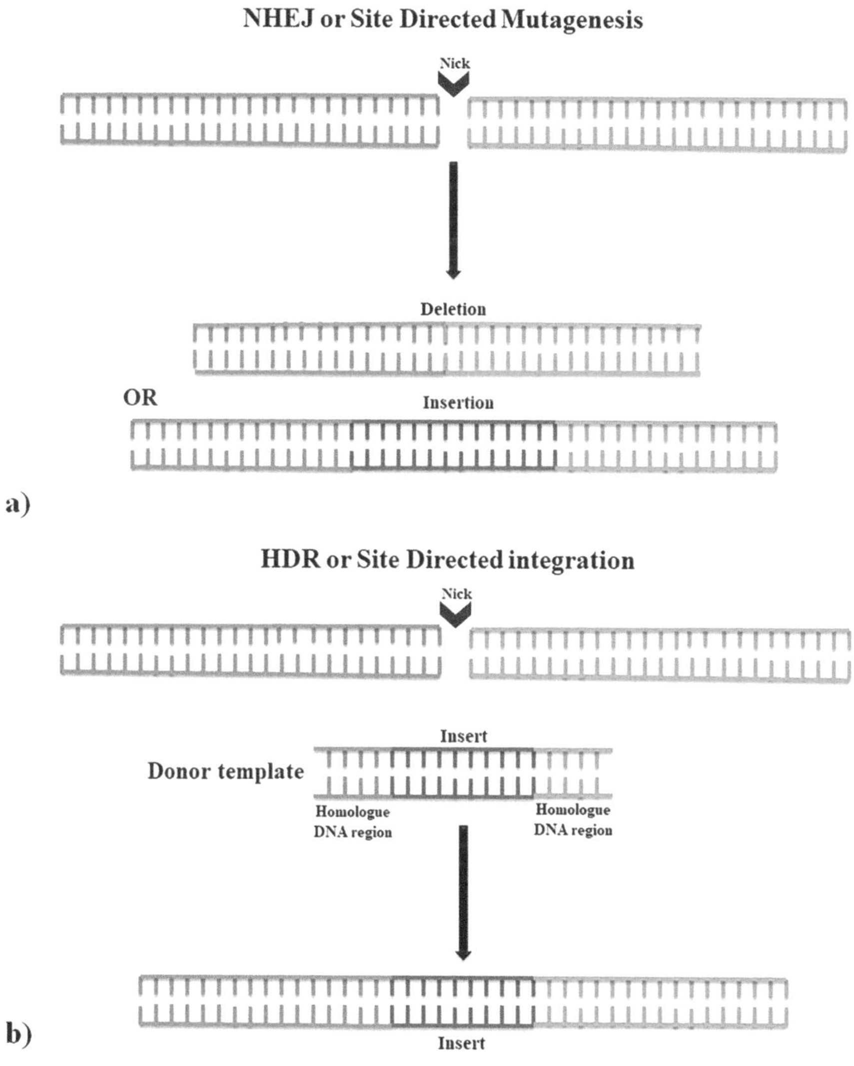

Fig. 7.1 Schematic representation of DNA double-stranded break repair with a natural DNA repair mechanism. (**a**) The NHEJ mechanism produces indels at the site of the cut/nick. (**b**) The HDR mechanism uses a donor DNA template with homologous ends and hybridizes this DNA insert to the template at the site of the cut/nick

et al., 1995; Dreier et al., 2005). Therefore, the biotechnologist reengineered the ZFs with novel binding specificity to the target in different applications.

ZFN contains a Cys_2-His_2 zinc-finger DNA-binding domain and a FokI cleavage domain (Mani et al., 2005). FokI (from *Flavobacterium okeanokoites*) is a type II restriction endonuclease that cuts nonspecifically from the DNA recognition site (Pruett-Miller, 2014). The nuclease domain of the FokI enzyme is redirected by ZFs, forming ZFNs. The DNA recognition domain is designed to bind DNA at a specific sequence, but the nuclease domain does not have a target specificity and drive-by ZFs to produce a sticky ending break on double-stranded DNA. The ZF monomer consists of nearly 30 amino acid residues arranged in the ββα motif, and the residues at the α-helix recognize triplet base pairs or codons at the targeted site (Pavletich & Pabo, 1991; Wolfe et al., 2000). However, three to six ZFs are used in pairs to recognize 18–36 base pairs of targeted sites with 5–7 base pairs of spacer DNA sequences. ZFNs dimerize at the recognition site, and at the spacer sequence, FokI breaks (Samanta et al., 2016). The FokI-restriction endonuclease has two monomers that, upon dimerization, show catalytic activity (separately attached to ZF dimers at the C-terminus) only when the ZF array recognizes a right dictated sequence site (Fig. 7.2). ZFN mutants of different organisms have been successfully obtained. ZFN-based plant genome editing is used in crops such as maize, tobacco, *Arabidopsis*, petunia, and soybean (Table 7.1). The widespread adoption of ZFNs

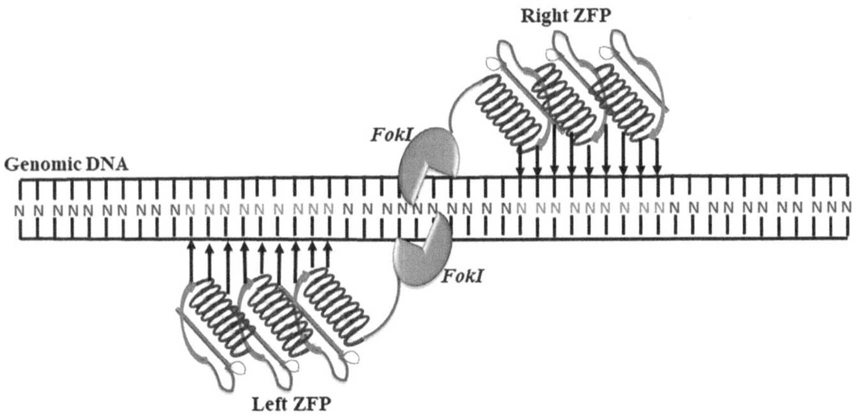

Fig. 7.2 Schematic representation of the ZFN mechanism. Two ZFN arrays (left and right) along with FokI subunits (cleavage domain) bind to the targeted region of gDNA and produce a cut

Table 7.1 List of genome-edited crops with their targeted loci using different genome editing systems

Genome editing system	Edited crop	Targeted locus	Reference
ZFN	*Zea mays*	*IPK1*	Shukla et al. (2009)
	Nicotiana tabacum	*SuRB*	Townsend et al. (2009)
	Arabidopsis thaliana	*AtABI4* *ADH1* and *TT4* *ABI4*	Osakabe et al. (2010) Zhang et al. (2010) Osakabe et al. (2010)
	Glycine max	*DCL1a, DCL1b, DCL4a, DCL4b, RDR6a, RDR6b, and HEN1a*	Curtin et al. (2011)
TALEN	*Arabidopsis thaliana*	*At16S rRNA*, *AtrpoC1*, and *AtpsbA* *Atatp1*	Nakazato et al. (2021, 2022)
	Solanum tuberosum	*StGBSS* *StSSR2* *StSSR2* *StSSR2* *StALS* *StVInv*	Kusano et al. (2018), Sawai et al. (2014), Yasumoto et al. (2019, 2020), Nicolia et al. (2015), Clasen et al. (2016)
	Chrysanthemum morifolium	*CmDMC1*	Shinoyama et al. (2020)

Table 7.1 (continued)

Genome editing system	Edited crop	Targeted locus	Reference
CRISPR–Cas9	*Arabidopsis thaliana*	*AtOST2* *AtPDS3*, *AtAG*, and *AtDUO1* *AtCLV3*, *AtAS1* *AtPDS*	Osakabe et al. (2018), Tsutsui and Higashiyama (2017), Yamamoto et al. (2019), Odahara et al. (2021)
	Oryza sativa	*OsPDS* *DsRed* *OsDL1*, *OsCYP72A33*, *OsCYP72A32*, OsVIP1, and *OsALS* *OsDMC1A* and *OsCDKB2* *OsDL*, *OsDMC1A*, *OsYSA*, and *OsPDS* *EGFP*, *OsFTIP1e*, and *OsALS* *OsEPSPS*, *OsALS*, and *OsDL* *OsTos17 LTRs* *DsRed2*, *OsDL*, and *OsGW7* *OsGCS1* and *OsPRR37* *OsAAO2* and *OsNCED1* *OsALS* and *OsCly1* *OsLig1* *OsPDS*	Mikami et al. (2015a, 2015b), Kaya et al. (2016), Endo et al. (2016a, 2016b), Mikami et al. (2016), Mikami et al. (2017), Shimatani et al. (2017), Endo et al. (2021), Saika et al. (2019), Toda et al. (2019), Negishi et al. (2020), Nishizawa-Yokoi et al. (2020), Nishizawa-Yokoi and Toki (2021), Endo et al. (2021)
	Solanum tuberosum	*StGBSSI* *StALS1* *StIAA2 gene* *GBSS* *St16DOX* *StERF3*	Kusano et al. (2018), Butler et al. (2015), Wang et al. (2015), Andersson et al. (2017, 2018), Nakayasu et al. (2018), Razzaq et al. (2022)
	Solanum lycopersicum	*SlGAD2*, *SlGAD3* *SlDELLA* and *SlETR1* *SlIAA9* *SlNADK2A*, *SlIAA9* *SlDDB1*, *SlDET1*, *SlCYC-B* *SlIAA9* *SlHWS* *SlDELLA*, *SlETR1*, *SlETR2*, *SlHWS* *DDTRF10/A*	Nonaka et al. (2017), Shimatani et al. (2017), Ueta et al. (2017), Hashimoto et al. (2018), Hunziker et al. (2020), Abe-Hara et al. (2021), Yuan et al. (2021), Kashojiya et al. (2022), Ijaz et al. (2022)
	Triticum aestivum	*TaGASR7* *TaQsd1* *TaQsd1*	Hamada et al. (2018), Abe et al. (2019), Liu et al. (2021)
	Nicotiana benthamiana	*NbPDS* *NbPDS* and *NbTOM1* *NbPDS3*	Kaya et al. (2017), Ariga et al. (2020), Nagahara et al. (2021)
	Glycine max	Gly-m Bd 28 K, Gly-m Bd 30 K loci	Sugano et al. (2020)
	Nicotiana tabacum	*NtPDS* and *NtFT4* *NtPDS*	Kaya et al. (2016, 2017)
	Vitis vinifera	*VvPDS* *VvPDS*	Nakajima et al. (2017), Osakabe et al. (2018)
	Malus pumila	*MpPDS* *MpPDS*	Nishitani et al. (2016), Osakabe et al. (2018)
	Vaccinium corymbosum	*VcCEN*	Omori et al. (2021)
	Ipomoea nil	*InDFR-B* *InCCD* *InEPH1*	Watanabe et al. (2017, 2018), Shibuya et al. (2018)
	Chrysanthemum indicum	*YGFP*	Kishi-Kaboshi et al. (2017)
	Cryptomeria japonica	*GFP*, *CjChlI*	Nanasato et al. (2021)

(continued)

Table 7.1 (continued)

Genome editing system	Edited crop	Targeted locus	Reference
CRISPR–Cas12a	*Oryza sativa*	*OsDL, OsALS, OsNCED1,* and *OsAO1* *OsDL, OsALS, OsLCD, OsAAO2,* and *OsNCED1*	Endo et al. (2016a, 2016b), Negishi et al. (2020)
	Nicotiana tabacum	*NtPDS* and *NtSTF1*	Endo et al. (2016a, 2016b)
	Glycine max	*FAD2-1A* and *FAD2-1B*	Kim et al. (2017)
CRISPR–Cas13a	*Nicotiana benthamiana*	*NbPDS*	Sharma et al. (2022)
	Solanum lycopersicum	*SlPDS*	Sharma et al. (2022)
	Arabidopsis thaliana	*AtTTG1*	Sharma et al. (2022)

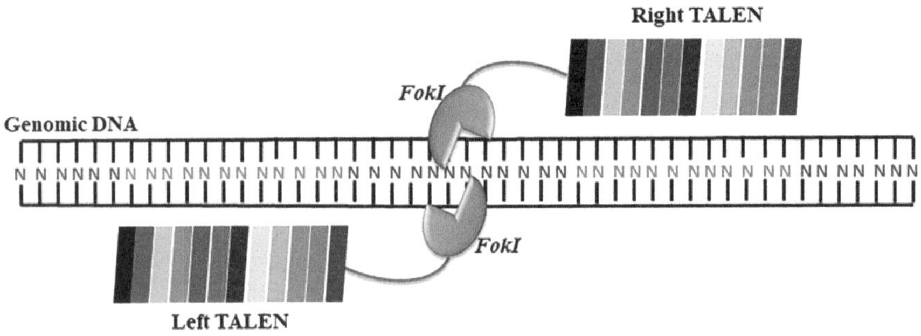

Fig. 7.3 Schematic representation of the TALEN mechanism. Two TALE arrays (left and right), known as the DNA-binding domain, along with FokI subunits (cleavage domain), bind to the targeted region of gDNA and produce a cut

has been hindered due to lower success, higher off-targeting, and complex or tedious engineering (Sander & Joung, 2014; Zhou et al., 2014; Gaj et al., 2016; Hameed et al., 2018).

Transcription Activator-Like Effector Kinases (TALENs)

TALEs are effector proteins of bacteria derived from *Xanthomonas* spp. However, in 2009, the amino acids identifying the target DNA were decoded (Boch et al., 2009). After months, their ability to modify any gene has also been customized. Similar to ZFN, TALEN is also a chimeric protein consisting of a TALE modular DNA-binding domain and a nonspecific FokI nuclease domain (Bogdanove & Voytas, 2011). Similarly, TALEN protein dimerization is facilitated by the FokI nuclease, which produces a cut at a 12–19 base pair spacer sequence separating the TALE array (Miller et al., 2011) (Fig. 7.3). Each TALE-binding domain consists of nearly 34 amino acids and recognizes a single base pair of DNA with variable 12- and 13-amino acid residues, known as repeat variable di-residues (abbreviated as RVD) (Moscou & Bogdanove, 2009). Therefore, a series of TALE domains are used to recognize approximately 12–20 base pairs of DNA, and as a result, increased genome editing specificity has been observed (Guilinger et al., 2014). The advantages of TALEN over ZFN include that it is less expensive, less time-consuming, and more specific and has reduced base mismatch. However, their larger length and codon degeneracy make it challenging to make them ideal for genome editing (Gaj et al., 2016).

CRISPR–Cas System

Microbes also possess strategic controls against invading organisms. These controls are known to their immune system. The key player in these controls is the genome of the invading entity and receptor entity. The ability of a receptor entity to battle with an attacking entity defines its fate. This is how viruses and microbes routinely survive in this diverse environment. The battle between microbes (bacteria/archaea) and viruses has been known for many years. Bacteriophages or bacterial eaters are commonly discussed viruses that occupy the bacterial control center and force them to produce their copies. These phages also infect other single-cell organisms, such as archaea. The foreign genetic elements enter into microbes and trigger an array of defense strategies, including preventing virus adsorption, blocking DNA injection, abortive infection, blocking virus replication-transcription-translation, and digestion of virus DNA by restriction-modification systems or using sugar-nonspecific nucleases. Foreign DNA is first distinguished by the host cell for survival from this invasive attack. In addition, conserving this exogenous DNA into host DNA maintains genetic integrity. Clustered regulatory interspaced palindromic repeats (CRISPRs) have been newly identified in regions known as acquired immunity genetic elements against viral attack (Horvath & Barrangou, 2010).

CRISPR is a DNA repeat family found in ~90% of archaeal and ~ 40% of bacterial host genomes (Jansen et al., 2002; van der Oost et al., 2009). The CRISPRs are palindromic, and hypervariable spacer regions separate direct DNA repeats. These regions were first identified in the bacterial (*Escherichia coli*) genome in 1983 and were abbreviated in 2002 (Jansen et al., 2002). The size of CRISPR repeats and adjacent spacers ranges between 23–47 bp and 21–72 bp, respectively. These repeats formed highly conserved and stable secondary structures. Approximately 375 spacer repeats are documented, but most loci contain fewer than 50 units (Horvath et al., 2009).

CRISPRs are typically found on chromosomal DNA, but some reports suggest their presence on plasmid DNA (Mojica et al., 2005; Lillestol et al., 2009). Several putative roles of CRISPRs involve neighboring gene expression modulation, chromosomal rearrangement, directly targeted DNA binding, repair, and replicon partitioning (Makarova et al., 2002). In 2005, homology among extrachromosomal elements (i.e., viruses and plasmids) and spacer sequences were identified in three different computer-based studies (Mojica et al., 2005; Bolotin et al., 2005), which provided clues for their role in the immune response (Makarova et al., 2006). Adjacent to the CRISPR region, the "cas" genes were identified. These genes encode proteins of different domains that typically function as polymerases, nucleases, helicases, and polynucleotide-binding proteins (Haft et al., 2005). However, these two types of loci combined formed the CRISPR–Cas system.

CRISPR–Cas systems can be categorized into two classes. Class I consists of multiple complexes of effector proteins, while class II includes single multidomain effector proteins (Makarova et al., 2015). The evolutionary features of ubiquitous Cas1–Cas2 suggest that CRISPR–Cas systems can also be characterized into six types (types I–VI); class I consists of type I, type III, and type IV, while class II consists of type II, type V, and type VI (Makarova et al., 2018, 2020). The type I systems of class I are the most widespread, containing Cas3 nuclease as the signature protein. This system has been divided into seven subtypes, types I–E, which are well characterized and used as models to understand the type I CRISPR–Cas system. Similarly, the type III system showed high similarity to type I based on its structural composition. This system has been divided into four subtypes. Type IV is categorized as class I, but sufficient information about this system has not yet been collected. The type II system of class II with three subtypes includes the most promising genome editing system, CRISPR–Cas9. Even the major research studies using the CRISPR–Cas system are based on the CRISPR–Cas9 prototype for genome editing in different organisms (Banan, 2020). Type V (three subtypes) and type VI systems of class I utilize Cas12 and Cas13 cleavage proteins, respectively (Hille et al., 2018). Genome editing using CRISPR–Cas12a and CRISPR–Cas13 has also been reported.

The CRISPR–Cas9 machinery consists of 20 nucleotides, gene-specific guide RNA (gRNA) (CRISPR RNA (crRNA) used to bind to the complementary target + *trans*-activating crRNA (tracrRNA) used for binding scaffold for the Cas9), and an endonuclease (Cas9) (Gao & Zhao, 2014). The single-guided RNA (sgRNA) binds to the targeted strand following Watson and Crick's (1953) base pairing. Primarily, the gRNA binds to its complementary DNA sequence, where the protospacer adjacent motif (PAM) sequence (most commonly NGG) is found downstream from the targeted site. Cas9 recognizes the gRNA-DNA duplex near the PAM, and its HNH (McrA-like, characterized by histidine and asparagine residues) and RuvC (RNase H fold) nuclease domains produce cuts at complementary and noncomplementary DNA strands to sgRNA, respectively (Fig. 7.4). Cas9 performed cleavage at three base pairs upstream of the PAM region (Ran et al., 2013). Notably, in addition to the 20 nucleotides of gRNA (seed sequence), there are ~80 nucleotides of gRNA scaffold sequence at the 3′ end, which is essential for Cas9 endonuclease binding (Jiang & Doudna, 2017). The DNA repair mechanism is activated when a cut is produced by Cas9 endonuclease. As a result, insertions, deletions, or frameshift mutations occur. The translation process is blocked, and the gene's function is lost (Zhang & Showalter, 2020). The versatility, ease of experimentation, reliability or stability of results, and multiplexing choice make the CRISPR–Cas9 approach fruitful.

The CRISPR–Cas12a (Cpf1) system does not need tracrRNA (usually used for binding scaffolds for Cas9 and Cas12b). After PAM recognition, the crRNA binds to the complementary target site, and Cas12 with the RuvC domain breaks the double strand, leaving overhangs of 5 and 7 nt (Fonfara et al., 2016) (Fig. 7.5). Moreover, Cas12a can recognize T-rich PAM sites on both target and nontarget strands (Zetsche et al., 2015). Cas12b does not have a PAM recognition domain. CRISPR–Cas12a

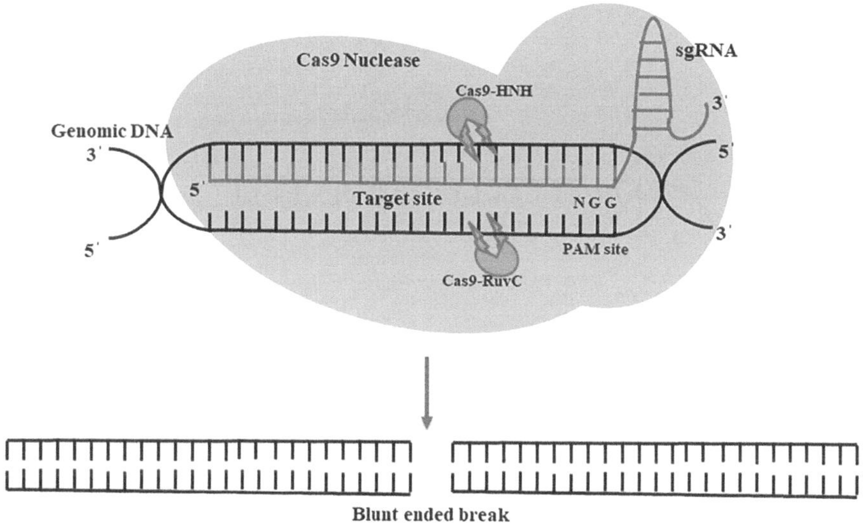

Fig. 7.4 Schematic representation of the CRISPR–Cas9 mechanism. The sgRNA binds to the Cas9 protein and starts searching for the target site on gDNA. Where the target site is encountered, a break 3 bp upstream of the PAM is produced

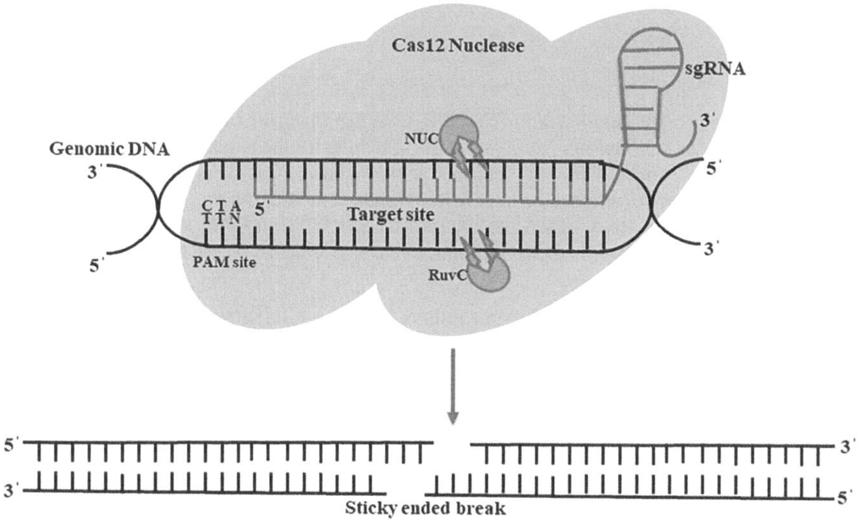

Fig. 7.5 Schematic representation of the CRISPR–Cas12 mechanism. The sgRNA (lacking the tracr region) binds to the Cas12 protein and starts searching for the target site on gDNA. Where the target site is encountered, the R-loop structure is formed. Then, the Cas12 nuclease produces a staggered cut on the targeted site

and CRISPR–Cas12b use seed sequences of 18 nt, making them highly specific for genome editing (Liu et al., 2017a, 2017b). However, the structural and functional composition of Cas12c needs to be elucidated.

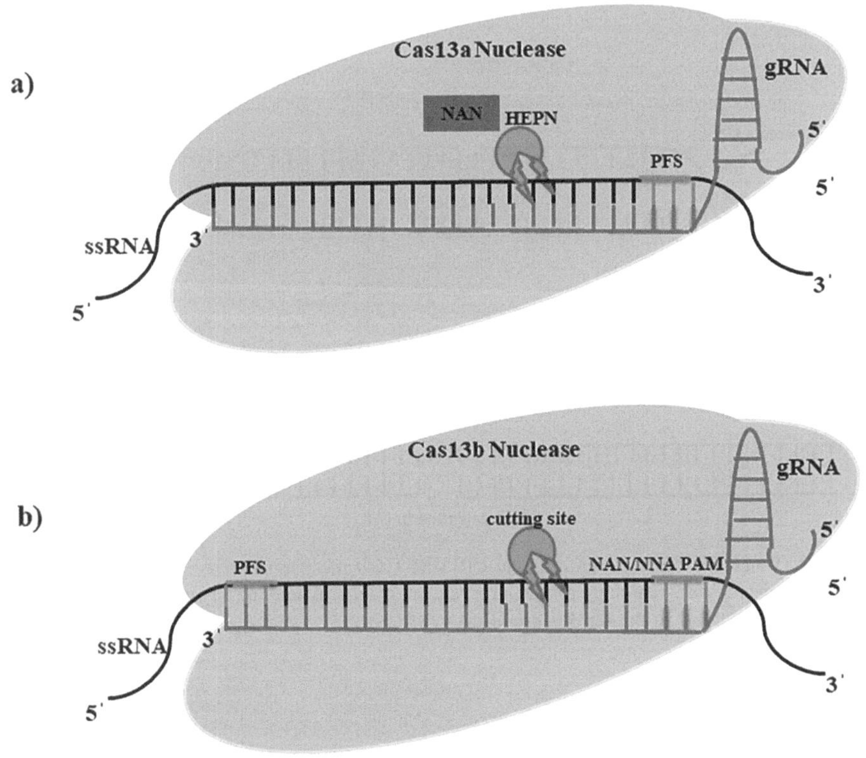

Fig. 7.6 Schematic representation of the CRISPR–Cas13 mechanism. (**a**) The gRNA (without the tracr region) binds to the Cas13a protein and starts searching for the target site on ssRNA. Where the target site was encountered, the PFS sequence at the 3′ end and Cas13a nuclease produced a cut on ssRNA. (**b**) The gRNA (without the tracr region) binds to the Cas13b protein and starts searching for the target site on ssRNA. When encountering a target site with a PFS sequence at the 5′ end and a PAM at the 3′ end, Cas13b nuclease produced a nonspecific cut on ssRNA

Cas13 (type VI) is a new addition to class II of the CRISPR–Cas system. Unlike other systems of this class, Cas13 recognizes ssRNA (Abudayyeh et al., 2016). The crRNA binds to the targeted ssRNA, and Cas13 cleaves the RNA–RNA duplex. Similar to Cas12a, Cas13 does not require tracrRNA and requires a seed sequence at the center for nuclease activity (Liu et al., 2017a, 2017b). Cas13a (C2c2) preferentially recognizes nonguanine protospacer flanking sites (Fig. 7.6a) (abbreviated as PFS), which could differ for different species (East-Seletsky et al., 2017). Cas13b required PFS at both sides: noncytosine PFS at the 5′ end and preferably NAN/NNA PFS at the 3′ end (Smargon et al., 2017) (Fig. 7.6b).

Advances in the CRISPR–Cas system bring closer to a comprehensive understanding of its complexity, diversity, and role in genome editing. This system has been efficiently utilized to edit various loci of different crops (Table 7.1). The CRISPR–Cas9 system is simple in design and efficient in working with a multiplex genome editing facility, which can be attained by designing multiple sequence-specific gRNA arrays (Jinek et al., 2012). However, the associated disadvantage of the system is off-target edits, which can be minimized using null cas9 (nCas9), which produces single-strand breaks (Rodriguez-Rodriguez et al., 2019). However, nCas9 could not completely minimize the risk of off-target effects. The CRISPR–Cas12 system has an advantage over CRISPR–Cas9, as it facilitates staggered double-stranded breaks, which utilize only the HDR system for repair (Hillary & Ceasar, 2022). However, the effectiveness of this system is cell type dependent, which is the main flaw in its application. Similar to the CRISPR–Cas9 and CRISPR–Cas12 systems, the CRISPR–Cas13 system is also a robust addition to the genome editing world. Unlike RNAi (RNA interference), it facilitates RNA targeting only cytoplasmic transcripts. The simple design of this system includes a single effector protein and gRNA (Lotfi & Rezaei, 2020). It serves as a molecular diagnostic tool for the identification of viruses. However, the major drawback of the system is also the off-targeting effect (Rodriguez-Rodriguez et al., 2019).

References

Abe, F., Haque, E., Hisano, H., Tanaka, T., Kamiya, Y., Mikami, M., Kawaura, K., Endo, M., Onishi, K., Hayashi, T., & Sato, K. (2019). Genome-edited triple-recessive mutation alters seed dormancy in wheat. *Cell Reports, 28*(5), 1362–1369.

Abe-Hara, C., Yamada, K., Wada, N., Ueta, R., Hashimoto, R., Osakabe, K., & Osakabe, Y. (2021). Effects of the sliaa9 mutation on shoot elongation growth of tomato cultivars. *Frontiers in Plant Science, 12*, 627832.

Abudayyeh, O. O., Gootenberg, J. S., Konermann, S., Joung, J., Slaymaker, I. M., Cox, D. B., Shmakov, S., Makarova, K. S., Semenova, E., Minakhin, L., & Severinov, K. (2016). C2c2 is a single-component programmable RNA-guided RNA-targeting CRISPR effector. *Science, 353*(6299), aaf5573.

Andersson, M., Turesson, H., Nicolia, A., Fält, A. S., Samuelsson, M., & Hofvander, P. (2017). Efficient targeted multiallelic mutagenesis in tetraploid potato (Solanum tuberosum) by transient CRISPR-Cas9 expression in protoplasts. *Plant Cell Reports, 36*, 117–128.

Andersson, M., Turesson, H., Olsson, N., Fält, A. S., Ohlsson, P., Gonzalez, M. N., Samuelsson, M., & Hofvander, P. (2018). Genome editing in potato via CRISPR-Cas9 ribonucleoprotein delivery. *Physiologia Plantarum, 164*(4), 378–384.

Ariga, H., Toki, S., & Ishibashi, K. (2020). Potato virus X vector-mediated DNA-free genome editing in plants. *Plant & Cell Physiology, 61*(11), 1946–1953.

Banan, M. (2020). Recent advances in CRISPR/Cas9-mediated knock-ins in mammalian cells. *Journal of Biotechnology, 308*, 1–9.

Boch, J., Scholze, H., Schornack, S., Landgraf, A., Hahn, S., Kay, S., Lahaye, T., Nickstadt, A., & Bonas, U. (2009). Breaking the code of DNA binding specificity of TAL-type III effectors. *Science, 326*(5959), 1509–1512.

Bogdanove, A. J., & Voytas, D. F. (2011). TAL effectors: Customizable proteins for DNA targeting. *Science, 333*(6051), 1843–1846.

Bolotin, A., Quinquis, B., Sorokin, A., & Ehrlich, S. D. (2005). Clustered regularly interspaced short palindrome repeats (CRISPRs) have spacers of extrachromosomal origin. *Microbiology, 151*(8), 2551–2561.

Butler, N. M., Atkins, P. A., Voytas, D. F., & Douches, D. S. (2015). Generation and inheritance of targeted mutations in potato (Solanum tuberosum L.) using the CRISPR/Cas system. *PLoS One, 10*(12), e0144591.

Chevalier, B. S., Kortemme, T., Chadsey, M. S., Baker, D., Monnat, R. J., & Stoddard, B. L. (2002). Design, activity, and structure of a highly specific artificial endonuclease. *Molecular Cell, 10*(4), 895–905.

Christian, M., Cermak, T., Doyle, E. L., Schmidt, C., Zhang, F., Hummel, A., Bogdanove, A. J., & Voytas, D. F. (2010). Targeting DNA double-strand breaks with TAL effector nucleases. *Genetics, 186*(2), 757–761.

Clasen, B. M., Stoddard, T. J., Luo, S., Demorest, Z. L., Li, J., Cedrone, F., Tibebu, R., Davison, S., Ray, E. E., Daulhac, A., & Coffman, A. (2016). Improving cold storage and processing traits in potato through targeted gene knockout. *Plant Biotechnology Journal, 14*(1), 169–176.

Curtin, S. J., Zhang, F., Sander, J. D., Haun, W. J., Starker, C., Baltes, N. J., Reyon, D., Dahlborg, E. J., Goodwin, M. J., Coffman, A. P., & Dobbs, D. (2011). Targeted mutagenesis of duplicated genes in soybean with zinc-finger nucleases. *Plant Physiology, 156*(2), 466–473.

Dreier, B., Fuller, R. P., Segal, D. J., Lund, C. V., Blancafort, P., Huber, A., Koksch, B., & Barbas, C. F. (2005). Development of zinc finger domains for recognition of the 5′-CNN-3′ family DNA sequences and their use in the construction of artificial transcription factors. *The Journal of Biological Chemistry, 280*(42), 35588–35597.

East-Seletsky, A., O'Connell, M. R., Burstein, D., Knott, G. J., & Doudna, J. A. (2017). RNA targeting by functionally orthogonal type VI-A CRISPR-Cas enzymes. *Molecular Cell, 66*(3), 373–383.

Endo, A., Masafumi, M., Kaya, H., & Toki, S. (2016a). Efficient targeted mutagenesis of rice and tobacco genomes using Cpf1 from Francisella novicida. *Scientific Reports, 6*(1), 38169.

Endo, M., Mikami, M., & Toki, S. (2016b). Biallelic gene targeting in rice. *Plant Physiology, 170*(2), 667–677.

Endo, M., Iwakami, S., & Toki, S. (2021). Precision genome editing in plants via gene targeting and subsequent break-induced single-strand annealing. *Plant Biotechnology Journal, 19*(3), 563–574.

Faruqi, A. F., Egholm, M., & Glazer, P. M. (1998). Peptide nucleic acid-targeted mutagenesis of a chromosomal gene in mouse cells. *Proceedings of the National Academy of Sciences, 95*(4), 1398–1403.

Fonfara, I., Richter, H., Bratovič, M., Le Rhun, A., & Charpentier, E. (2016). The CRISPR-associated DNA-cleaving enzyme Cpf1 also processes precursor CRISPR RNA. *Nature, 532*(7600), 517–521.

Gaj, T., Sirk, S. J., Shui, S. L., & Liu, J. (2016). Genome-editing technologies: Principles and applications. *Cold Spring Harbor Perspectives in Biology, 8*(12), a023754.

Gao, Y., & Zhao, Y. (2014). Self-processing of ribozyme-flanked RNAs into guide RNAs in vitro and in vivo for CRISPR-mediated genome editing. *Journal of Integrative Plant Biology, 56*(4), 343–349.

Guilinger, J. P., Pattanayak, V., Reyon, D., Tsai, S. Q., Sander, J. D., Joung, J. K., & Liu, D. R. (2014). Broad specificity profiling of TALENs results in engineered nucleases with improved DNA-cleavage specificity. *Nature Methods, 11*(4), 429–435.

Haft, D. H., Selengut, J., Mongodin, E. F., & Nelson, K. E. (2005). A guild of 45 CRISPR-associated (Cas) protein families and multiple CRISPR/Cas subtypes exist in prokaryotic genomes. *PLoS Computational Biology, 1*(6), e60.

Hamada, H., Liu, Y., Nagira, Y., Miki, R., Taoka, N., & Imai, R. (2018). Biolistic-delivery-based transient CRISPR/Cas9 expression enables in planta genome editing in wheat. *Scientific Reports, 8*(1), 1–7.

Hameed, A., Zaidi, S. S. E. A., Shakir, S., & Mansoor, S. (2018). Applications of new breeding technologies for potato improvement. *Frontiers in Plant Science, 9*, 925.

Hashimoto, R., Ueta, R., Abe, C., Osakabe, Y., & Osakabe, K. (2018). Efficient multiplex genome editing induces precise, and self-ligated type mutations in tomato plants. *Frontiers in Plant Science, 9*, 916.

Hendel, A., Fine, E. J., Bao, G., & Porteus, M. H. (2015). Quantifying on-and off-target genome editing. *Trends in Biotechnology, 33*(2), 132–140.

Hillary, V. E., & Ceasar, S. A. (2022). Prime editing in plants and mammalian cells: Mechanism, achievements, limitations, and future prospects. *BioEssays, 44*(9), 2200032.
Hille, F., Richter, H., Wong, S. P., Bratovič, M., Ressel, S., & Charpentier, E. (2018). The biology of CRISPR-Cas: Backward and forward. *Cell, 172*(6), 1239–1259.
Holmes, R. K., & Jobling, M. G. (1996). Genetics. In S. Baron (Ed.), *Medical microbiology* (4th ed.). University of Texas Medical Branch at Galveston.
Horvath, P., & Barrangou, R. (2010). CRISPR/Cas, the immune system of bacteria and archaea. *Science, 327*(5962), 167–170.
Horvath, P., Coûté-Monvoisin, A. C., Romero, D. A., Boyaval, P., Fremaux, C., & Barrangou, R. (2009). Comparative analysis of CRISPR loci in lactic acid bacteria genomes. *International Journal of Food Microbiology, 131*(1), 62–70.
Hunziker, J., Nishida, K., Kondo, A., Kishimoto, S., Ariizumi, T., & Ezura, H. (2020). Multiple gene substitution by Target-AID base-editing technology in tomato. *Scientific Reports, 10*(1), 20471.
Ijaz, S., Haq, I. U., & Razzaq, H. A. (2022). Mutation introduced in DDTFR10/A gene of ethylene response element-binding protein (EREBP) family through CRISPR/Cas9 genome editing confers increased Fusarium wilt tolerance in tomato. *Physiology and Molecular Biology of Plants, 29*(1), 1–10.
Isalan, M. (2013). Zinc fingers. In W. J. Lennarz & M. D. Lane (Eds.), *Encyclopedia of biological chemistry* (2nd ed., pp. 575–579). Academic Press.
Jacquier, A., & Dujon, B. (1985). An intron-encoded protein is active in a gene conversion process that spreads an intron into a mitochondrial gene. *Cell, 41*(2), 383–394.
Jansen, R., Embden, J. D. V., Gaastra, W., & Schouls, L. M. (2002). Identification of genes that are associated with DNA repeats in prokaryotes. *Molecular Microbiology, 43*(6), 1565–1575.
Jiang, F., & Doudna, J. A. (2017). CRISPR–Cas9 structures and mechanisms. *Annual Review of Biophysics, 46*, 505–529.
Jinek, M., Chylinski, K., Fonfara, I., Hauer, M., Doudna, J. A., & Charpentier, E. (2012). A programmable dual-RNA–guided DNA endonuclease in adaptive bacterial immunity. *Science, 337*(6096), 816–821.
Kashojiya, S., Lu, Y., Takayama, M., Komatsu, H., Minh, L. H. T., Nishida, K., Shirasawa, K., Miura, K., Nonaka, S., Masuda, J. I., & Kondo, A. (2022). Modification of tomato breeding traits and plant hormone signaling by Target-AID, the genome-editing system inducing efficient nucleotide substitution. *Horticulture Research, 9*.
Kaya, H., Mikami, M., Endo, A., Endo, M., & Toki, S. (2016). Highly specific targeted mutagenesis in plants using Staphylococcus aureus Cas9. *Scientific Reports, 6*(1), 26871.
Kaya, H., Ishibashi, K., & Toki, S. (2017). A split Staphylococcus aureus Cas9 as a compact genome-editing tool in plants. *Plant & Cell Physiology, 58*(4), 643–649.
Kim, H., Kim, S. T., Ryu, J., Kang, B. C., Kim, J. S., & Kim, S. G. (2017). CRISPR/Cpf1-mediated DNA-free plant genome editing. *Nature Communications, 8*(1), 14406.
Kishi-Kaboshi, M., Aida, R., & Sasaki, K. (2017). Generation of gene-edited Chrysanthemum morifolium using multicopy transgenes as targets and markers. *Plant & Cell Physiology, 58*(2), 216–226.
Kusano, H., Ohnuma, M., Mutsuro-Aoki, H., Asahi, T., Ichinosawa, D., Onodera, H., Asano, K., Noda, T., Horie, T., Fukumoto, K., & Kihira, M. (2018). Establishment of a modified CRISPR/Cas9 system with increased mutagenesis frequency using the translational enhancer dMac3 and multiple guide RNAs in potato. *Scientific Reports, 8*(1), 13753.
Lillestøl, R. K., Shah, S. A., Brügger, K., Redder, P., Phan, H., Christiansen, J., & Garrett, R. A. (2009). CRISPR families of the crenarchaeal genus Sulfolobus: Bidirectional transcription and dynamic properties. *Molecular Microbiology, 72*(1), 259–272.
Liu, L., Chen, P., Wang, M., Li, X., Wang, J., Yin, M., & Wang, Y. (2017a). C2c1-sgRNA complex structure reveals RNA-guided DNA cleavage mechanism. *Molecular Cell, 65*(2), 310–322.
Liu, L., Li, X., Wang, J., Wang, M., Chen, P., Yin, M., Li, J., Sheng, G., & Wang, Y. (2017b). Two distant catalytic sites are responsible for C2c2 RNase activities. *Cell, 168*(1–2), 121–134.

Liu, Y., Luo, W., Linghu, Q., Abe, F., Hisano, H., Sato, K., Kamiya, Y., Kawaura, K., Onishi, K., Endo, M., & Toki, S. (2021). In planta genome editing in commercial wheat varieties. *Frontiers in Plant Science, 12*, 648841.

Lotfi, M., & Rezaei, N. (2020). CRISPR/Cas13: A potential therapeutic option of COVID-19. *Biomedicine & Pharmacotherapy, 131*, 110738.

Makarova, K. S., Aravind, L., Grishin, N. V., Rogozin, I. B., & Koonin, E. V. (2002). A DNA repair system specific for thermophilic Archaea and bacteria predicted by genomic context analysis. *Nucleic Acids Research, 30*(2), 482–496.

Makarova, K. S., Grishin, N. V., Shabalina, S. A., Wolf, Y. I., & Koonin, E. V. (2006). A putative RNA-interference-based immune system in prokaryotes: Computational analysis of the predicted enzymatic machinery, functional analogies with eukaryotic RNAi, and hypothetical mechanisms of action. *Biology Direct, 1*(1), 1–26.

Makarova, K. S., Wolf, Y. I., Alkhnbashi, O. S., Costa, F., Shah, S. A., Saunders, S. J., Barrangou, R., Brouns, S. J., Charpentier, E., Haft, D. H., & Horvath, P. (2015). An updated evolutionary classification of CRISPR–Cas systems. *Nature Reviews. Microbiology, 13*(11), 722–736.

Makarova, K. S., Wolf, Y. I., & Koonin, E. V. (2018). Classification and nomenclature of CRISPR-Cas systems: Where from here? *The CRISPR Journal, 1*(5), 325–336.

Makarova, K. S., Wolf, Y. I., Iranzo, J., Shmakov, S. A., Alkhnbashi, O. S., Brouns, S. J., Charpentier, E., Cheng, D., Haft, D. H., Horvath, P., & Moineau, S. (2020). Evolutionary classification of CRISPR–Cas systems: A burst of class 2 and derived variants. *Nature Reviews. Microbiology, 18*(2), 67–83.

Mani, M., Smith, J., Kandavelou, K., Berg, J. M., & Chandrasegaran, S. (2005). Binding of two zinc finger nuclease monomers to two specific sites is required for effective double-strand DNA cleavage. *Biochemical and Biophysical Research Communications, 334*(4), 1191–1197.

Mikami, M., Toki, S., & Endo, M. (2015a). Comparison of CRISPR/Cas9 expression constructs for efficient targeted mutagenesis in rice. *Plant Molecular Biology, 88*(6), 561–572.

Mikami, M., Toki, S., & Endo, M. (2015b). Parameters affecting frequency of CRISPR/Cas9 mediated targeted mutagenesis in rice. *Plant Cell Reports, 34*, 1807–1815.

Mikami, M., Toki, S., & Endo, M. (2016). Precision targeted mutagenesis via Cas9 paired nickases in rice. *Plant & Cell Physiology, 57*(5), 1058–1068.

Mikami, M., Toki, S., & Endo, M. (2017). In planta processing of the SpCas9–gRNA complex. *Plant & Cell Physiology, 58*(11), 1857–1867.

Miller, J., McLachlan, A. D., & Klug, A. (1985). Repetitive zinc-binding domains in the protein transcription factor IIIA from Xenopus oocytes. *The EMBO Journal, 4*(6), 1609–1614.

Miller, J. C., Tan, S., Qiao, G., Barlow, K. A., Wang, J., Xia, D. F., Meng, X., Paschon, D. E., Leung, E., Hinkley, S. J., & Dulay, G. P. (2011). A TALE nuclease architecture for efficient genome editing. *Nature Biotechnology, 29*(2), 143–148.

Mojica, F. J., Díez-Villaseñor, C. S., García-Martínez, J., & Soria, E. (2005). Intervening sequences of regularly spaced prokaryotic repeats derive from foreign genetic elements. *Journal of Molecular Evolution, 60*, 174–182.

Moscou, M. J., & Bogdanove, A. J. (2009). A simple cipher governs DNA recognition by TAL effectors. *Science, 326*(5959), 1501–1501.

Nagahara, S., Higashiyama, T., & Mizuta, Y. (2021). Detection of a biolistic delivery of fluorescent markers and CRISPR/Cas9 to the pollen tube. *Plant Reproduction, 34*(3), 191–205.

Nakajima, I., Ban, Y., Azuma, A., Onoue, N., Moriguchi, T., Yamamoto, T., Toki, S., & Endo, M. (2017). CRISPR/Cas9-mediated targeted mutagenesis in grape. *PLoS One, 12*, e0177966.

Nakayasu, M., Akiyama, R., Lee, H. J., Osakabe, K., Osakabe, Y., Watanabe, B., Sugimoto, Y., Umemoto, N., Saito, K., Muranaka, T., & Mizutani, M. (2018). Generation of α-solanine-free hairy roots of potato by CRISPR/Cas9 mediated genome editing of the St16DOX gene. *Plant Physiology and Biochemistry, 131*, 70–77.

Nakazato, I., Okuno, M., Yamamoto, H., Tamura, Y., Itoh, T., Shikanai, T., Takanashi, H., Tsutsumi, N., & Arimura, S. I. (2021). Targeted base editing in the plastid genome of Arabidopsis thaliana. *Nature Plants, 7*(7), 906–913.

Nakazato, I., Okuno, M., Zhou, C., Itoh, T., Tsutsumi, N., Takenaka, M., & Arimura, S. I. (2022). Targeted base editing in the mitochondrial genome of Arabidopsis thaliana. *Proceedings of the National Academy of Sciences, 119*(20), e2121177119.

Nanasato, Y., Mikami, M., Futamura, N., Endo, M., Nishiguchi, M., Ohmiya, Y., Konagaya, K. I., & Taniguchi, T. (2021). CRISPR/Cas9-mediated targeted mutagenesis in Japanese cedar (Cryptomeria japonica D Don). *Scientific Reports, 11*(1), 16186.

Negishi, K., Mikami, M., Toki, S., & Endo, M. (2020). Enhanced FnCas12a-mediated targeted mutagenesis using crRNA With altered target length in rice. *Frontiers in Genome Editing, 2*, 608563.

Nicolia, A., Proux-Wéra, E., Åhman, I., Onkokesung, N., Andersson, M., Andreasson, E., & Zhu, L. H. (2015). Targeted gene mutation in tetraploid potato through transient TALEN expression in protoplasts. *Journal of Biotechnology, 204*, 17–24.

Nishitani, C., Hirai, N., Komori, S., Wada, M., Okada, K., Osakabe, K., Yamamoto, T., & Osakabe, Y. (2016). Efficient genome editing in apple using a CRISPR/Cas9 system. *Scientific Reports, 6*(1), 31481.

Nishizawa-Yokoi, A., & Toki, S. (2021). A piggyBac-mediated transgenesis system for the temporary expression of CRISPR/Cas9 in rice. *Plant Biotechnology Journal, 19*(7), 1386–1395.

Nishizawa-Yokoi, A., Mikami, M., & Toki, S. (2020). A universal system of CRISPR/Cas9-mediated gene targeting using all-in-one vector in plants. *Frontiers in Genome Editing, 2*, 604289.

Nonaka, S., Arai, C., Takayama, M., Matsukura, C., & Ezura, H. (2017). Efficient increase of ɣ-aminobutyric acid (GABA) content in tomato fruits by targeted mutagenesis. *Scientific Reports, 7*(1), 1–14.

Odahara, M., Watanabe, K., Kawasaki, R., Tsuchiya, K., Tateishi, A., Motoda, Y., Kigawa, T., Kodama, Y., & Numata, K. (2021). Nanoscale polyion complex vesicles for delivery of cargo proteins and Cas9 ribonucleoprotein complexes to plant cells. *ACS Applied Nano Materials, 4*(6), 5630–5635.

Omori, M., Yamane, H., Osakabe, K., Osakabe, Y., & Tao, R. (2021). Targeted mutagenesis of CENTRORADIALIS using CRISPR/Cas9 system through the improvement of genetic transformation efficiency of tetraploid highbush blueberry. *The Journal of Horticultural Science and Biotechnology, 96*(2), 153–161.

Osakabe, K., Osakabe, Y., & Toki, S. (2010). Site-directed mutagenesis in Arabidopsis using custom-designed zinc finger nucleases. *Proceedings of the National Academy of Sciences, 107*(26), 12034–12039.

Osakabe, Y., Liang, Z., Ren, C., Nishitani, C., Osakabe, K., Wada, M., Komori, S., Malnoy, M., Velasco, R., Poli, M., & Jung, M. H. (2018). CRISPR–Cas9-mediated genome editing in apple and grapevine. *Nature Protocols, 13*(12), 2844–2863.

Pavletich, N. P., & Pabo, C. O. (1991). Zinc finger-DNA recognition: Crystal structure of a Zif268-DNA complex at 2.1 Å. *Science, 252*(5007), 809–817.

Pruett-Miller, S. M. (2014). Chapter 29: Genome editing in somatic cells using zinc finger nucleases and transcription activator-like effector nucleases. In J. Cibelli, J. Gurdon, I. Wilmut, R. Jaenisch, R. Lanza, M. D. West, & K. H. S. Campbell (Eds.), *Principles of cloning* (2nd ed., pp. 369–378). Academic Press.

Ran, F. A., Hsu, P. D., Wright, J., Agarwala, V., Scott, D. A., & Zhang, F. (2013). Genome engineering using the CRISPR-Cas9 system. *Nature Protocols, 8*(11), 2281–2308.

Razzaq, H. A., Ijaz, S., Haq, I. U., & Khan, I. A. (2022). Functional inhibition of the StERF3 gene by dual targeting through CRISPR/Cas9 enhances resistance to the late blight disease in Solanum Tuberosum L. *Molecular Biology Reports, 49*(12), 11675–11684.

Rodríguez-Rodríguez, D. R., Ramírez-Solís, R., Garza-Elizondo, M. A., Garza-Rodríguez, M. D. L., & Barrera-Saldaña, H. A. (2019). Genome editing: A perspective on the application of CRISPR/Cas9 to study human diseases. *International Journal of Molecular Medicine, 43*(4), 1559–1574.

Rouet, P., Smih, F., & Jasin, M. (1994). Introduction of double-strand breaks into the genome of mouse cells by expression of a rare-cutting endonuclease. *Molecular and Cellular Biology, 14*(12), 8096–8106.

Saika, H., Mori, A., Endo, M., & Toki, S. (2019). Targeted deletion of rice retrotransposon Tos17 via CRISPR/Cas9. *Plant Cell Reports, 38*, 455–458.

Samanta, M. K., Dey, A., & Gayen, S. (2016). CRISPR/Cas9: An advanced tool for editing plant genomes. *Transgenic Research, 25*, 561–573.

Sander, J. D., & Joung, J. K. (2014). CRISPR-Cas systems for editing, regulating and targeting genomes. *Nature Biotechnology, 32*(4), 347–355.

Sawai, S., Ohyama, K., Yasumoto, S., Seki, H., Sakuma, T., Yamamoto, T., Takebayashi, Y., Kojima, M., Sakakibara, H., Aoki, T., & Muranaka, T. (2014). Sterol side chain reductase 2 is a key enzyme in the biosynthesis of cholesterol, the common precursor of toxic steroidal glycoalkaloids in potato. *Plant Cell, 26*(9), 3763–3774.

Sharma, V. K., Marla, S., Zheng, W., Mishra, D., Huang, J., Zhang, W., Morris, G. P., & Cook, D. E. (2022). CRISPR guides induce gene silencing in plants in the absence of Cas. *Genome Biology, 23*, 1–24.

Shibuya, K., Watanabe, K., & Ono, M. (2018). CRISPR/Cas9-mediated mutagenesis of the EPHEMERAL1 locus that regulates petal senescence in Japanese morning glory. *Plant Physiology and Biochemistry, 131*, 53–57.

Shim, G., Kim, D., Park, G. T., Jin, H., Suh, S. K., & Oh, Y. K. (2017). Therapeutic gene editing: Delivery and regulatory perspectives. *Acta Pharmacologica Sinica, 38*(6), 738–753.

Shimatani, Z., Kashojiya, S., Takayama, M., Terada, R., Arazoe, T., Ishii, H., Teramura, H., Yamamoto, T., Komatsu, H., Miura, K., & Ezura, H. (2017). Targeted base editing in rice and tomato using a CRISPR-Cas9 cytidine deaminase fusion. *Nature Biotechnology, 35*(5), 441–443.

Shinoyama, H., Ichikawa, H., Nishizawa-Yokoi, A., Skaptsov, M., & Toki, S. (2020). Simultaneous TALEN-mediated knockout of chrysanthemum DMC1 genes confers male and female sterility. *Scientific Reports, 10*(1), 1–14.

Shukla, V. K., Doyon, Y., Miller, J. C., DeKelver, R. C., Moehle, E. A., Worden, S. E., Mitchell, J. C., Arnold, N. L., Gopalan, S., Meng, X., & Choi, V. M. (2009). Precise genome modification in the crop species Zea mays using zinc-finger nucleases. *Nature, 459*(7245), 437–441.

Smargon, A. A., Cox, D. B., Pyzocha, N. K., Zheng, K., Slaymaker, I. M., Gootenberg, J. S., Abudayyeh, O. A., Essletzbichler, P., Shmakov, S., Makarova, K. S., & Koonin, E. V. (2017). Cas13b is a type VI-B CRISPR-associated RNA-guided RNase differentially regulated by accessory proteins Csx27 and Csx28. *Molecular Cell, 65*(4), 618–630.

Sugano, S., Hirose, A., Kanazashi, Y., Adachi, K., Hibara, M., Itoh, T., Mikami, M., Endo, M., Hirose, S., Maruyama, N., & Abe, J. (2020). Simultaneous induction of mutant alleles of two allergenic genes in soybean by using site-directed mutagenesis. *BMC Plant Biology, 20*(1), 1–15.

Toda, E., Koiso, N., Takebayashi, A., Ichikawa, M., Kiba, T., Osakabe, K., Osakabe, Y., Sakakibara, H., Kato, N., & Okamoto, T. (2019). An efficient DNA-and selectable-marker-free genome-editing system using zygotes in rice. *Nature Plants, 5*(4), 363–368.

Townsend, J. A., Wright, D. A., Winfrey, R. J., Fu, F., Maeder, M. L., Joung, J. K., & Voytas, D. F. (2009). High-frequency modification of plant genes using engineered zinc-finger nucleases. *Nature, 459*(7245), 442–445.

Tsutsui, H., & Higashiyama, T. (2017). pKAMA-ITACHI vectors for highly efficient CRISPR/Cas9-mediated gene knockout in Arabidopsis thaliana. *Plant & Cell Physiology, 58*(1), 46–56.

Ueta, R., Abe, C., Watanabe, T., Sugano, S. S., Ishihara, R., Ezura, H., Osakabe, Y., & Osakabe, K. (2017). Rapid breeding of parthenocarpic tomato plants using CRISPR/Cas9. *Scientific Reports, 7*(1), 507.

Van der Oost, J., Jore, M. M., Westra, E. R., Lundgren, M., & Brouns, S. J. (2009). CRISPR-based adaptive and heritable immunity in prokaryotes. *Trends in Biochemical Sciences, 34*(8), 401–407.

Wang, S., Zhang, S., Wang, W., Xiong, X., Meng, F., & Cui, X. (2015). Efficient targeted mutagenesis in potato by the CRISPR/Cas9 system. *Plant Cell Reports, 34*, 1473–1476.

Watson, J. D., & Crick, F. H. (1953, January). The structure of DNA. In Cold Spring Harbor symposia on quantitative biology (Vol. 18, pp. 123–131). Cold Spring Harbor Laboratory Press.

Watanabe, K., Kobayashi, A., Endo, M., Sage-Ono, K., Toki, S., & Ono, M. (2017). CRISPR/Cas9-mediated mutagenesis of the dihydroflavonol-4-reductase-B (DFR-B) locus in the Japanese morning glory Ipomoea (Pharbitis) nil. *Scientific Reports, 7*(1), 10028.

Watanabe, K., Oda-Yamamizo, C., Sage-Ono, K., Ohmiya, A., & Ono, M. (2018). Alteration of flower colour in Ipomoea nil through CRISPR/Cas9-mediated mutagenesis of carotenoid cleavage dioxygenase 4. *Transgenic Research, 27*, 25–38.

Wolfe, S. A., Nekludova, L., & Pabo, C. O. (2000). DNA recognition by Cys2His2 zinc finger proteins. *Annual Review of Biophysics and Biomolecular Structure, 29*(1), 183–212.

Wu, H., Yang, W. P., & Barbas, C. F., 3rd. (1995). Building zinc fingers by selection: Toward a therapeutic application. *Proceedings of the National Academy of Sciences, 92*(2), 344–348.

Yamamoto, A., Ishida, T., Yoshimura, M., Kimura, Y., & Sawa, S. (2019). Developing heritable mutations in Arabidopsis thaliana using a modified CRISPR/Cas9 toolkit comprising PAM-altered Cas9 variants and gRNAs. *Plant & Cell Physiology, 60*(10), 2255–2262.

Yang, J., Zimmerly, S., Perlman, P. S., & Lambowitz, A. M. (1996). Efficient integration of an intron RNA into double-stranded DNA by reverse splicing. *Nature, 381*, 332–335.

Yanik, M., Müller, B., Song, F., Gall, J., Wagner, F., Wende, W., Lorenz, B., & Stieger, K. (2017). In vivo genome editing as a potential treatment strategy for inherited retinal dystrophies. *Progress in Retinal and Eye Research, 56*, 1–18.

Yasumoto, S., Umemoto, N., Lee, H. J., Nakayasu, M., Sawai, S., Sakuma, T., Yamamoto, T., Mizutani, M., Saito, K., & Muranaka, T. (2019). Efficient genome engineering using Platinum TALEN in potato. *Plant Biotechnology, 36*(3), 167–173.

Yasumoto, S., Sawai, S., Lee, H. J., Mizutani, M., Saito, K., Umemoto, N., & Muranaka, T. (2020). Targeted genome editing in tetraploid potato through transient TALEN expression by Agrobacterium infection. *Plant Biotechnology, 37*(2), 205–211.

Yuan, S., Kawasaki, S., Abdellatif, I. M., Nishida, K., Kondo, A., Ariizumi, T., Ezura, H., & Miura, K. (2021). Efficient base editing in tomato using a highly expressed transient system. *Plant Cell Reports, 40*, 667–676.

Zetsche, B., Gootenberg, J. S., Abudayyeh, O. O., Slaymaker, I. M., Makarova, K. S., Essletzbichler, P., Volz, S. E., Joung, J., Van Der Oost, J., Regev, A., & Koonin, E. V. (2015). Cpf1 is a single RNA-guided endonuclease of a class 2 CRISPR-Cas system. *Cell, 163*(3), 759–771.

Zhang, Y., & Showalter, A. M. (2020). CRISPR/Cas9 genome editing technology: A valuable tool for understanding plant cell wall biosynthesis and function. *Frontiers in Plant Science, 11*, 589517.

Zhang, F., Maeder, M. L., Unger-Wallace, E., Hoshaw, J. P., Reyon, D., Christian, M., Li, X., Pierick, C. J., Dobbs, D., Peterson, T., & Joung, J. K. (2010). High frequency targeted mutagenesis in Arabidopsis thaliana using zinc finger nucleases. *Proceedings of the National Academy of Sciences, 107*(26), 12028–12033.

Zhou, H., Liu, B., Weeks, D. P., Spalding, M. H., & Yang, B. (2014). Large chromosomal deletions and heritable small genetic changes induced by CRISPR/Cas9 in rice. *Nucleic Acids Research, 42*(17), 10903–10914.

Chapter 8
Application of CRISPR–Cas Technology in Drug Development

Sidra Altaf, Faiza Saleem, Hafiza Maimoona Ikram, Arslan Iftikhar, and Haseeb Anwar

Abstract The discovery of clustered regularly interspaced short palindromic repeat (CRISPR)-Cas technology has transformed the science of genome editing, providing exemplary accuracy and efficiency in changing DNA sequences. It has generated a resurgence in genetic engineering that offers benefits in various sectors, including basic scientific research and translational medicine. By allowing us to alter, discover, and identify certain gene sequences, these genome editing techniques have expanded the possibilities for scientific investigation. The CRISPR–Cas9 system has gained significant acceptance throughout the research community in a very short period due to its extraordinary simplicity, resilience, and versatility. Its uses span various activities, including the construction of transgenic animals and plants, the clarification of disease processes, the discovery of disease targets, the generation of model cell lines, and the control of gene expression. This technology also emphasizes how crucial it is to monitor its development closely. Starting with its invention and emphasizing significant developments in the realm of drug discovery, this review article seeks to offer a thorough overview of the present status of CRISPR–Cas technology in a tailored style, future trends, their opportunities, challenges, and ethical issues that could arise from the continuous advancement and use of this technology.

Keywords Pharmacy · Drugs · Genome editing · Gene therapy

Introduction

CRISPR–Cas has emerged as a revolutionary gene editing technique in recent years, revolutionizing the biotechnology industry and providing unprecedented accuracy and efficiency in DNA manipulation. CRISPR and CRISPR-associated proteins,

---

S. Altaf (✉) · F. Saleem · H. M. Ikram · A. Iftikhar · H. Anwar
Department of Pharmacy, University of Agriculture, Faisalabad, Pakistan
e-mail: sidraaltaf@uaf.edu.pk

S. Ijaz et al. (eds.), *Trends in Plant Biotechnology*,
https://doi.org/10.1007/978-981-97-0814-7_8

known as CRISPR–Cas, were first found in bacteria as an adaptive immunological defense against viral infections. This cutting-edge gene editing technique has the potential to completely transform the way we cure illnesses, enhance crops, and even bring in new life-forms. CRISPR uses a guide RNA–Cas to target a particular DNA sequence that is subsequently cut by the Cas9 enzyme. This enables precise alteration of the genome, which may be used to fix genetic flaws, develop novel medications, or even create new animal breeds (Xu & Li, 2020; Wang & Cui, 2020; Zhu et al., 2020).

Identifying viable targets for a therapeutic approach is one of the key challenges in drug development. This groundbreaking finding is based on bacteria's immunological mechanisms to defend themselves against viral invasion (Mosterd et al., 2021). Researchers now have an unrivaled ability to comprehensively examine genes and understand their functions in illnesses because of CRISPR–Cas. Using genome-wide CRISPR–Cas knockout screens, researchers have effectively discovered genes associated with several illnesses, including cancer, cardiovascular disease, and neurological disorders. These studies have uncovered novel pharmacological targets presenting new prospects for therapeutic intervention (Chandrasekaran et al., 2022; Shim, 2022; Zhang, 2021).

History

The CRISPR–Cas journey began long before it became known as a gene editing technology in 2012. Early in the 1990s, Francisco J. Mojica laid the groundwork for further investigation by describing the repetitive DNA sequences in bacteria (Kratzer et al., 2021; Halder et al., 2022). CRISPRs were first discovered in bacteria in 1987 when analyzing the nucleotide sequence of the iap gene in *Escherichia coli*, which oversees alkaline phosphatase isozyme conversion (Bhattacharjee et al., 2022; Lee & Sashital, 2022; Tukur, 2021). The 3′ end of the iap gene lies in proximity to five repeated palindromic nucleotide sequences identified in this study (Golubov, 2021; Hillary & Ceasar, 2023). In between these repetitions were varied sequences of a fixed length that showed no resemblance to one another. Six years later, scientists observed an identical pattern of nucleotide sequences in Archaea (Shmakova et al., 2022; Mosterd et al., 2021). Confusion in the field resulted from these perplexing repetitions being studied by numerous research teams at different times, each giving them a different name or abbreviation. The first CRISPR-associated (cas) genes were identified, and the word CRISPR was coined in 2002 (Kumar et al., 2022; Roy Chowdhury & Gargate, 2021; El ouar & Djekoun, 2021; Rasool et al., 2017). These genes were discovered to flank CRISPR loci and were frequently missing in CRISPR-negative strains (Wu et al., 2022).

Before 2005, it was unclear what the variable sequences (known as spacers) between the repetitions were. Then, homology was found between certain spacers within a few CRISPR loci and viral and plasmid sequences. Several theories have suggested that CRISPRs may be used to protect against foreign genetic material

(Martínez Arbas et al., 2021; Kamruzzaman & Iredell, 2020). A cross-talk mechanism between CRISPRs and Cas proteins in defense against plasmids and phages, similar to eukaryotic RNA interference systems, was revealed in 2006 by bioinformatics (Golubov, 2021).

In 2007, an experiment using the dairy bacterial species *Streptococcus thermophilus* marked a crucial turning point in our knowledge of CRISPR–Cas. Most of the phage-infected cells died from infection when exposed to virulent streptococcal phages, but a tiny percentage managed to survive (Chavez-Granados et al., 2022; Safari et al., 2019; Markakiou et al., 2020; Watson et al., 2021). The incorporation of a new spacer and the insertion of a repeat were discovered when the CRISPR array of the surviving bacterial cells was examined. These additional spacers came from the genome of the dangerous phage utilized in the experiment. They were named "protospacers" to distinguish them from bacterial spacers. The phage bearing that sequence gained resistance when a similar spacer was added and regained resistance when the spacer was removed. The phage bearing that sequence was rendered resistant by including an identical spacer, and sensitivity to the phage was restored by deleting the spacer (Golubov, 2021). Plasmid interference was also demonstrated in 2008. Cas9 was shown to be the only protein necessary for DNA cleavage in *S. thermophilus* and *Streptococcus pyogenes*, whereas numerous Cas proteins contribute to the interference stage in *E. coli* (Shmakova et al., 2022).

Further research in *Escherichia coli* demonstrated that the Cas proteins process the CRISPR locus during transcription, synthesizing small guide RNAs (sgRNAs). These sgRNAs enable interference with known foreign DNA invasions, made possible by a combination of Cas proteins and CRISPR-RNA (crRNA) (Xu & Tsang, 2022). In 2010, the *S. thermophilus* CRISPR–Cas system was discovered to precisely cleave target plasmid and phage DNA inside the protospacer region, explaining plasmid interference and phage resistance (Munawar et al., 2021).

The molecular mechanism of the gene editing tool CRISPR was later discovered by Jennifer A. Doudna, Emmanuelle M. Charpentier, and their teams in 2012, allowing for accurate and precise targeted cuts in the genome. The related proteins were named Cas, and the repeating elements were named CRISPR (Shmakova et al., 2022). This amazing discovery piqued experts' curiosity and highlighted this system's potential for genetic modification (Zhang et al., 2021a, 2021b; Agarwal & Gupta, 2021; Ghorbani et al., 2021).

Classification

The CRISPR–Cas system has been divided into two classes based on effector protein properties, functions, and architecture. Class 1 systems comprise several effector complexes with 4–7 Cas proteins, whereas class 2 systems only include one Cas protein with various subdomains. Types I, III, and IV are the three subtypes found in the class 1 system, while types II, V, and VI are found in the class 2 system, with each type having a unique gene. Type VI targets RNA, while types I, II, IV, and V

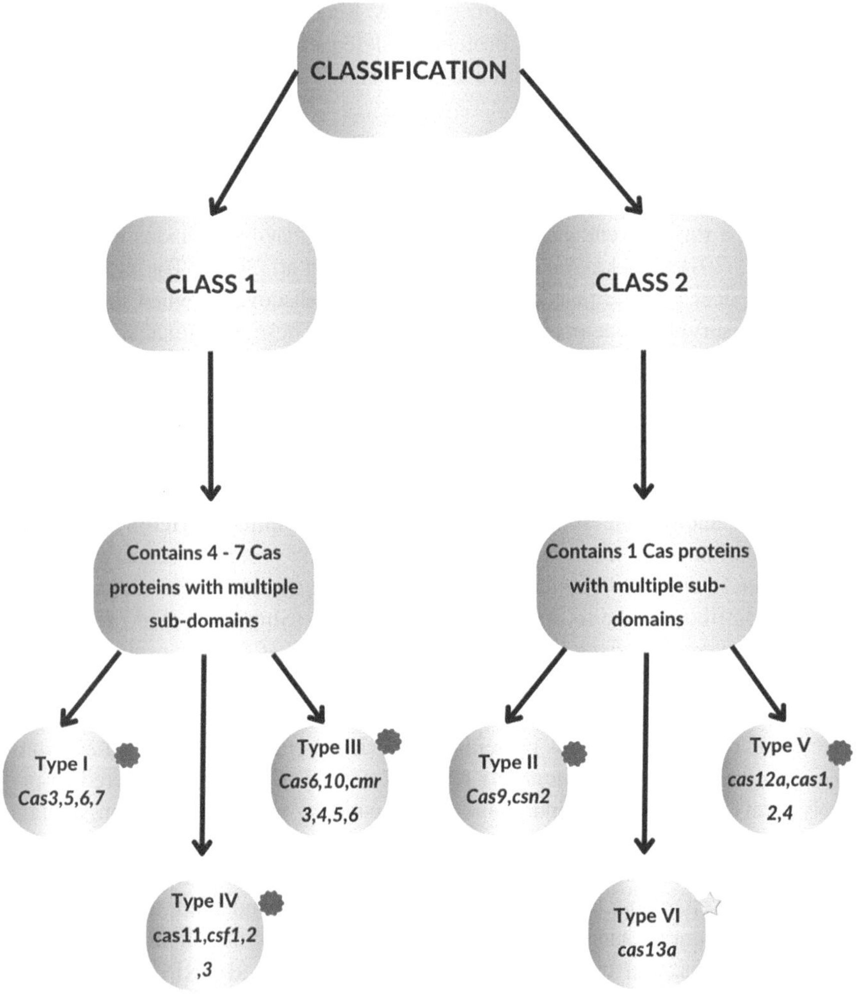

Fig. 8.1 Classification of the CRISPR–Cas system based on the properties, functions, and architecture of Cas endonucleases with CRISPR. The CRISPR–Cas system is divided into two classes. Class 1 contains 4–7 signature proteins with multiple subdomains, while class 2 contains only one Cas protein, i.e., Cas9, the most important multidomain protein for its function to produce crRNA. Type VI target RNA is shown as a yellow star, while types I, II, IV, and V target DNA is shown as a red star

target DNA, as shown in Fig. 8.1. Class 1 systems regulate interference using complexes of many Cas proteins, while class 2 systems cut nucleic acids using a single Cas multidomain effector protein. Type I contains the Cas3 gene, which codes a protein with nuclease and helicase domains. The Cas10 gene is the signature gene for type III, which encodes a multidomain protein with nuclease activity responsible for the interference stage. csf1 has been identified as the signature gene for type

IV. Types I and III utilize effector complexes comprising several Cas proteins and crRNA. The Cascade complex is the effector complex in type I systems, whereas the Cmr complex is in type III systems. These effector complexes have different protein compositions and minimal similarity in their sequences, but their structures and functions are almost identical since they have a common ancestor. Class 2 CRISPR–Cas systems have gained popularity for the development of an innovative genome editing technology because of the straightforward structural design of effector complexes (Zhang et al., 2020a, 2020b, 2020c; Sofos et al., 2020; Niu et al., 2020; Li & Bondy-Denomy, 2021; Gong et al., 2020).

CRISPR–Cas System and Mechanism

This breakthrough discovery was made possible by an immune strategy adapted from bacteria's defenses against viral invasions. The CRISPR array serves as a genetic memory store for previous viral interactions. The CRISPR array, which comprises repetitive DNA sequences, and the Cas proteins, which carry out the gene editing process, are the two main parts of the CRISPR–Cas system. One genome might include 1–18 CRISPR arrays; each array could have 2–374 repeat units. Cas proteins can identify and specifically target viral DNA for eradication when a bacterium is encountered by a viral intruder again. There are three stages in the CRISPR–Cas mechanism: adaptation, expression, and interference (Watson et al., 2021; Gong et al., 2020). During acquisition, the bacteria take a brief section of the foreign viral DNA protospacer sequences and add it as a new spacer to its own CRISPR array (Shivram et al., 2021). These protospacers are identified and chosen based on certain downstream features in the foreign DNA strand known as protospacer adjacent motifs (PAMs). Bacteria recognize the PAM sequence when a similar virus invades again. The CRISPR array is transcribed into tiny RNA molecules known as CRISPR RNAs (crRNAs) during the expression step. These crRNAs and the Cas proteins work together to create a complex that can identify and bind to complementary DNA sequences (Kurihara et al., 2022). The crRNA-Cas complex finally directs Cas proteins to the invading viral DNA during the interference stage, where they cleave and neutralize it and establish a double-strand break at the targeted location, causing the cell's inbuilt DNA repair pathways to kick in (Li & Bondy-Denomy, 2021; Takeda et al., 2021). This is accomplished by the bacteria copying the sequence close to the PAM sequence and integrating it in the CRISPR array. PAMs are distinct sequences in invader DNA that Cas can identify and that help the protein attach to the invading DNA. The Cas of *Streptococcus pyogenes*, frequently utilized in genome editing, recognizes the pattern 5′-NGG-3′ (Hu et al., 2021). The Cas9 protein's Arg1333 and Arg1335 residues interact with the two guanine bases of the PAM motif, leading to this recognition. This interaction helps the DNA strand unwind so that the crRNA may bind to the foreign DNA's complementary sequence (Khurana et al., 2022).

The Cas9 HNH nuclease domain (Cas9 nuclear domain) cuts the complementary crRNA-binding DNA strand and cleaves it, resulting in open-end cleavage of the foreign DNA strand (Liu et al., 2020; Kato et al., 2022). The nonhomologous end joining repair mechanism, the most prevalent repair mechanism and a highly inaccurate repair process that may result in indel mutations, is one of the two repair processes activated by this DNA damage. The second repair mechanism is homology-directed repair, which aims to fix the damage in DNA by using homologous DNA as a template (Ghaemi et al., 2021; Schenke & Cai, 2020; Beying et al., 2020; Yu & Marchisio, 2020). CRISPR–Cas9 is an effective method for gene editing because it can be used to add precise alterations to the genetic code.

Significance of CRISPR–Cas in Drug Development

This technology has the potential to completely alter the medication development process. The precise targeting and editing of genes in cells using CRISPR–Cas might be utilized to create novel treatments for various ailments. CRISPR–Cas technology has drawn more attention in recent years as a potential tool for medication development (Yang et al., 2021; El ouar & Djekoun, 2021; Khurana et al., 2022). Several clinical trials are being conducted, and some companies have begun researching CRISPR-based treatments (Ding et al., 2021; Dey, 2021).

Developing new cancer medications using CRISPR–Cas technology is one of its most promising uses. Cancer-causing genetic mutations can be fixed with CRISPR–Cas, or cancer cells can be targeted and eliminated using this technique (Chavez-Granados et al., 2022) (Yang et al., 2021). CRISPR–Cas has been utilized for leukemia treatment by targeting and killing leukemia cells. The treatment is presently being evaluated in clinical trials after being shown to be effective for leukemia patients (Lee et al., 2020; Munawar et al., 2021). HIV, cystic fibrosis, COVID-19, antimicrobial resistance, and sickle cell anemia are among the various illnesses that can be treated with CRISPR–Cas technology (Ding et al., 2021; Gholizadeh et al., 2020). CRISPR–Cas can be employed in any of these situations to pinpoint and treat the genetic abnormalities that cause the disease (Kumar et al., 2020; Bayarsaikhan et al., 2021).

Target Identification and Validation

Some techniques may be utilized to identify targets when using CRISPR–Cas9. CRISPR–Cas systems come in various forms but function similarly. Finding the desired DNA sequence is the first step. There are several ways to accomplish this, including bioinformatics or DNA sequencing. Upon determining the target sequence, a guide RNA is developed. The Cas9 protein is subsequently added to the guide RNA, which can cut DNA at the target region. A variety of methods can then repair the cut or damaged DNA. One method is a repair template, a DNA fragment with the desired sequence alteration. The Cas9 protein aids in directing the repair machinery to the proper location, and the repair template is attached to the cut DNA. A nonhomologous end joining (NHEJ) repair mechanism is a second option. The cut ends of the DNA are linked in this mechanism, which might result in minute insertions or deletions (Moore, 2015; Rasul et al., 2022a, 2022b, 2022c).

Using CRISPR–Cas9 to knock down a gene of interest in a cell line or an animal model is a common method. A gene may be a potential therapeutic target if knocking it out causes an apparent change in the cell line phenotype or animal model (Gullapalli, 2021). Introducing mutations into a gene of interest using CRISPR–Cas9 is another way to use the technology to identify targets (Rasul et al., 2022a, 2022b, 2022c; Chandrasekaran et al., 2022; Khurana et al., 2022; Grotz & Deswal, 2021; Sreedurgalakshmi et al., 2021).

Pros and Cons of Target Validation

Regarding target identification, CRISPR–Cas9 offers many benefits over other approaches. Since CRISPR–Cas9 is an extremely precise gene editing tool, it may delete or add mutations to certain genes. This makes CRISPR–Cas9 the best tool for locating genes crucial for illness. However, this implies that CRISPR–Cas9 occasionally deletes or modifies genes that are not the focus of interest (Kostyusheva et al., 2022; Wu et al., 2021). Another drawback of CRISPR–Cas9 is that it might be challenging to utilize it to delete or modify genes necessary for cell survival. It can be challenging to demonstrate the safety of deleting or changing a gene of interest, which presents another difficulty in validating therapeutic targets with CRISPR–Cas9. This is because CRISPR–Cas9 might occasionally result in cell death (Chen et al., 2020; Modell et al., 2022). With CRISPR–Cas9, therapeutic target validation can be difficult. One issue is that it might be challenging to demonstrate that deleting or changing a gene of interest truly has a therapeutic impact. This is because various other stimuli can influence a cell line's or animal model's phenotype (Rasul et al., 2022a, 2022b, 2022c; Ghorbani et al., 2021).

Development of Animal Models

Animal models are essential for investigating illness pathophysiology and assessing the effectiveness of treatments in preclinical drug development. The creation of precise genetically modified animal models using CRISPR–Cas has considerably improved. This method enables the introduction of disease-relevant mutations, the knockout of certain genes, or the insertion of reporter genes to track the development of illness. Clinical trial success rates have increased due to the accelerated assessment of drug candidates made possible by the more effective and quick development of complex animal models (Khurana et al., 2022).

Drug Screening

Drug screening is one of the most intriguing uses of CRISPR–Cas9. The process of identifying substances that can successfully cure an illness is known as drug screening. The search for effective medicine is a laborious and expensive process that sometimes takes years. By enabling researchers to test hundreds of chemicals swiftly and efficiently for their capacity to target certain genes, CRISPR–Cas9 can hasten the drug development process (Wang et al., 2020a, 2020b, 2020c; Sreedurgalakshmi et al., 2021; Ding et al., 2021; Dilip Kumar et al., 2021; Yang & Zhang, 2023). CRISPR–Cas9 may be used in a variety of ways for drug testing. One such method uses a library of CRISPR–Cas9 guide RNAs targeting a distinct gene. The cells are exposed to various chemicals after editing the cell genomes using guide RNAs. The target gene of the CRISPR–Cas9 guide RNA is then identified by analyzing the cells most responsive to a certain substance (Sharma et al., 2021; Yagoubat et al., 2020). A high-throughput screening (HTS) platform is an additional method for CRISPR–Cas9 drug screening (Xing & Meng, 2020; Bhat et al., 2022; Awwad et al., 2023). Robots are used in HTS platforms to automate the evaluation of a compound's capacity to target a particular gene. Finding novel therapeutic targets and accelerating drug discovery are possible benefits now through this technology.

New therapeutic targets for several illnesses have already been found using CRISPR–Cas (Selvakumar et al., 2022). Due to the intricacy of their mechanisms, current cancer treatments are constrained, which emphasizes the need for additional therapeutic modalities. The discovery, validation, and targeting of high-value targets may be accelerated by integrating the CRISPR–Cas system with next-generation sequencing (NGS). For instance, CRISPR–Cas9 was employed to find a novel cancer therapeutic target (Xing & Meng, 2020; Aghamiri et al., 2020; Selvakumar et al., 2022). The protein MYC, known to be involved in the development of cancer cells, is the drug's target. A substance inhibiting MYC activity effectively eliminated cancer cells in the lab (Awwad et al., 2023; Xu et al., 2021). According to another study, CRISPR–Cas9 was used to find novel cancer treatment targets, and it was determined that a substance that inhibits the action of the protein BRD9 was successful in destroying cancer cells (Webb et al., 2022; Mu et al., 2021; Gier et al., 2020). Additionally, CRISPR–Cas9 has been utilized since 2016 to create novel HIV medications. Scientists pursuing alternate HIV infection treatment methods now have a new field of study to explore. Using CRISPR–Cas9, a novel HIV therapeutic target, protein CCR5, which HIV uses to enter cells, was discovered and genetically modified. HIV infection was successfully avoided by a substance that inhibits CCR5 function in the lab (Jena et al., 2022; Herskovitz et al., 2021).

CRISPR–Cas technology has accelerated the development of personalized medicine. Induced pluripotent stem cell (iPSC) technology and CRISPR–Cas technology can create patient-specific cell lines that mimic disease phenotypes (Wang et al., 2021a, 2021b). These cell lines may be used to conduct high-throughput drug screening experiments, determine personalized therapeutic targets, and explore underlying disease factors (Dilip Kumar et al., 2021; Geng et al., 2020). Additionally, CRISPR–Cas9 makes it possible to fix genes that cause illnesses in cells obtained from patients, which raises the prospect of employing gene editing as a therapeutic strategy (Stellon et al., 2022; Kitano et al., 2022; Lahm et al., 2023).

These are only a handful of the several studies that have recently been published and show the immense potential of CRISPR–Cas9 for drug screening. Although CRISPR–Cas9 is still a young technology, it has the power to alter how we treat illness completely. In addition to the research above, several other studies have been conducted that showed the great potential of CRISPR–Cas9 for drug screening.

CRISPR–Cas for Gene Therapy

Genetic engineers can use the physiological mechanism to cause DSBs at particular sites in the genome purposefully. They can then introduce a modified template to substitute the damaged DNA sequence using DNA repair. During this procedure, called genome editing, the incorrect sequence linked to a specific human illness is changed or replaced (Zhou & Wang, 2021) (Yu & Wu, 2021). The development of the human genome project has substantially accelerated gene therapy research. We now know more than ever about the precise mutations and deletions that cause pathological disorders. Scientists find a complementary sequence in the genome and choose a 20-nucleotide target that is as near as feasible to the PAM site to target the disorders successfully. The target's distance from the PAM site is negatively correlated with the effectiveness of creating the targeted mutation (Tan et al., 2020).

Additionally, the guide RNA sequence affects the efficacy of inducing precise DSBs. Currently, researchers are investigating different guide RNA settings to enhance its efficacy. Finding nucleases that can accurately produce DSBs at specified places without producing off-target alterations has piqued scientists' interest. They made an enormous advancement in their quest when they identified Cas9 (Tran et al., 2022; Aghamiri et al., 2020).

Using two distinct Cas9 proteins that target both ends of the target sequence is one way to improve the cutting precision of the Cas9 enzyme. It is possible to achieve this by introducing the D10A mutation, which renders the RuvC catalytic domain inactive. This redesigned version of Cas9 is called NickaseCas9 (nCas9). nCas9 cuts the DNA at a complementary location to the crRNA by keeping just one functioning HNH catalytic domain. Two nCas9 proteins and two unique sgRNAs are used to achieve a DSB at the targeted location (Pattharaprachayakul et al., 2020; Molla et al., 2020; Dey, 2021).

Researchers can create endonuclease-deficient Cas9 (dCas9) by altering Cas9 catalytic domains while maintaining its DNA recognition capabilities. With this modified version of Cas9, precision targeting of particular genomic regions is possible without leading to DNA strand breakage. Since dCas9 has been researched for its capacity to modify gene expression, its use goes beyond gene editing. Control over the activity of target genes is made possible by combining the dCas9 protein with a transcriptional repressor or activator. Additionally, the value of dCas9 in molecular manipulations can be increased by manipulating it to trigger enzymatic activity at particular gene sites (Gao et al., 2020; Zhou & Wang, 2021; Yu & Wu, 2021; Slaymaker & Gaudelli, 2021).

Fusion of the FokI nuclease with dCas9 is another alteration that enhances the accuracy of DNA cutting (Munawar et al., 2021). Comparing this change to paired Cas9 nickase, the efficiency is increased up to four times. The increased effectiveness is mostly related to the FokI nuclease's inherent capacity to break double-stranded DNA. As a result, nicking occurs when RNA–DNA dimerization occurs, which lowers the likelihood of off-target cleavage (Breier & Peer, 2023; Liu et al., 2021). Various computational methods are available to produce highly efficient sgRNAs for genome editing. These instruments include CRISPRseek, E-CRISP, CRISPR direct, CHOPCHOP, and sgRNA designer (Sledzinski et al., 2020; Gerashchenkov et al., 2020) (Bhat et al., 2020; Choudhary et al., 2020; Zirin et al., 2022).

By permitting the precise regulation of gene expression, CRISPR–Cas technology has made it easier to characterize the functional properties of therapeutic targets. Researchers may examine the consequences of gene disruption on disease-related mechanisms and processes by using CRISPR-based gene activation or repression. This method has advanced our understanding of disease pathways and provided significant new insights into the functional impact of gene dysregulation, which will help us design more effective targeted treatments (Sreedurgalakshmi et al., 2021; Dunbar, 2021).

CRISPR–Cas9 Is Revolutionizing Biological Research in Utilizing Organoids

The use of CRISPR–Cas9 has significantly changed biological research, especially in the field of organoids. Organoids derived from three-dimensional (3D) culture technologies have become an important link between in vivo and in vitro research settings. As a 3D multicellular tissue construct grown in an in vitro environment, they represent a tiny representation of an organ. Scientists have realized through years of study and development that organ systems may be successfully investigated in vitro utilizing 3D culture platforms ("CRISPR–Cas Tools and Their Application in Genetic Engineering of Human Stem Cells and Organoids," 2020) (Ramakrishna et al., 2021). RNAi technology was used for physiological and pathological research prior to the advent of CRISPR. Although this method was helpful, researchers admitted that it was best suited for verifying transcriptome-level studies and lacked the resilience to explore the ramifications of genetic malfunction in other biological investigations. Scientists worked with nucleases such as zinc finger nucleases (ZFNs) and transcription activator-type effector nucleases (TALENs) before the development of CRISPR (Chavez-Granados et al., 2022; Khurana et al., 2022). The understanding of biological systems has substantially improved due to the exceptional capability of the CRISPR–Cas9 tool to target and alter DNA strands accurately, surpassing the efficiency of its competitors, and has also revealed potential therapeutic targets.

Adult stem cells, induced pluripotent stem cells, and embryonic stem cells can all make organoid cultures. These organoids may be grown and genetically modified using CRISPR technology to explore their biological characteristics (Menche & Farin, 2021; Kim et al., 2020). The amazing capacity of adult stem cell-derived organoids to self-organize allows them to closely mimic the cellular stability, genomic stability, and functional properties of the different epithelial tissues through which they are produced. CRISPR–Cas9 technology opens the door to various strategies, including nonhomologous end joining-mediated gene alterations, which inject brief insertions or deletions (indels) into the organoid's endogenous locus. When concurrently delivered by CRISPR–Cas9 into organoid cells, the homology-dependent repair pathway may be used for targeted gene insertion, enabling the introduction of a particular gene at the location of a DNA break (Khurana et al., 2022).

Dekkers et al. conducted a successful study on the genetic modifications underlying the neoplastic transformation of breast epithelial tissue. They used CRISPR–Cas9 to specifically knock down genes, including pten, nf1, rb1, and p53, in breast epithelial organoids produced from reduced mammoplasties in cancer patients (Dekkers et al., 2020). Our knowledge and ability to treat various breast cancer subtypes have been considerably enhanced due to this manipulation of breast epithelial organoids.

CRISPR–Cas9 technology was used by Takeda et al. to identify genes linked to tumorigenicity in colorectal cancer (Takeda et al., 2019). CRISPR–Cas technology has also been employed to develop disease-specific models, while brain organoids have been used to study the brain's pathology, evolution, and physiology (He et al., 2021). Organoids offer a great platform for testing therapy efficacy, and base editing has emerged as a potential technique for treating monogenic brain illnesses such as Alzheimer's disease (Ferrari et al., 2020; Khoshandam et al., 2023). The treatment of familial Alzheimer's disease and other monogenic diseases that affect various brain regions may be possible through genome editing.

Rijn et al. used intestinal organoids from stem cells found in the crypt bottom to research congenital diarrheal diseases. They proved that a deficit in the dgat1 gene contributes to intestinal failure by using CRISPR. Triacylglycerol metabolism is altered by DGAT1 absence, making patients more vulnerable to lipotoxicity and consequent intestinal failure (van Rijn et al., 2018). Using CRISPR–Cas9, Hendrik et al. successfully constructed multigene knock-in and knockout fetal liver and human adult organoids. These investigations have established a solid basis for a more thorough investigation of cellular differentiation and gene activities. Likewise, human liver organoids transformed using CRISPR were used to explore the potential of BAP1 to decrease tumor growth in cholangiocarcinoma (Wang et al., 2022a, 2022b; Hendriks et al., 2020a, 2020b).

Chen et al. developed a 3D model to study lung development and disease pathophysiology using lung organoids derived from a human pluripotent stem cell (Chen et al., 2017). These examples demonstrate how CRISPR technology is widely applied across multiple tissue systems, utilizing organoids to study disease mechanisms, comprehend illnesses, and identify novel therapeutic targets.

Safety and Efficacy

The safety and effectiveness of already available medications can also be increased using CRISPR–Cas9. It can be used for fixing mutations that render cells drug resistant. This can be accomplished by attaching a CRISPR–Cas9 system to a guide RNA specifically targeting the mutation. The mutation in the cells can then be eliminated using the CRISPR–Cas9 system. According to a recent study, this technology was used to eliminate a mutation that makes cells resistant to a drug called imatinib. This shows that CRISPR–Cas9 might be utilized to increase imatinib's effectiveness in leukemia treatment.

Strategies Used in CRISPR–Cas Technology

CRISPR–Cas may introduce specific mutations in the DNA of living cells. However, the selectivity of CRISPR–Cas-induced mutations may be imperfect, resulting in off-target effects, such as inadvertent cleavage of DNA at places with mismatches with the guide RNA. Off-target consequences can result in undesired mutations, chromosomal rearrangements, or changes in gene expression, reducing editing accuracy and safety. As a result, optimizing the design and modification of sgRNA (Doench et al., 2014), utilizing Cas variants with higher specificity (Kleinstiver et al., 2016), or using anti-CRISPR proteins to limit undesirable Cas activity can improve editing outcomes and lessen unwanted effects. Off-target impacts can be reduced in a variety of ways, such as:

1. Improved sgRNA design and modification:

   The nucleic acid that binds to the target DNA and directs the Cas protein to cleave it is known as sgRNA. sgRNA design and alteration can influence its binding affinity and specificity to target DNA and its stability and folding in the cell. Improving the design and modification of sgRNA can aid in the reduction of off-target effects by increasing the number of matched base pairs in the target site, decreasing the number of potential off-target sites, improving sgRNA stability and folding, or introducing chemical modifications or mismatches in sgRNA (Wang et al., 2020a, 2020b, 2020c).

2. Cas variants with higher specificity:

   Cas proteins are the enzymes that break the guide RNA-guided target DNA. Cas proteins have varied cleavage processes and recognition sequences, which alter their selectivity and sensitivity to target DNA mismatches. By improving the stringency of target identification, reducing the number of potential off-target sites, or introducing mutations or fusions in Cas proteins, Cas variants with higher specificity can aid in limiting off-target effects (Ghorbani et al., 2021; Cebrian-Serrano & Davies, 2017; Modrzejewski et al., 2020).

3. Anti-CRISPR proteins to inhibit undesirable Cas activity:

   Anti-CRISPR proteins are naturally occurring inhibitors of CRISPR–Cas systems produced by some bacteria or phages to avoid CRISPR-mediated immunity. They can attach to Cas proteins or guide RNAs, preventing them from interacting with the target DNA. Anti-CRISPR proteins can aid in decreasing off-target effects by decreasing the duration of Cas activity, limiting its geographic distribution, or modulating its timing (Yu & Marchisio, 2020).

   The distribution of CRISPR–Cas components (Cas protein and guide RNA) to target cells or tissues remains a critical and unresolved issue that affects editing efficiency and safety. Improving delivery efficiency by selecting optimal vectors, methods, and conditions for introducing CRISPR–Cas components into target cells or tissues might thereby improve editing outcomes while reducing undesired consequences (Feng et al., 2022; Tyumentseva et al., 2021).

4. CRISPR–Cas as a ribonucleoprotein complex (RNP):

   RNPs are CRISPR–Cas components preassembled in vitro and consist of a Cas protein and a guide RNA. Using CRISPR–Cas as an RNP has various advantages, including high editing efficiency, minimal nonspecific activity, instantaneous editing after delivery, rapid screening of guide RNAs, and reduced immunogenicity due to the presence of CRISPR–Cas elements in the target cell for a short period (Banakar et al., 2021).

5. Various vectors to distribute CRISPR–Cas components:

   Vectors are carriers that can transfer CRISPR–Cas components to target cells or tissues. Different vectors have varied qualities, such as size, stability, specificity, immunogenicity, and biosafety. Using multiple vectors to carry CRISPR–Cas components can help to optimize delivery efficiency based on target cell type and tissue. Viral vectors have great transduction efficiency and long-term expression, but they may also be immunogenic and capable of insertional mutagenesis. However, nanoparticles, polymers, liposomes, and plasmids are nonviral vectors and show minimal immunogenicity with a good biosafety profile, but these nonviral vectors have inadequate transfection effectiveness and show short-term expression (Dong & Kantor, 2021; Huang et al., 2022).

6. Various methods to supply CRISPR–Cas components:

Cost, speed, scalability, invasiveness, and applicability are all pros and cons of different approaches. Considering several different strategies and techniques to deliver CRISPR–Cas components may enhance delivery efficiency. For example, electroporation involves electric pulses to form pores in the plasma membrane of target cells, thus allowing CRISPR–Cas components to enter the cell. This technique is empowered with high delivery efficiency and sufficient capability to target a wide range of cell types and tissues; however, it may also cause cell damage and cell death (Alghadban et al., 2020; Duan et al., 2021a, 2021b; Tröder et al., 2018). Another technique is microinjection, which involves the insertion of CRISPR–Cas components directly into the cytoplasm or nucleus of the cell via a small needle. It is a highly efficient technique for delivery as well as highly specific for delivery of components to a single cell or embryo, although it may take a longer duration and more labor than other techniques (Tyumentseva et al., 2021).

7. Variable conditions to deliver CRISPR–Cas components:

   Several variable conditions may influence the delivery efficiency of CRISPR–Cas components, including pH, ratio, timing, buffer, concentration, or temperature. Using multiple delivery settings for CRISPR–Cas components can help to fine-tune the delivery effectiveness based on the target cell type and gene target. Increasing the concentration or ratio of CRISPR–Cas components, for example, may improve editing efficiency and increase off-target effects or toxicity. Changing the delivery time or temperature may influence the availability or activity of CRISPR–Cas components in target cells. Choosing a proper pH or buffer may increase CRISPR–Cas component stability or solubility (Duan et al., 2021a, 2021b).

   The efficacy of CRISPR–Cas-induced mutations varies depending on cell type and gene target, making consistent and optimal editing outcomes difficult to obtain. As a result, optimizing editing time and temperature by varying the duration and degree of exposure to CRISPR–Cas components based on cell type and gene target can improve editing efficiency and limit undesired effects (S. Feng et al., 2022; Malzahn et al., 2019; Prasad et al., 2021).

8. Cas variants with varying temperature:

   Cas proteins are the enzymes that break the guide RNA-guided target DNA. The ideal temperature for Cas protein activity varies by Cas protein, ranging from 37 °C to 55 °C. For example, Cas9 from *Streptococcus pyogenes* (SpCas9) prefers a temperature of 37 °C, but Cas12a from *Acidaminococcus* sp. (AsCas12a) prefers a temperature of 45 °C. Using Cas variants with varying temperature sensitivities can aid in adjusting editing efficiency based on cell type and gene target (Banakar et al., 2021; Malzahn et al., 2019).

9. Temperature-sensitive guide RNAs:

   Nucleic acids known as guide RNAs bind to the target DNA and direct the Cas protein to break it. Different guide RNAs have variable stability and folding at different temperatures, altering their binding affinity and specificity to the target DNA. By adjusting their activity at different temperatures, temperature-sensitive guide RNAs can aid in managing editing efficiency (Poultney et al., 2011).

10. Time-controlled delivery methods:

    The methods or technologies that transport CRISPR–Cas components (Cas protein and guide RNA) to target cells are called delivery systems. The kinetics and dynamics of delivering CRISPR–Cas components vary among delivery systems, altering their availability and duration in target cells. By adjusting the exposure duration of the CRISPR–Cas components to the target cells, time-controlled delivery methods can aid in optimizing editing efficiency (Sinclair et al., 2023; Tyumentseva et al., 2021).

    Because not all cells exposed to CRISPR–Cas will be effectively edited, separating the altered cells from the nonedited ones is necessary. Several methods for accomplishing this enrichment include fluorescent markers, antibiotic resistance genes, and mutant enrichment approaches (Feng et al., 2022; Kim et al., 2022).

11. Fluorescent markers:

    Fluorescent markers are proteins that produce light of a specific color when stimulated by a light source. They can mark altered cells by introducing them into the DNA alongside the desired mutation, utilizing a DNA template and the homology-directed repair (HDR) mechanism. The altered cells can be spotted and sorted using fluorescence-activated cell sorting (FACS) or other methods. This technique is frequently utilized in CRISPR–Cas systems due to the benefits of visualization, time savings, and lower labor (Koch et al., 2018; Petersen et al., 2019).

12. Antibiotic resistance genes:

    These genes provide cells with resistance to antibiotics that would otherwise kill them. A DNA template and the HDR method can also mark the altered cells by inserting them into the DNA along with the intended mutation. The modified cells can be selected by growing them in antibiotic-containing media, while the nonedited cells die. This approach is easy and inexpensive, but it has significant downsides, including antibiotic toxicity, unintended incorporation of the resistance gene into the genome, and ethical problems (Duan et al., 2021a, 2021b; Kamruzzaman & Iredell, 2020; Tao et al., 2022a, 2022b).

13. Mutant enrichment techniques:

    Mutant enrichment approaches utilize molecular variations between edited and nonedited cells, such as DNA sequencing, RNA expression, or protein function. They can enrich modified cells using specialized circumstances or reagents that favor or detect edited cells over nonedited cells. Some strategies, for example, use Cas variations, anti-CRISPR proteins, or mutant-specific PCR primers to amplify or suppress the altered cells. This technique is more adaptable and diverse than the others, although it may necessitate additional optimization and validation for each target (Feng et al., 2022; Kim et al., 2022).

    Some targets may be present at very low levels or have similar sequences to other molecules, making precise and sensitive detection difficult. As a result, combining amplification strategies based on diverse nucleic acid amplification techniques can improve the sensitivity and specificity of CRISPR–Cas systems for these targets (Dubey et al., 2022; Feng et al., 2021; Qiu et al., 2022).

14. Integration of nucleic acid amplification: CRISPR–Cas system:

    The target nucleic acid is first amplified using technologies such as polymerase chain reaction (PCR), reverse transcription polymerase amplification (RT-PCR), recombinase polymerase amplification (RPA), or loop-mediated isothermal amplification (LAMP). After that, a CRISPR–Cas system, such as Cas9, Cas12, or Cas13, recognizes and cleaves the amplified product. The cleavage produces a signal that can be detected using various techniques, such as fluorescence, colorimetry, or electrochemistry. Because the CRISPR–Cas system can identify the target sequence from other similar sequences in the amplified product, this technique improves detection specificity (Dubey et al., 2022).

15. Incorporation of nucleic acid amplification: CRISPR–Cas system:

    The CRISPR–Cas system, such as Cas9 or Cas12, first recognizes and cleaves the target nucleic acid. The cleaved product is then amplified using a method such as PCR, RT-PCR, RPA, or LAMP. The amplification produces a signal that can be detected using various methods, including fluorescence, colorimetry, and electrochemistry. Because the CRISPR–Cas system may enrich the target sequence from other abundant sequences in the sample, this method improves detection sensitivity (Feng et al., 2021; Xu & Li, 2020).

16. Facilitation of nucleic acid amplification: CRISPR–Cas system:

    The CRISPR–Cas system, such as Cas9 or Cas12a, unwounds and nicks the target nucleic acid. The unwinding and nicked result is then amplified at a low temperature using a process such as RPA or LAMP. The amplification produces a signal that can be detected using various methods, including fluorescence, colorimetry, and electrochemistry. Because the CRISPR–Cas system can eliminate the need for high temperature or sophisticated enzymes for nucleic acid amplification, this technique enhances detection efficiency (Feng et al., 2021).

17. Combination of functional nucleic acids and molecular translators: CRISPR–Cas system:

    Functional nucleic acids (FNAs), such as aptamers or DNAzymes, first recognize and bind target proteins or small molecules. FNA then activates a molecular translator, such as exonuclease III (Exo III) or RNA polymerase (RNAP), forming a DNA or RNA reporter. The reporter is then recognized and cleaved by a CRISPR–Cas system, such as Cas12 or Cas13. The cleavage generates a signal, which can be detected using a variety of techniques, including fluorescence, colorimetry, and electrochemistry. This approach recognizes non-nucleic acid targets by converting them into nucleic acid signals that the CRISPR–Cas system can amplify (Su et al., 2023; Feng et al., 2021; Chen et al., 2022).

    Some targets may lack nucleic acid sequences that the CRISPR–Cas system can recognize and cleave, making them difficult to identify and alter directly. Combining CRISPR–Cas systems with functional nucleic acids (FNAs) and molecular translators can detect and manipulate non-nucleic acid targets (Feng et al., 2021).

18. FNAs to recognize and bind non-nucleic acid targets:

    FNAs are nucleic acids with specific activities, such as attaching to a specific molecule, catalyzing a chemical, or changing structure in response to a stimulus. Aptamers, DNAzymes, ribozymes, and molecular beacons are examples. They have a high affinity and specificity for recognizing and binding non-nucleic acid substrates such as proteins, metal ions, and tiny compounds (Micura & Höbartner, 2020; Zhu et al., 2021).
19. Converting non-nucleic acid signals into nucleic acid signals via molecular translators:

    Molecular translators are enzymes or substances that can translate FNA binding or activity into the production or degradation of a DNA or RNA reporter. Exonucleases, polymerases, ligases, nickases, and strand displacement probes are among them. They can be utilized to transform FNA recognition or binding of non-nucleic acid targets into the production or degradation of a DNA or RNA reporter that the CRISPR–Cas system can detect or manipulate (Feng et al., 2021; He et al., 2023; Javaid & Choi, 2021).
20. Detecting and manipulating nucleic acid signal: CRISPR–Cas systems:

    Cas proteins and guide RNAs in CRISPR–Cas systems can recognize and cleave specific nucleic acid sequences. Cas9, Cas12, Cas13, and their variations are among them (Table 8.1). They can be employed to detect or alter the DNA or RNA reporter that the molecular translator generates or degrades in response to the non-nucleic acid target recognized or bound (Jeong et al., 2023; Li et al., 2022; Su et al., 2023).

Applications of the CRISPR–Cas System

CRISPR–Cas is the most accurate, efficient, and effective genome editing tool for living cells. It has many potential uses in many applied disciplines (Jeong et al., 2023; Zhang et al., 2019). Here are various practical examples and sources for CRISPR–Cas applications:

Applications in Agriculture

Making crops resistant to pathogens such as bacteria, fungi, viruses, or insects is one of the most challenging tasks in plant breeding. Utilizing CRISPR–Cas, it is possible to introduce genes that protect against disease and alter or delete genes that make plants susceptible to illness. For instance, by removing a gene encoding a vulnerability, CRISPR–Cas has been used to develop rice plants resistant to bacterial blight (Wolter et al., 2019). To defend plants against viral infections, CRISPR–Cas has also been utilized to insert antiviral proteins or RNA interference molecules into plants.

Table 8.1 Strategies used in CRISPR–Cas technology and their fates

Strategies	Goals	Methods	Advantages	Limitations	Clinical trials	Future prospective	Reference
Enriching the edited cells	To select for desired mutations from the nonedited cells	Using fluorescent markers, antibiotic resistance genes, or mutant enrichment techniques	To increase the purity and yield of edited cells	To require additional steps or reagents, to introduce unwanted genes or sequences, or to need more optimization and validation	Several trials for employing fluorescent markers or antibiotic resistance genes to enrich modified cells for treating blood diseases (such as sickle cell disease or beta-thalassemia) or cancers (such as multiple myeloma or leukemia) are underway or planned. No published trials using mutant enrichment approaches	To increase the effectiveness and safety of enrichment procedures, to investigate other applications in other diseases or tissues, and to create more sensitive and specific detection methods for altered cells	Kim et al. (2022) Feng et al. (2021)

Incorporating amplification strategies	To improve the sensitivity and specificity of CRISPR–Cas systems for nucleic acid, protein, and small molecule detection	Using several nucleic acid amplification strategies to boost the signal produced by CRISPR–Cas systems	To detect low-abundance or comparable targets	Increasing the assay's complexity and cost, introducing false positives or negatives, or affecting the precision or specificity of CRISPR–Cas systems	Several trials are currently underway or planned to identify nucleic acids (such as SARS-CoV-2 RNA or HPV DNA) or proteins (such as SARS-CoV-2 antigen) in various samples (such as saliva, blood, urine, or swabs) using CRISPR–Cas systems with amplification methods. There are no published trials for detecting small molecules	To increase the performance and robustness of CRISPR–Cas systems through amplification techniques To investigate further applications in disease diagnosis, prognosis, or monitoring To develop more portable and user-friendly devices for point-of-care testing	Feng et al. (2021) Liu et al. (2023) Ghorbani et al. (2021)

(continued)

Table 8.1 (continued)

Strategies	Goals	Methods	Advantages	Limitations	Clinical trials	Future prospective	Reference
Combination with functional nucleic acids and molecular translators	To detect and manipulate non-nucleic acid targets	Employing functional nucleic acids to recognize and bind non-nucleic acid targets and then employing molecular translators to turn them into nucleic acid signals that CRISPR–Cas systems can detect or edit	To broaden the variety of targets that CRISPR–Cas systems can detect or alter	To rely on the availability and quality of functional nucleic acids and molecular translators, to add new procedures or reagents, or to require extra optimization and validation	No documented trials involving the use of CRISPR–Cas systems with functional nucleic acids and molecular translators	To improve the design and manufacture of functional nucleic acids and molecular translators; to investigate new applications in disease diagnostics, therapy, or biosensing; and to create more adaptable and modular CRISPR–Cas systems To improve the characterization and engineering of Cas variants and guide RNAs, to investigate new uses in various cell types and tissues, and to create more precise and dynamic delivery methods	Feng et al. (2021), Modrzejewski et al. (2020), Rasul et al. (2022a, 2022b, 2022c)

Editing time and temperature optimization	To tailor editing efficiency based on cell type and gene target	Using Cas variants with varying temperature sensitivity, temperature-sensitive guide RNAs, time-controlled delivery devices, or varied delivery circumstances	To improve editing efficiency and minimize undesirable consequences	Depending on the availability and quality of Cas variants and guide RNAs, introducing extra processes or reagents or requiring additional optimization and validation	Several clinical trials are underway to use Cas variants with varying temperature sensitivities (such as AsCas12a) to treat malignancies (such as glioblastoma multiforme) No published trials using temperature-sensitive guide RNAs, time-controlled delivery methods, or various delivery settings	To improve the characterization and engineering of Cas variants and guide RNAs, to investigate further applications in diverse cell types and tissues, and to build more precise and dynamic delivery systems	Banakar et al. (2021), Malzahn et al. (2019), Feng et al. (2022)

(continued)

Table 8.1 (continued)

Strategies	Goals	Methods	Advantages	Limitations	Clinical trials	Future prospective	Reference
Improving delivery efficiency	To effectively and securely introduce CRISPR–Cas components into target cells or tissues	Using CRISPR–Cas as a ribonucleoprotein complex, alternative vectors for delivery, other delivery strategies, or different delivery conditions	To improve editing results while reducing undesired effects	To rely on the availability and quality of CRISPR–Cas components and delivery systems, to introduce extra procedures or reagents, or to require additional optimization and validation	Several trials are underway or planned to use CRISPR–Cas as a ribonucleoprotein complex, different vectors (such as viral or nonviral vectors), or different methods of delivery (such as electroporation, microinjection, or nanoparticles) to treat various diseases (such as blood disorders, cancers, inherited eye diseases, diabetes, or infectious diseases). No trials have been reported for using alternative conditions for delivery	To improve the design and synthesis of CRISPR–Cas components and delivery systems, to investigate new uses in various cell types and tissues, and to create more efficient and safe delivery methods	Petersen et al. (2019), Dong and Kantor (2021), Zhang et al. (2020a, 2020b, 2020c), Tyumentseva et al. (2021)

Decreasing off-target effects	To avoid inadvertent cleavage of DNA at locations where the sequences differ from the guide RNA	Improving the design and modification of sgRNA, utilizing more selective Cas variants, or using anti-CRISPR proteins to limit undesired Cas activity	To boost editing accuracy and safety	Depending on the availability and quality of sgRNA, Cas variants, and anti-CRISPR proteins, introducing extra processes or reagents or requiring additional optimization and validation	Several clinical trials are either underway or planned to use improved design and modification of sgRNA (such as shortened sgRNA) or Cas variations with higher specificity (such as SpCas9-HF1) to treat a variety of diseases (including blood disorders, malignancies, hereditary eye diseases, diabetes, and infectious diseases). Anti-CRISPR protein experiments have not yet been reported	To improve the design and manufacture of sgRNA, Cas variants, and anti-CRISPR proteins; to investigate further applications in diverse cell types and tissues; and to create more accurate and sensitive methods for off-target detection	Cho et al. (2014), Fu et al. (2013, 2014) Maity et al. (2022)

Plants can alter their ability to generate specific substances such as oils, carbohydrates, sugars, vitamins, colors, flavors, or medications. CRISPR–Cas can change the biosynthesis pathways of these chemicals by activating or inactivating genes that regulate their production or control. For example, CRISPR–Cas has been used to increase the oil content of soybean seeds by knocking out a gene that suppresses oil formation (Wolter et al., 2019). CRISPR–Cas has also been used to reduce caffeine concentrations in coffee plants by targeting genes encoding essential enzymes involved in caffeine production (Touzdjian Pinheiro Kohlrausch Távora et al., 2022).

Numerous morphological and physiological traits that plants display impact their growth, development, reproduction, and capacity for adaptation. CRISPR–Cas can alter these traits by editing genes that control plant hormones, transcription factors, signaling pathways, or structural elements. For instance, CRISPR–Cas has been used to alter the hue of petunia flowers by deleting a gene that codes for an enzyme essential for the anthocyanin coloring process (Lin et al., 2022). By focusing on genes that regulate cell proliferation and growth, CRISPR–Cas has also been used to alter tomato fruit shape and size (Tiwari & Singh, 2023).

Applications in Animals

Animal genomes can be edited with the help of the potent CRISPR–Cas system for various purposes. Animals are essential sources of food, fiber, and other goods used by humans (Ur Rahman & Mohsin, 2019). They can also be used to study human illnesses and develop new remedies. Genes that impact animal health, production, or research suitability can be modified or deleted advantageously. This is possible with CRISPR–Cas. By eliminating a gene that encodes a virus receptor, CRISPR–Cas, for instance, has been used to create pigs resistant to the porcine reproductive and respiratory syndrome virus (Khwatenge & Nahashon, 2021). Targeting genes that affect the growth of hair follicles using CRISPR–Cas has also been used to increase the amount of wool produced by sheep (Kalds et al., 2022).

Numerous human diseases are caused by genetic anomalies or environmental variables that interfere with the function of particular genes or metabolic pathways. Researchers can benefit from animal models that mimic these illnesses to learn more about their underlying causes and test potential treatments. Using CRISPR–Cas, disease-related genes can be altered or deleted to create animal models of human diseases. By introducing mutations in genes involved in the development of amyloid plaques and neurodegeneration, CRISPR–Cas, for instance, has been used to create mice that imitate Alzheimer's disease. By eliminating a gene involved in social behavior and brain development, CRISPR–Cas was also utilized to create monkeys that imitate autism spectrum disorder (Yang et al., 2019).

Infectious diseases affecting people or other animals can act as vectors or reservoirs in animals (Rehman et al., 2021). Particularly in developing countries, these diseases can cause substantial morbidity and mortality (Afzal et al., 2022). Animals can be altered using CRISPR–Cas to change their susceptibility to or capacity for transmitting diseases such as malaria, rabies, or Zika virus (Wang et al., 2021a, 2021b). For instance, CRISPR–Cas has been used to reduce mosquito-borne malaria transmission by changing genes vital for parasite development or female fertility. CRISPR–Cas has also been utilized to minimize mouse susceptibility to Zika virus infection [165] by deleting a virus receptor gene (Dukhovny et al., 2019).

Applications in Humans

CRISPR–Cas can alter human cells in vitro (outside the body) or in vivo (within the body) to achieve various aims (Batista & Flotte, 2021). CRISPR–Cas, for example, can be used to repair defective genes in afflicted cells to address genetic flaws that underlie inherited disorders such as cystic fibrosis (Maule et al., 2020), hemophilia (Segurado et al., 2022), or muscular dystrophy. CRISPR–Cas can improve capabilities such as the immune response, metabolism, or cell regeneration by activating or introducing genes that give these characteristics. CRISPR–Cas can also be used to confer resistance to diseases such as HIV, malaria, or cancer by deleting or changing genes that make cells susceptible to these pathogens or tumors (Uddin et al., 2020).

Gene therapy is a cutting-edge new approach for treating conditions brought on by genetic anomalies or environmental variables that prevent the proper activation of particular genes or signaling pathways. Gene therapy uses various methods, such as viral vectors, nanoparticles, or electroporation, to deliver therapeutic genes or gene editing tools, such as CRISPR–Cas, to target cells or tissues (Uddin et al., 2020). By correcting the mutation that results in improper hemoglobin production in hematopoietic stem cells, CRISPR–Cas has been used to provide gene treatments for inherited blood illnesses such as sickle cell anemia and thalassemia. By deleting larger CAG repeats that cause neurotoxicity in brain cells, CRISPR–Cas has also been utilized to create gene treatments for neurodegenerative diseases, including Huntington's disease. By reprogramming immune cells such as T cells and natural killer cells to express chimeric antigen receptors (CARs) that target cancer cells, CRISPR–Cas has also been utilized to generate cancer gene treatments (Liu et al., 2023).

By methodically knocking out (Shalem et al., 2014) or knocking down every gene in the genome and analyzing the phenotypic changes, genome-wide screening is a potent technique for identifying genes involved in diverse biological processes or disease pathways. CRISPR–Cas can be used to perform genome-wide screening by delivering libraries of guide RNAs (Tyumentseva et al., 2021) (gRNAs) that target every gene in the genome to the cells of interest and Cas proteins. By stimulating the NHEJ pathway, gRNAs can cause mutations or deletions in target genes, or they can cause knockdowns by interfering with transcription or translation. Cell viability, fluorescence, reporter assays, and sequencing can all be used to measure phenotypic changes. CRISPR–Cas, for example, has been used to undertake genome-wide screening to find genes required for cell survival, growth, differentiation, or treatment response (Table 8.2). CRISPR–Cas has also been employed in

Table 8.2 Applications of CRISPR–Cas Technology

Example	Description	References
Applications in plants		
Mushrooms	Browning can be avoided by turning off a gene that generates an oxidative enzyme	Chen et al. (2018)
Rice	Drought tolerance can be improved by adding a gene that modulates water-use efficiency	Zafar et al. (2020)
Tomato	Editing genes that influence blooming time and branching to increase output and fruit size	Tiwari and Singh (2023)
Application in animals		
Pigs	Removing retroviruses and immunogenic substances from their organs to make them more suitable for transplantation into humans	Ross and Coates (2018)
Mice	Developing Alzheimer's disease models by introducing mutations in genes linked to amyloid plaques and neurodegeneration	Lu et al. (2021)
Mosquitos	Disrupting genes involved in parasite development or female fertility to reduce their potential to transmit malaria	Hammond et al. (2016)
Application in humans		
Blood cells	Correcting the sickle cell anemia mutation by replacing it with the normal version of the gene	Christen et al. (2022)
Immune cells	T cells are engineered to express chimeric antigen receptors (CARs) targeting cancer cells	Parnas et al. (2015)
Embryonic stem cells	Genes are knocked out or knocked in to explore their functions in development, differentiation, or illness	Zhu et al. (2015)

genome-wide screening to uncover genes involved in disease pathways such as viral infection (Li et al., 2020), inflammation, or cancer (Maity et al., 2022).

CRISPR–Cas in Therapeutics of Various Diseases

Therapeutics for Malignant Tumors

CRISPR–Cas9 can modify immune cells to enhance their capacity to identify and eliminate cancer cells or to insert genes that cause cancer cells to self-destruct or limit tumor growth in various ways (Shojaei Baghini et al., 2022; Wang et al., 2022a, 2022b; Xiao-Jie et al., 2015; Zhang et al., 2021a, 2021b).

Gene Knockout

Genes that affect immune cell function or survival, such as PD-1, CTLA-4, and LAG-3, can be knocked out using CRISPR–Cas9. This can increase T-cell or natural killer cell antitumor activity through immune checkpoint suppression (Liu et al.,

2023; Min et al., 2021). Genes such as EGFR, KRAS, BCR-ABL1, and others that support cancer cell growth or survival can also be knocked out using CRISPR–Cas9. This may impact the development, invasion, or resistance of cancer cells (Liu et al., 2023).

Gene Editing

CRISPR–Cas9 can be used to change the genes that produce receptors or ligands, such as CARs, TCRs, CD19, CD20, CD22, and others, that aid in recognizing or eliminating cancer cells. This can increase the activity and selectivity of immune cells against cancerous cells (Liu et al., 2023; Min et al., 2021). EGFRvIII, KRASG12D, TP53R175H, and other genes encoding oncogenic variants or mutations confer resistance or aggressiveness to cancer cells and can be modified using CRISPR–Cas9. This may assist cancer cells in regaining their regular sensitivity or functionality (Liu et al., 2023).

Gene Repression

Genes that are overexpressed or active in cancer cells and support their survival, proliferation, or evasion of immune surveillance, including PD-L1, MYC, and BCL2, can be inhibited using CRISPR–Cas9. The expression or activity of these genes may be reduced, rendering cancer cells more susceptible to immune system attack or apoptosis (Liu et al., 2023).

Gene Base Editing

Single-nucleotide changes can be made in cancer-related genes such as NOTCH, EGFR, KRAS, and others using the CRISPR–Cas9 system. Modifying or limiting the activity of specific genes can alter the phenotype or behavior of cancer cells (Liu et al., 2023).

Gene Insertion

Genes encoding tumor suppressor or suicide proteins, such as p53, BAX, HSV-TK, and others, can be inserted into cancer cells using CRISPR–Cas9. This may result in cell cycle arrest, apoptosis, or necrosis of cancer cells (Liu et al., 2023). Table 8.3 summarizes some of the instances from the literature.

Table 8.3 CRISPR–Cas in therapeutics of malignant tumors

Cancer type	Target gene	Strategy	Clinical trial phase	Reference
Breast cancer	KRAS, EGFR, NOTCH, HER2, BRCA1/2	Knockout, editing, impression, base editing	Preclinical	Karn et al. (2022)
Glioblastoma	EGFRvIII, PDL-1	Knockout, editing	Phase 1/2	Rasul et al. (2022a, 2022b, 2022c)
Leukemia	BRC-ABL1, CRLF2-JAK2	Knockout, editing	Phase 1/2	Ou et al. (2021)
Lung cancer	KRAS, DP-1, TP53, EGFR	Knockout, editing, base editing	Preclinical	Rasul et al. (2022a, 2022b, 2022c), Ou et al. (2021)
Lymphoma	CD19, CD20, CD22	CAR-T cells, knockout	Phase 1/2	González-Romero et al. (2019)
Melanoma	BRAF, NRAS, PTEN	Knockout, editing	Preclinical	Rasul et al. (2022a, 2022b, 2022c)
Multiple myeloma	CD38, CS1, BCMA	CAR-T cells, knockout	Phase 1/2	Rasul et al. (2022a, 2022b, 2022c)
Pancreatic cancer	KRAS, CDKN2A, TP53	Knockout, editing	Preclinical	Ou et al. (2021)
Prostate cancer	AR, TP53, PTEN	Knockout, editing	Preclinical	Ou et al. (2021)

Therapeutics of Hematological Diseases

Sickle Cell Anemia

A mutation in the HBB gene, which codes for the beta-globin component of hemoglobin, causes sickle cell anemia. Due to this mutation, hemoglobin molecules clump together and form sickle-shaped red blood cells, which can block blood vessels and cause discomfort, organ damage, and infections. CRISPR can correct the HBB gene mutation or activate the expression of fetal hemoglobin (HbF), preventing red blood cell sickling. Several clinical trials are currently being conducted to assess the safety and efficacy of CRISPR-based therapies for sickle cell anemia (González-Romero et al., 2019).

Beta-Thalassemia

Beta-thalassemia is a disorder caused by mutations in the HBB gene that limit or eliminate beta-globin production, resulting in low hemoglobin levels and severe anemia. CRISPR can correct HBB gene mutations or increase HbF expression, which can compensate for the lack of beta-globin. Currently, numerous clinical trials are in progress to assess the effectiveness and safety of CRISPR-based treatments for beta-thalassemia (González-Romero et al., 2019).

Hemophilia

Mutations in the F8 or F9 genes cause hemophilia by disrupting the production of clotting factors VIII and IX, which are crucial for blood coagulation and preventing excessive bleeding. Through CRISPR, genetic defects can be corrected, and functional copies of these genes can be incorporated into other areas of the genome, such as the liver. In preclinical research, CRISPR was found to restore clotting factor levels and reduce bleeding in hemophilia animal models (Segurado et al., 2022; Tang et al., 2023; Zhang et al., 2019).

Transthyretin Amyloidosis

Transthyretin mutations in the TTR gene, which codes for transthyretin, a blood protein that transports thyroid hormones and vitamin A, cause amyloidosis. Transthyretin misfolds and accumulates amyloid in several organs, including the heart and nerves, eventually leading to organ failure and death. CRISPR can be used to silence the TTR gene in the liver, which produces the majority of transthyretin, and reduce blood amyloidogenic protein levels (Table 8.4). A clinical trial of CRISPR-based treatment for transthyretin amyloidosis yielded promising results (Gillmore et al., 2021).

Table 8.4 CRISPR–Cas in therapeutics of hematological diseases

Disease	Targeted genes	Strategies	Clinical trial phase	References
Sickle cell anemia	HBB or BCL11A	HBB mutation correction or HbF activation	Phase ½	Frangoul et al. (2021)
Beta-thalassemia	HBB or BCL11A	HBB mutation correction or HbF activation	Phase ½	Frangoul et al. (2021), Gamage et al. (2023)
Hemophilia	F8 or F9	F8 or F9 mutation correction or insert functional gene copy	Phase 1	Tang et al. (2023), Zhang et al. (2019)
Transthyretin amyloidosis	TTR	Liver TTR gene suppression	Phase 1	Batista and Flotte (2021), Gillmore et al. (2021)

Therapeutics of Infectious Diseases

AIDS

CRISPR–Cas can modify the CCR5 or CXCR4 gene, which encodes a receptor for HIV entry into host cells, rendering them immune to HIV infection. CRISPR can also identify and eliminate HIV genomes that have entered host cells. Several clinical trials are underway to assess the safety and efficacy of CRISPR-based HIV treatments (Ding et al., 2021; Lin et al., 2021).

Hepatitis B and C

In hepatitis B and C, CRISPR–Cas can target and break the HBV or HCV genome in infected cells, lowering viral burden and halting viral reproduction. CRISPR has been proven effective in altering the genes responsible for viral entry, replication, and pathogenesis, such as NTCP, APOBEC3G, and IFNAR1. Extensive studies on animals and laboratory cells have demonstrated the potential of CRISPR in reducing HBV or HCV infection both in vitro and in vivo (Shenge & Osiowy, 2021).

COVID-19

It is possible to detect and eliminate viral RNA associated with COVID-19 caused by SARS-CoV-2 using CRISPR enzymes designed for that purpose. This might result in quick, accurate, and affordable COVID-19 identification. CRISPR can also modify host genes involved in viral entry, such as ACE2 or TMPRSS2, and target and remove the SARS-CoV-2 genome from infected cells. In preclinical studies, CRISPR has been demonstrated to reduce SARS-CoV-2 infection both in vitro and in vivo (Kumar et al., 2020; Zhang et al., 2020a, 2020b, 2020c).

Influenza

CRISPR can modify host genes required for viral infection, such as ANP32A and TMPRSS2, and target and kill the influenza virus genome in infected cells. CRISPR can modify human B cells to make antibodies that neutralize the influenza virus. CRISPR has been shown in preclinical trials to protect mice from deadly influenza virus attacks (Li et al., 2020; Lin et al., 2021; Song et al., 2021).

Malaria

CRISPR can be used to modify the genomes of malaria parasites such as *Plasmodium falciparum* or *Plasmodium vivax* to understand their biology and medication resistance. CRISPR can also be used to change the genes of mosquitos such as *Anopheles gambiae* and *Anopheles stephensi* to diminish their ability to transmit malaria parasites or support population management (Table 8.5). CRISPR has been shown in preclinical experiments to minimize mosquito-borne malaria transmission (Dong et al., 2018).

Therapeutics of Cardiovascular Diseases

Duchenne Muscular Dystrophy (DMD)

Duchenne muscular dystrophy (DMD) is a hereditary disorder that causes progressive muscle atrophy and weakness. Mutations in the DMD gene cause it, which creates the protein dystrophin, which supports muscle fibers. Dystrophin production and function can be restored by employing CRISPR to fix DMD mutations or insert functional gene copies into muscle cells. According to a recent study, CRISPR-mediated gene editing restored cardiac and muscle function in DMD dogs (Erkut & Yokota, 2022).

Table 8.5 CRISPR–Cas in therapeutics of infectious diseases

Disease	Targeted genes	Strategies	Clinical trial phase	References
AIDS	CCR5 or CXCR4	To make CD4+ T cells resistant to HIV infection, disable the CCR5 or CXCR4 gene	Phase ½	Ding et al. (2021), Lin et al. (2021)
Hepatitis B and C	HBV and HCV genome	Targeted and cleaved HBV genome in infected cells, lowering viral load and halting viral replication	Preclinical	Shenge and Osiowy (2021)
COVID-19	SARS-CoV-2 RNA, ACE2, or TMPRSS2	Target and destroy the SARS-CoV-2, modification in host gene	Preclinical	Kumar et al. (2020), Zhang et al. (2020a, 2020b, 2020c)
Influenza	ANP32A or TMPRSS2 or influenza virus genome	Target and destroy the influenza virus genome, modification in host gene	Preclinical	Li et al. (2020), Lin et al. (2021), Song et al. (2021)
Malaria	*Anopheles* or *Plasmodium* genes	Gene editing	Preclinical	Dong et al. (2018)

Familial Hypercholesterolemia

A hereditary disorder called familial hypercholesterolemia raises blood LDL cholesterol levels and the risk of heart attack and stroke. Mutations in the PCSK9 gene, which encodes an enzyme that breaks down LDLR, or the LDLR gene, which produces a receptor for LDL cholesterol, can contribute to it. Improved LDL cholesterol clearance can be achieved by using CRISPR to upregulate LDLR expression or downregulate PCSK9 expression in the liver. In a preliminary investigation, mice with familial hypercholesterolemia had LDL cholesterol levels lowered by CRISPR-mediated gene editing by up to 64% (Xu & Song, 2021).

Marfan Syndrome

Marfan syndrome is an inheritable disease that damages connective tissue, resulting in heart, blood vessel, eye, bone, and joint disorders. Mutations in the FBN1 gene, which codes for fibrillin-1, a protein that creates elastic fibers in connective tissue, cause it. CRISPR can restore fibrillin-1 expression and function by correcting FBN1 mutations or inserting functional gene copies into damaged cells. A preclinical study found that CRISPR-mediated gene editing improved heart function and reduced aortic aneurysms in mice with Marfan syndrome (Yin et al., 2021).

Therapeutics of Inherited Diseases

Cystic Fibrosis (CF)

In CF, the most common cause of mutation is a loss of three nucleotides in the CFTR gene, which encodes a chloride channel that regulates fluid balance in numerous organs. The creation of a defective protein that cannot reach the cell membrane or function correctly results in thick and sticky mucus that clogs the airways, digestive tract, and other organs. CRISPR can repair this mutation by adding the missing nucleotides or replacing the complete CFTR gene with a normal copy. CRISPR can modify the CFTR gene in cultured cells or organoids produced from CF patients. There are still obstacles to overcome before CRISPR may be used in humans, such as delivering CRISPR components to damaged cells, preventing off-target effects or immunological reactions, and ensuring long-term safety and efficacy (Maule et al., 2020).

Huntington's Disease (HD)

The HTT gene, which codes for the huntingtin protein, contains an enlarged CAG repeat, which causes Huntington's disease (HD). Although the typical purpose of huntingtin is unknown, it is thought to have a part in various cellular processes, such as transport, signaling, and metabolism. As a result of the extended CAG repeat, a mutant huntingtin protein with an excessively long polyglutamine tract is created, making it more prone to form toxic aggregates that can harm brain neurons. To prevent the production of the mutant huntingtin protein, CRISPR can target and remove the enlarged CAG repeat in the HTT gene. Numerous studies have proven that CRISPR may successfully modify the HTT gene in HD animal models or cultured cells (Seo et al., 2023).

Blindness

Leber congenital amaurosis (LCA) and retinitis pigmentosa (RP) are hereditary illnesses that cause significant vision loss or blindness due to abnormalities in retinal cells. These diseases can be caused by mutations in several genes involved in phototransduction, vitamin A metabolism, retinal development, or retinal pigment epithelium (RPE) function. Damaged genes in photoreceptor or RPE cells can be repaired or replaced using CRISPR, restoring function (Ruan et al., 2016).

Challenges and Limitations for CRISPR–Cas

CRISPR–Cas may occasionally bind and cut DNA or RNA sequences that are similar but not identical to the intended target, resulting in accidental mutations or modifications in other genes or sections of the genome. Off-target effects, such as function loss, toxicity, or gene flow, could have unintended repercussions for the organism or the environment (Jiang et al., 2023; Mali et al., 2013). To overcome this issue, scientists are developing techniques to improve CRISPR specificity and accuracy, such as using high-fidelity Cas variants, optimizing guide RNA design, and screening for probable off-target regions (Uddin et al., 2020).

Delivery of CRISPR–Cas components to target cells or tissues requires quick and safe transportation, which varies depending on the organism, tissue, or cell type. Various methods are available, including viral vectors, nanoparticles, liposomes, electroporation, microinjection, and biolistics. However, these delivery approaches can have potential issues such as low efficiency, high cost, immunogenicity, toxicity, and instability. Researchers are exploring strategies to enhance CRISPR delivery and stability to overcome these challenges. These include using nonviral vectors, encapsulating CRISPR–Cas components in biocompatible materials, and developing tissue-specific promoters (Wang et al., 2022a, 2022b).

Conclusion and Future Perspectives

Despite the potential benefits of CRISPR–Cas technology for drug development, numerous issues still require attention. CRISPR–Cas9, a powerful tool for drug testing, is not without its limitations. While generating and deploying CRISPR–Cas9 libraries may be relatively inexpensive for some scientists, the use of CRISPR–Cas9 itself can be quite costly. Moreover, targeting specific genes with CRISPR–Cas9 can be difficult, as this technology tends to target undesired genes despite its overall accuracy. This could result in false positives and negatives in drug screening investigations, a significant concern. Delivery methods, off-target effects, and ethical considerations are all critical concerns that necessitate further research and development. Clinical trials of CRISPR–Cas-based medications must also include comprehensive safety and efficacy evaluations.

The scientific community and society must pay due attention to CRISPR technology's ethical, social, legal, and regulatory implications. Failure to do so would be a grave mistake. For example, CRISPR could be used in bioterrorism, eugenics, or gene doping. CRISPR may pose concerns regarding who owns, patents, and licenses the technology and its derivatives. CRISPR may also necessitate particular safety, quality control, and oversight rules. To address this issue, researchers are working with diverse stakeholders and policymakers to develop a transparent and responsible framework for CRISPR technology governance and regulation.

References

Afzal, S., et al. (2022). Multifactorial complexity and zoonotic aspects of dog-mediated human rabies in Pakistan. *Pakistan Veterinary Journal, 42*, 445–452.

Agarwal, N., & Gupta, R. (2021). History, evolution and classification of CRISPR-Cas associated systems. *Progress in Molecular Biology and Translational Science, 179*, 11–76.

Aghamiri, S., et al. (2020). Nanoparticles-mediated CRISPR/Cas9 delivery: Recent advances in cancer treatment. *Journal of Drug Delivery Science and Technology, 56*, 101533.

Alghadban, S., et al. (2020). Electroporation and genetic supply of Cas9 increase the generation efficiency of CRISPR/Cas9 knock-in alleles in C57BL/6J mouse zygotes. *Scientific Reports, 10*, 1–12.

Awwad, S. W., Serrano-Benitez, A., Thomas, J. C., Gupta, V., & Jackson, S. P. (2023). Revolutionizing DNA repair research and cancer therapy with CRISPR–Cas screens. *Nature Reviews. Molecular Cell Biology*, 1–18. https://doi.org/10.1038/s41580-022-00571-x

Banakar, R., et al. (2021). Efficiency, specificity and temperature sensitivity of Cas9 and Cas12a RNPs for DNA-free genome editing in plants. *Frontiers in Genome Editing, 3*, 1–11.

Batista, A. R., & Flotte, T. R. (2021). In vivo gene editing works in humans: Results of a phase 1 clinical trial for TTR amyloidosis. *Molecular Therapy, 29*, 2633–2634.

Bayarsaikhan, D., Bayarsaikhan, G., & Lee, B. (2021). Recent advances in stem cells and gene editing: Drug discovery and therapeutics. *Progress in Molecular Biology and Translational Science, 181*, 231–269.

Beying, N., Schmidt, C., Pacher, M., Houben, A., & Puchta, H. (2020). CRISPR–Cas9-mediated induction of heritable chromosomal translocations in Arabidopsis. *Nature Plants, 6*, 638–645.

Bhat, M. A., et al. (2020). The era of editing plant genomes using CRISPR/Cas: A critical appraisal. *Journal of Biotechnology, 324*, 34–60.

Bhat, A. A., et al. (2022). Integration of CRISPR/Cas9 with artificial intelligence for improved cancer therapeutics. *Journal of Translational Medicine, 201*(20), 1–18.

Bhattacharjee, R., Das Roy, L., & Choudhury, A. (2022). Understanding on CRISPR/Cas9 mediated cutting-edge approaches for cancer therapeutics. Discov. *Oncologia, 13*, 1–15.

Breier, D., & Peer, D. (2023). Genome editing in cancer: Challenges and potential opportunities. *Bioactive Materials, 21*, 394–402.

Cebrian-Serrano, A., & Davies, B. (2017). CRISPR-Cas orthologues and variants: Optimizing the repertoire, specificity and delivery of genome engineering tools. *Mammalian Genome, 28*, 247–261.

Chandrasekaran, A. P., Karapurkar, J. K., Chung, H. Y., & Ramakrishna, S. (2022). The role of the CRISPR-Cas system in cancer drug development: Mechanisms of action and therapy. *Biotechnology Journal, 17*, 2100468.

Chavez-Granados, P. A., Manisekaran, R., Acosta-Torres, L. S., & Garcia-Contreras, R. (2022). CRISPR/Cas gene-editing technology and its advances in dentistry. *Biochimie, 194*, 96–107.

Chen, Y. W., et al. (2017). A three-dimensional model of human lung development and disease from pluripotent stem cells. *Nature Cell Biology, 195*(19), 542–549.

Chen, B. X., et al. (2018). Efficient CRISPR-Cas9 gene disruption system in edible-medicinal mushroom Cordyceps militaris. *Frontiers in Microbiology, 9*, 1–11.

Chen, S., Yao, Y., Zhang, Y., & Fan, G. (2020). CRISPR system: Discovery, development and off-target detection. *Cellular Signalling, 70*, 109577.

Chen, K., et al. (2022). Research progress of CRISPR-based biosensors and bioassays for molecular diagnosis. *Frontiers in Bioengineering and Biotechnology, 10*, 1–18.

Cho, S. W., et al. (2014). Analysis of off-target effects of CRISPR/Cas-derived RNA-guided endonucleases and nickases. *Genome Research, 24*, 132–141.

Choudhary, S., Ubale, A., Padiya, J., & Mikkilineni, V. (2020). Application of bioinformatics tools in CRISPR/Cas. *CRISPR/Cas Genome Editing*, 31–52. https://doi.org/10.1007/978-3-030-42022-2_2

Christen, F., et al. (2022). Modeling clonal hematopoiesis in umbilical cord blood cells by CRISPR/Cas9. *Leukemia, 36*, 1102–1110.

Dekkers, J. F., et al. (2020). Modeling breast cancer using CRISPR-Cas9–mediated engineering of human breast organoids. *JNCI Journal of the National Cancer Institute, 112*, 540–544.

Dey, A. (2021). CRISPR/Cas genome editing to optimize pharmacologically active plant natural products. *Pharmacological Research, 164*, 105359.

Dilip Kumar, S., Aashabharathi, M., KarthigaDevi, G., Subbaiya, R., & Saravanan, M. (2021). Insights of CRISPR-Cas systems in stem cells: Progress in regenerative medicine. *Molecular Biology Reports, 49*(1), 657–673.

Ding, R., et al. (2021). CRISPR/Cas system: A potential technology for the prevention and control of COVID-19 and emerging infectious diseases. *Frontiers in Cellular and Infection Microbiology, 11*, 342.

Doench, J. G., et al. (2014). Rational design of highly active sgRNAs for CRISPR-Cas9–mediated gene inactivation. *Nature Biotechnology, 32*(12), 1262–1267.

Dong, W., & Kantor, B. (2021). Lentiviral vectors for delivery of gene-editing systems based on crispr/cas: Current state and perspectives. *Viruses, 13*(7), 1288.

Dong, Y., Simões, M. L., Marois, E., & Dimopoulos, G. (2018). CRISPR/Cas9-mediated gene knockout of Anopheles gambiae FREP1 suppresses malaria parasite infection. *PLoS Pathogens, 14*, 1–16.

Duan, L., et al. (2021a). Nanoparticle delivery of CRISPR/Cas9 for genome editing. *Frontiers in Genetics, 12*, 673286.

Duan, C., Cao, H., Zhang, L. H., & Xu, Z. (2021b). Harnessing the CRISPR-Cas systems to Combat antimicrobial resistance. *Frontiers in Microbiology, 12*, 716064.

Dubey, A. K., et al. (2022). Exploring nano-enabled CRISPR-Cas-powered strategies for efficient diagnostics and treatment of infectious diseases. *Journal of Nanostructure in Chemistry, 12*(5), 833–864. (Springer Berlin, Heidelberg).

Dukhovny, A., et al. (2019). A CRISPR activation screen identifies genes that protect against Zika Virus infection. *Journal of Virology, 93*, 1–16.

Dunbar, C. E. (2021). A plethora of gene therapies for hemoglobinopathies. *Nature Medicine, 27*(2), 202–204.

El ouar, I., & Djekoun, A. (2021). Therapeutic and diagnostic relevance of Crispr technology. *Biomedicine & Pharmacotherapy, 138*, 111487.

Erkut, E., & Yokota, T. (2022). CRISPR therapeutics for Duchenne muscular dystrophy. *International Journal of Molecular Sciences, 23*, 1–18.

Feng, W., et al. (2021). CRISPR technology incorporating amplification strategies: Molecular assays for nucleic acids, proteins, and small molecules. *Chemical Science, 12*, 4683–4698.

Feng, S., et al. (2022). Strategies for high-efficiency mutation Using the CRISPR/Cas system. *Frontiers in Cell and Development Biology, 9*, 1–20.

Ferrari, G., Thrasher, A. J., & Aiuti, A. (2020). Gene therapy using haematopoietic stem and progenitor cells. *Nature Reviews. Genetics, 22*(4), 216–234.

Frangoul, H., et al. (2021). CRISPR-Cas9 gene editing for sickle cell disease and β-thalassemia. *The New England Journal of Medicine, 384*, 252–260.

Fu, Y., et al. (2013). High-frequency off-target mutagenesis induced by CRISPR-Cas nucleases in human cells. *Nature Biotechnology, 31*, 822–826.

Fu, Y., Sander, J. D., Reyon, D., Cascio, V. M., & Joung, J. K. (2014). Improving CRISPR-Cas nuclease specificity using truncated guide RNAs. *Nature Biotechnology, 32*, 279–284.

Gamage, U., Warnakulasuriya, K., Hansika, S., & Silva, G. N. (2023). CRISPR Gene therapy: A promising one-time therapeutic approach for transfusion-dependent β-Thalassemia—CRISPR-Cas9 gene editing for β-thalassemia. *Thalass Reports, 13*, 51–69.

Gao, J., Luo, T., Lin, N., Zhang, S., & Wang, J. (2020). A new tool for CRISPR-Cas13a-based cancer gene therapy. *Molecular Therapy - Oncolytics, 19*, 79–92.

Geng, B. C., et al. (2020). A simple, quick, and efficient CRISPR/Cas9 genome editing method for human induced pluripotent stem cells. *Acta Pharmacologica Sinica, 41*(11), 1427–1432.

Gerashchenkov, G. A., et al. (2020). Design of guide RNA for CRISPR/Cas plant genome editing. *Molecular Biology, 54*(1), 24–42.

Ghaemi, A., et al. (2021). CRISPR-cas9 genome editing delivery systems for targeted cancer therapy. *Life Sciences, 267*, 118969.

Gholizadeh, P., et al. (2020). How CRISPR-Cas system could be used to combat antimicrobial resistance. *Infection and Drug Resistance, 13*, 1111.

Ghorbani, A., et al. (2021). A short overview of CRISPR-Cas technology and its application in viral disease control. *Transgenic Research, 30*, 221–238.

Gier, R. A., et al. (2020). High-performance CRISPR-Cas12a genome editing for combinatorial genetic screening. *Nature Communications, 11*(1), 1–9.

Gillmore, J. D., et al. (2021). CRISPR-Cas9 in vivo gene editing for transthyretin amyloidosis. *The New England Journal of Medicine, 385*, 493–502.

Golubov, A. (2021). CRISPR—Bacterial immune system. In *Genome stab. From virus to hum application* (pp. 91–105). Academic Press. https://doi.org/10.1016/B978-0-323-85679-9.00006-4

Gong, T., Zeng, J., Tang, B., Zhou, X., & Li, Y. (2020). CRISPR-Cas systems in oral microbiome: From immune defense to physiological regulation. *Molecular Oral Microbiology, 35*, 41–48.

González-Romero, E., et al. (2019). CRISPR to fix bad blood: A new tool in basic and clinical hematology. *Haematologica, 104*, 881–893.

Grotz, A., & Deswal, S. (2021). Pooled CRISPR KO screens for target identification. *Genome Editing in Drug Discovery*, 113–125. https://doi.org/10.1002/9781119671404.CH8

Gullapalli, S. (2021). Drug target identification and validation. *Drug Discovery and Development*, 235–249. https://doi.org/10.1007/978-981-15-5534-3_8/TABLES/5

Halder, K., Chaudhuri, A., Abdin, M. Z., Majee, M., & Datta, A. (2022). RNA interference for improving disease resistance in plants and its relevance in this clustered regularly interspaced short palindromic repeats-dominated era in terms of dsRNA-based biopesticides. *Frontiers in Plant Science, 13*, 1434.

Hammond, A., et al. (2016). A CRISPR-Cas9 gene drive system targeting female reproduction in the malaria mosquito vector Anopheles gambiae. *Nature Biotechnology, 34*, 78–83.

He, Z., et al. (2021). Lineage recording in human cerebral organoids. *Nature Methods, 191*(19), 90–99.

He, Y., et al. (2023). The CRISPR/Cas system: A customizable toolbox for molecular detection. *Genes (Basel), 14*, 850.

Hendriks, D., Clevers, H., & Artegiani, B. (2020a). CRISPR-Cas tools and their application in genetic engineering of human stem cells and organoids. *Cell Stem Cell, 27*(5), 705–731.

Hendriks, D., Artegiani, B., Hu, H., de Sousa, C., Lopes, S., & Clevers, H. (2020b). Establishment of human fetal hepatocyte organoids and CRISPR–Cas9-based gene knockin and knockout in organoid cultures from human liver. *Nature Protocols, 16*(1), 182–217.

Herskovitz, J., et al. (2021). CRISPR-Cas9 mediated exonic disruption for HIV-1 elimination. *eBioMedicine, 73*, 103678.

Hillary, V. E., & Ceasar, S. A. (2023). A review on the mechanism and applications of CRISPR/Cas9/Cas12/Cas13/Cas14 proteins utilized for genome engineering. *Molecular Biotechnology, 65*, 311–325.

Hu, C., et al. (2021). Mechanism for Cas4-assisted directional spacer acquisition in CRISPR–Cas. *Nature, 598*, 515–520.

Huang, J., Zhou, Y., Li, J., Lu, A., & Liang, C. (2022). CRISPR/Cas systems: Delivery and application in gene therapy. *Frontiers in Bioengineering and Biotechnology, 10*, 1–21.

Javaid, N., & Choi, S. (2021). CRISPR/Cas system and factors affecting its precision and efficiency. *Frontiers in Cell and Development Biology, 9*, 1–25.

Jena, R., et al. (2022). Treatment strategies for HIV infection with emphasis on role of CRISPR/Cas9 gene: Success so far and road ahead. *European Journal of Pharmacology, 931*, 175173.

Jeong, S. H., Lee, H. J., & Lee, S. J. (2023). Recent advances in CRISPR-Cas technologies for synthetic biology. *Journal of Microbiology, 61*, 13–36.

Jiang, K., et al. (2023). Next-generation diagnostic test for dengue virus detection using an ultrafast plasmonic colorimetric RT-PCR strategy. *Analytica Chimica Acta, 1274*, 341565.

Kalds, P., et al. (2022). Genetics of the phenotypic evolution in sheep: A molecular look at diversity-driving genes. *Genetics, Selection, Evolution, 54*, 1–27.

Kamruzzaman, M., & Iredell, J. R. (2020). CRISPR-Cas system in antibiotic resistance plasmids in Klebsiella pneumoniae. *Frontiers in Microbiology, 10*, 2934.

Karn, V., et al. (2022). CRISPR/Cas9 system in breast cancer therapy: Advancement, limitations and future scope. *Cancer Cell International, 22*, 1–14.

Kato, K., et al. (2022). Structure of the IscB–ωRNA ribonucleoprotein complex, the likely ancestor of CRISPR-Cas9. *Nature Communications, 13*, 1–10.

Khoshandam, M., Soltaninejad, H., Mousazadeh, M., Hamidieh, A. A., & Hosseinkhani, S. (2023). Clinical applications of the CRISPR/Cas9 genome-editing system: Delivery options and challenges in precision medicine. *Genes & Diseases*. https://doi.org/10.1016/J.GENDIS.2023.02.027

Khurana, A., et al. (2022). A comprehensive overview of CRISPR/Cas 9 technology and application thereof in drug discovery. *Journal of Cellular Biochemistry, 123*, 1674–1698.

Khwatenge, C. N., & Nahashon, S. N. (2021). Recent advances in the application of CRISPR/Cas9 gene editing system in poultry species. *Frontiers in Genetics, 12*, 627714.

Kim, J., Koo, B. K., & Knoblich, J. A. (2020). Human organoids: Model systems for human biology and medicine. *Nature Reviews Molecular Cell Biology, 21*(10), 571–584.
Kim, B., et al. (2022). Application of CRISPR/Cas9-based mutant enrichment technique to improve the clinical sensitivity of plasma EGFR testing in patients with non-small cell lung cancer. *Cancer Cell International, 22*, 1–11.
Kitano, Y., et al. (2022). Generation of hypoimmunogenic induced pluripotent stem cells by CRISPR-Cas9 system and detailed evaluation for clinical application. *Molecular Therapy: Methods & Clinical Development, 26*, 15–25.
Kleinstiver, B. P., et al. (2016). High-fidelity CRISPR–Cas9 nucleases with no detectable genome-wide off-target effects. *Nature, 529*(7587), 490–495.
Koch, B., et al. (2018). Generation and validation of homozygous fluorescent knock-in cells using CRISPR-Cas9 genome editing. *Nature Protocols, 13*, 1465–1487.
Kostyusheva, A., et al. (2022). CRISPR-Cas systems for diagnosing infectious diseases. *Methods, 203*, 431–446.
Kratzer, K., Getz, L. J., Peterlini, T., Masson, J. Y., & Dellaire, G. (2021). Addressing the dark matter of gene therapy: Technical and ethical barriers to clinical application. *Human Genetics, 141*, 1175–1193.
Kumar, P., et al. (2020). CRISPR-Cas system: An approach with potentials for COVID-19 diagnosis and therapeutics. *Frontiers in Cellular and Infection Microbiology, 10*, 639.
Kumar, R. R., et al. (2022). Iron and Zinc at a cross-road: A trade-off between micronutrients and anti-nutritional factors in pearl millet flour for enhancing the bioavailability. *Journal of Food Composition and Analysis, 111*, 104591.
Kurihara, N., et al. (2022). Structure of the type V-C CRISPR-Cas effector enzyme. *Molecular Cell, 82*, 1865–1877.
Lahm, H., et al. (2023). Correction of a deleterious TBX5 mutation in an induced pluripotent stem cell line (DHMi004-A-1) using a completely plasmid-free CRISPR/Cas 9 approach. *Stem Cell Research, 70*, 103126.
Lee, H., & Sashital, D. G. (2022). Creating memories: Molecular mechanisms of CRISPR adaptation. *Trends in Biochemical Sciences, 47*, 464–476.
Lee, J., et al. (2020). Recent advances in genome editing of stem cells for drug discovery and therapeutic application. *Pharmacology & Therapeutics, 209*, 107501. https://doi.org/10.1016/j.pharmthera.2020.107501
Li, Y., & Bondy-Denomy, J. (2021). Anti-CRISPRs go viral: The infection biology of CRISPR-Cas inhibitors. *Cell Host & Microbe, 29*, 704–714.
Li, B., et al. (2020). Genome-wide CRISPR screen identifies host dependency factors for influenza A virus infection. *Nature Communications, 11*.
Li, J., et al. (2022). Application of CRISPR/Cas systems in the nucleic acid detection of infectious diseases. *Diagnostics, 12*(10), 2455.
Lin, H., et al. (2021). The use of CRISPR/Cas9 as a tool to study human infectious viruses. *Frontiers in Cellular and Infection Microbiology, 11*, 1–14.
Lin, Y., Jones, M. L., Jones, M. L., & Gene, S. (2022). CRISPR/Cas9-mediated editing of autophagy Gene 6 in petunia decreases flower longevity, seed yield, and phosphorus remobilization by accelerating ethylene production and senescence-related gene expression. *Frontiers in Plant Science, 13*, 840218.
Liu, Q., Zhang, H., & Huang, X. (2020). Anti-CRISPR proteins targeting the CRISPR-Cas system enrich the toolkit for genetic engineering. *The FEBS Journal, 287*, 626–644.
Liu, R., Liang, L., Freed, E. F., & Gill, R. T. (2021). Directed evolution of CRISPR/Cas systems for precise gene editing. *Trends in Biotechnology, 39*, 262–273.
Liu, Z., et al. (2023). Recent advances and applications of CRISPR—Cas9 in cancer immunotherapy. *Molecular Cancer, 1–19*. https://doi.org/10.1186/s12943-023-01738-6
Lu, L., Yu, X., Cai, Y., Sun, M., & Yang, H. (2021). Application of CRISPR/Cas9 in Alzheimer's disease. *Frontiers in Neuroscience, 15*, 1–14.

Maity, S., Mukherjee, R., & Banerjee, S. (2022). Recent advances and therapeutic strategies using CRISPR genome editing technique for the treatment of cancer. *Molecular Biotechnology, 65*(2), 206–226.

Mali, P., et al. (2013). RNA-guided human genome engineering via Cas9. *Science (80-), 339*, 823–826.

Malzahn, A. A., et al. (2019). Application of CRISPR-Cas12a temperature sensitivity for improved genome editing in rice, maize, and Arabidopsis. *BMC Biology, 17*, 1–14.

Markakiou, S., Gaspar, P., Johansen, E., Zeidan, A. A., & Neves, A. R. (2020). Harnessing the metabolic potential of Streptococcus thermophilus for new biotechnological applications. *Current Opinion in Biotechnology, 61*, 142–152.

Martínez Arbas, S., et al. (2021). Roles of bacteriophages, plasmids and CRISPR immunity in microbial community dynamics revealed using time-series integrated meta-omics. *Nature Microbiology, 6*, 123–135.

Maule, G., Arosio, D., & Cereseto, A. (2020). Gene therapy for cystic fibrosis: Progress and challenges of genome editing. *International Journal of Molecular Sciences, 21*, 1–13.

Menche, C., & Farin, H. F. (2021). Strategies for genetic manipulation of adult stem cell-derived organoids. *Experimental & Molecular Medicine, 53*(10), 1483–1494.

Micura, R., & Höbartner, C. (2020). Fundamental studies of functional nucleic acids: Aptamers, riboswitches, ribozymes and DNAzymes. *Chemical Society Reviews, 49*, 7331–7353.

Min, C., et al. (2021). Abstract 1905: In vivo CRISPR screen identifies tumor suppressors as drivers of tumor cell-intrinsic immune evasion. *Cancer Research, 81*, 1905–1905.

Modell, A. E., Lim, D., Nguyen, T. M., Sreekanth, V., & Choudhary, A. (2022). CRISPR-based therapeutics: Current challenges and future applications. *Trends in Pharmacological Sciences, 43*, 151–161.

Modrzejewski, D., et al. (2020). Which factors affect the occurrence of off-target effects caused by the use of CRISPR/Cas: A systematic review in plants. *Frontiers in Plant Science, 11*, 574959.

Molla, K. A., Karmakar, S., & Islam, M. T. (2020). Wide horizons of CRISPR-Cas-derived technologies for basic biology, agriculture, and medicine. *CRISPR-Cas Methods*, 1–23. https://doi.org/10.1007/978-1-0716-0616-2_1

Moore, J. D. (2015). The impact of CRISPR–Cas9 on target identification and validation. *Drug Discovery Today, 20*, 450–457.

Mosterd, C., Rousseau, G. M., & Moineau, S. (2021). A short overview of the CRISPR-Cas adaptation stage 1. *Canadian Journal of Microbiology, 67*, 1–12.

Mu, J., Sun, X., Zhao, Z., Sun, H., & Sun, P. (2021). BRD9 inhibition promotes PUMA-dependent apoptosis and augments the effect of imatinib in gastrointestinal stromal tumors. *Cell Death & Disease, 12*(11), 1–10.

Munawar, N., Ahmad, A., Munawar, N., & Ahmad, A. (2021). CRISPR/Cas system: An introduction. *CRISPR Crops*, 1–35. https://doi.org/10.1007/978-981-15-7142-8_1

Niu, Y., et al. (2020). A Type I-F Anti-CRISPR protein inhibits the CRISPR-Cas surveillance complex by ADP-ribosylation. *Molecular Cell, 80*, 512–524.

Ou, X., Ma, Q., Yin, W., Ma, X., & He, Z. (2021). CRISPR/Cas9 gene-editing in cancer immunotherapy: Promoting the present revolution in cancer therapy and exploring more. *Frontiers in Cell and Development Biology, 9*, 1–12.

Parnas, O., et al. (2015). A genome-wide CRISPR screen in primary immune cells to dissect regulatory networks. *Cell, 162*, 675–686.

Pattharaprachayakul, N., Lee, M., Incharoensakdi, A., & Woo, H. M. (2020). Current understanding of the cyanobacterial CRISPR-Cas systems and development of the synthetic CRISPR-Cas systems for cyanobacteria. *Enzyme and Microbial Technology, 140*, 109619.

Petersen, B. L., et al. (2019). Improved CRISPR/Cas9 gene editing by fluorescence activated cell sorting of green fluorescence protein tagged protoplasts. *BMC Biotechnology, 19*, 1–12.

Poultney, C. S., et al. (2011). Rational design of temperature-sensitive alleles using computational structure prediction. *PLoS One, 6*(9), e23947.

Prasad, K., George, A., Ravi, N. S., & Mohankumar, K. M. (2021). CRISPR/Cas based gene editing: Marking a new era in medical science. *Molecular Biology Reports, 48*, 4879–4895.
Qiu, M., Zhou, X. M., & Liu, L. (2022). Improved strategies for CRISPR-Cas12-based nucleic acids detection. *Journal of Analysis and Testing, 6*, 44–52.
Ramakrishna, G., Babu, P. E., Singh, R., & Trehanpati, N. (2021). Application of CRISPR-Cas9 based gene editing to study the pathogenesis of colon and liver cancer using organoids. *Hepatology International, 15*(6), 1309–1317.
Rasool, S., Hussain, T., Zehra, A., Khan, S., & Khan, S. (2017). CRISPR: Genome—editing and beyond. *Current Trends In Biotechnology and Pharmacy, 11*, 181–189.
Rasul, A., et al. (2022a). Target identification approaches in drug discovery. *Drug Target Selection and Validation*, 41–59. https://doi.org/10.1007/978-3-030-95895-4_3
Rasul, M. F., et al. (2022b). Strategies to overcome the main challenges of the use of CRISPR/Cas9 as a replacement for cancer therapy. *Molecular Cancer, 211*(21), 1–30.
Rasul, M. F., et al. (2022c). Strategies to overcome the main challenges of the use of CRISPR/Cas9 as a replacement for cancer therapy. *Molecular Cancer, 21*, 1–30.
Rehman, A. U., et al. (2021). Monitoring the health status and herd-level risk factors of tuberculosis in water buffalo (Bubalus bubalis) dairy farms in Pakistan. *Pakistan Veterinary Journal, 41*, 552–556.
Ross, M. J., & Coates, P. T. (2018). Using CRISPR to inactivate endogenous retroviruses in pigs: An important step toward safe xenotransplantation? *Kidney International, 93*, 4–6.
Roy Chowdhury, A., & Gargate, G. A. (2021). The trends in CRISPR research: A patent and literature study with a focus on India. *World Patent Information, 65*, 102038.
Ruan, G., Barry, E., Lukason, M., Cheng, S., & Scaria, A. (2016). Using CRISPR/Cas9 as a therapeutic approach for leber congenital amaurosis 10 (LCA10). *Molecular Therapy, 24*, S131–S132.
Safari, F., et al. (2019). The interaction of phages and bacteria: The co-evolutionary arms race. *Critical Reviews in Biotechnology, 40*, 119–137. https://doi.org/10.1080/07388551.2019.1674774
Schenke, D., & Cai, D. (2020). Applications of CRISPR/Cas to improve crop disease resistance: Beyond inactivation of susceptibility factors. *iScience, 23*, 101478.
Segurado, O. G., Jiang, R., & Pipe, S. W. (2022). Challenges and opportunities when transitioning from in vivo gene replacement to in vivo CRISPR/Cas9 therapies—A spotlight on hemophilia. *Expert Opinion on Biological Therapy, 22*, 1091–1098.
Selvakumar, S. C., et al. (2022). CRISPR/Cas9 and next generation sequencing in the personalized treatment of cancer. *Molecular Cancer, 21*, 1–14.
Seo, J. H., et al. (2023). DNA double-strand break-free CRISPR interference delays Huntington's disease progression in mice. *Communications Biology, 6*(1), 466.
Shalem, O., et al. (2014). Genome-scale CRISPR-Cas9 knockout screening in human cells. *Science (80-), 343*, 84–87.
Sharma, G., Sharma, A. R., Bhattacharya, M., Lee, S. S., & Chakraborty, C. (2021). CRISPR-Cas9: A preclinical and clinical perspective for the treatment of human diseases. *Molecular Therapy, 29*, 571–586.
Shenge, J. A., & Osiowy, C. (2021). Rapid diagnostics for Hepatitis B and C viruses in low- and middle-income countries. *Frontiers in Virology, 1*, 1–12.
Shim, H. (2022). Investigating the genomic background of CRISPR-Cas genomes for CRISPR-based antimicrobials. *Evolutionary Bioinformatics, 18*, 11769343221103887.
Shivram, H., Cress, B. F., Knott, G. J., & Doudna, J. A. (2021). Controlling and enhancing CRISPR systems. *Nature Chemical Biology, 17*, 10–19.
Shmakova, A. A., Shmakova, O. P., Karpukhina, A. A., & Vassetzky, Y. S. (2022). CRISPR/Cas: History and perspectives. *Russian Journal of Developmental Biology, 53*, 272–282.
Shojaei Baghini, S., et al. (2022). CRISPR/Cas9 application in cancer therapy: A pioneering genome editing tool. *Cellular & Molecular Biology Letters, 27*(1), 35.

Sinclair, F., Begum, A. A., Dai, C. C., Toth, I., & Moyle, P. M. (2023). Recent advances in the delivery and applications of nonviral CRISPR/Cas9 gene editing. *Drug Delivery and Translational Research, 13*, 1500–1519.

Slaymaker, I. M., & Gaudelli, N. M. (2021). Engineering Cas9 for human genome editing. *Current Opinion in Structural Biology, 69*, 86–98.

Sledzinski, P., Nowaczyk, M., & Olejniczak, M. (2020). Computational tools and resources supporting CRISPR-Cas experiments. *Cells, 9*, 1288.

Sofos, N., et al. (2020). Structures of the Cmr-β complex reveal the regulation of the immunity mechanism of Type III-B CRISPR-Cas. *Molecular Cell, 79*, 741–757.e7.

Song, Y., et al. (2021). A genome-wide CRISPR/Cas9 gene knockout screen identifies immunoglobulin superfamily DCC subclass member 4 as a key host factor that promotes influenza virus endocytosis. *PLoS Pathogens, 17*, 1–24.

Sreedurgalakshmi, K., Srikar, R., & Rajkumari, R. (2021). CRISPR-Cas deployment in non-small cell lung cancer for target screening, validations, and discoveries. *Cancer Gene Therapy, 28*(6), 566–580.

Stellon, D., et al. (2022). CRISPR/Cas-mediated knock-in of genetically encoded fluorescent biosensors into the AAVS1 locus of human-induced pluripotent stem cells. *Methods in Molecular Biology, 2549*, 379–398.

Su, W., Li, J., Ji, C., Chen, C., Wang, Y., Dai, H., Li, F., & Liu, P. (2023). CRISPR/Cas systems for the detection of nucleic acid and non-nucleic acid targets. *Nano Research, 16*(7), 9940–9953.

Takeda, H., et al. (2019). CRISPR-Cas9-mediated gene knockout in intestinal tumor organoids provides functional validation for colorectal cancer driver genes. *Proceedings of the National Academy of Sciences of the United States of America, 116*, 15635–15644.

Takeda, S. N., et al. (2021). Structure of the miniature type V-F CRISPR-Cas effector enzyme. *Molecular Cell, 81*, 558–570.

Tan, J., Zhang, F., Karcher, D., & Bock, R. (2020). Expanding the genome-targeting scope and the site selectivity of high-precision base editors. *Nature Communications, 11*(1), 1–11.

Tang, Q., et al. (2023). CRISPR-mediated in situ introduction or integration of F9-Padua in human iPSCs for gene therapy of Hemophilia B. *International Journal of Molecular Sciences, 24*(10), 9013.

Tao, S., et al. (2022a). Association of CRISPR-Cas system with the antibiotic resistance and virulence genes in nosocomial isolates of enterococcus. *Infection and Drug Resistance, 15*, 6939–6949.

Tao, S., Chen, H., Li, N., & Liang, W. (2022b). The application of the CRISPR-Cas system in antibiotic resistance. *Infection and Drug Resistance, 15*, 4155–4168.

Tiwari, J. K., & Singh, A. K. (2023). CRISPR/Cas genome editing in tomato improvement: Advances and applications. *Frontiers in Plant Science, 14*, 1121209. https://doi.org/10.3389/fpls.2023.1121209

Touzdjian Pinheiro Kohlrausch Távora, F., et al. (2022). CRISPR/Cas- and topical RNAi-based technologies for crop management and improvement: Reviewing the risk assessment and challenges towards a more sustainable agriculture. *Frontiers in Bioengineering and Biotechnology, 10*, 1–20.

Tran, N. T., et al. (2022). Precise CRISPR-Cas–mediated gene repair with minimal off-target and unintended on-target mutations in human hematopoietic stem cells. *Science Advances, 8*, eabm9106.

Tröder, S. E., et al. (2018). An optimized electroporation approach for efficient CRISPR/Cas9 genome editing in murine zygotes. *PLoS One, 13*, 1–14.

Tukur, A. (2021). The CRISPR-Cas system: Its origin, functions and applications in biotechnology: A review. *Bayero Journal of Pure and Applied Science, 13*(2), 65–71.

Tyumentseva, M. A., Tyumentsev, A. I., & Akimkin, V. G. (2021). Protocol for assessment of the efficiency of CRISPR/Cas RNP delivery to different types of target cells. *PLoS One, 16*, 1–22.

Uddin, F., Rudin, C. M., & Sen, T. (2020). CRISPR gene therapy: Applications, limitations, and implications for the future. *Frontiers in Oncology, 10*, 1387.

Ur Rahman, S., & Mohsin, M. (2019). The under reported issue of antibiotic-resistance in food-producing animals in Pakistan. *Pakistan Veterinary Journal, 39*, 323–328.

van Rijn, J. M., et al. (2018). Intestinal failure and aberrant lipid metabolism in patients with DGAT1 deficiency. *Gastroenterology, 155*, 130–143.e15.

Wang, Z., & Cui, W. (2020). CRISPR-Cas system for biomedical diagnostic platforms. *Viewpoints, 1*, 20200008.

Wang, Y., Huang, G., Yang, H., & Zhang, X. N. (2020a). Advances on nonviral vectors of CRISPR/Cas9 system for genome editing. *Acta Pharmaceutica Sinica, 55*, 2606–2617.

Wang, D. X., et al. (2020b). CRISPR/Cas12a-based dual amplified biosensing system for sensitive and rapid detection of polynucleotide kinase/phosphatase. *Biosensors & Bioelectronics, 168*, 112556.

Wang, Y., et al. (2020c). SgBE: A structure-guided design of sgRNA architecture specifies base editing window and enables simultaneous conversion of cytosine and adenosine. *Genome Biology, 21*, 1–16.

Wang, Q., et al. (2021a). Use of CRISPR/Cas ribonucleoproteins for high throughput gene editing of induced pluripotent stem cells. *Methods, 194*, 18–29.

Wang, M., Wang, L., & Zhao, Q. (2021b). Construction of a Triple Gene-Deleted Live Vaccine Candidate against Pseudorabies virus using CRISPR/Cas9 and Cre/LoxP system-based strategy. *Pakistan Veterinary Journal, 41*, 493–498.

Wang, Q., Guo, F., Jin, Y., & Ma, Y. (2022a). Applications of human organoids in the personalized treatment for digestive diseases. *Signal Transduction and Targeted Therapy, 71*(7), 1–30.

Wang, L., Chen, Y., Liu, X., Li, Z., & Dai, X. (2022b). The application of CRISPR/Cas9 technology for cancer immunotherapy: Current status and problems. *Frontiers in Oncology, 11*, 1–11.

Watson, B. N. J., Steens, J. A., Staals, R. H. J., Westra, E. R., & van Houte, S. (2021). Coevolution between bacterial CRISPR-Cas systems and their bacteriophages. *Cell Host & Microbe, 29*, 715–725.

Webb, T., Craigon, C., & Ciulli, A. (2022). Targeting epigenetic modulators using PROTAC degraders: Current status and future perspective. *Bioorganic & Medicinal Chemistry Letters, 63*, 128653.

Wolter, F., Schindele, P., & Puchta, H. (2019). Plant breeding at the speed of light: The power of CRISPR/Cas to generate directed genetic diversity at multiple sites. *BMC Plant Biology, 19*, 1–8.

Wu, H., et al. (2021). Versatile detection with CRISPR/Cas system from applications to challenges. *TrAC Trends in Analytical Chemistry, 135*, 116150.

Wu, Q., et al. (2022). CRISPR-Cas systems target endogenous genes to impact bacterial physiology and alter mammalian immune responses. *Molecular Biomedicine, 3*, 1–16.

Xiao-Jie, L., Hui-Ying, X., Zun-Ping, K., Jin-Lian, C., & Li-Juan, J. (2015). CRISPR-Cas9: A new and promising player in gene therapy. *Journal of Medical Genetics, 52*, 289–296.

Xing, H., & Meng, L. M. (2020). CRISPR-cas9: A powerful tool towards precision medicine in cancer treatment. *Acta Pharmacologica Sinica, 41*(5), 583–587.

Xu, Y., & Li, Z. (2020). CRISPR-Cas systems: Overview, innovations and applications in human disease research and gene therapy. *Computational and Structural Biotechnology Journal, 18*, 2401–2415.

Xu, M., & Song, J. (2021). Targeted therapy in cardiovascular disease: A precision therapy era. *Frontiers in Pharmacology, 12*, 1–21.

Xu, C. L., & Tsang, S. H. (2022). *The history of CRISPR: From discovery to the present. CRISPR genome surgery in stem cells and disease tissues* (pp. 1–6). Academic Press. https://doi.org/10.1016/B978-0-12-817876-8.00007-3

Xu, X., et al. (2021). Nanotechnology-based delivery of CRISPR/Cas9 for cancer treatment. *Advanced Drug Delivery Reviews, 176*, 113891.

Yagoubat, A., Corrales, R. M., Bastien, P., Lévêque, M. F., & Sterkers, Y. (2020). Gene editing in trypanosomatids: Tips and tricks in the CRISPR-Cas9 era. *Trends in Parasitology, 36*, 745–760.

Yang, X., & Zhang, B. (2023). A review on CRISPR/Cas: A versatile tool for cancer screening, diagnosis, and clinic treatment. *Functional & Integrative Genomics, 23*(2), 1–26.

Yang, W., Li, S., & Li, X. J. (2019). A CRISPR monkey model unravels a unique function of PINK1 in primate brains. *Molecular Neurodegeneration, 14*, 1–4.

Yang, Y., Xu, J., Ge, S., & Lai, L. (2021). CRISPR/Cas: Advances, limitations, and applications for precision cancer research. *Frontiers in Medicine, 8*, 115.

Yin, X., Hao, J., & Yao, Y. (2021). CRISPR/Cas9 in zebrafish: An attractive model for FBN1 genetic defects in humans. *Molecular Genetics & Genomic Medicine, 9*, 1–13.

Yu, L., & Marchisio, M. A. (2020). Types I and V anti-CRISPR proteins: From phage defense to eukaryotic synthetic gene circuits. *Frontiers in Bioengineering and Biotechnology, 8*, 575393.

Yu, W., & Wu, Z. (2021). Ocular delivery of CRISPR/Cas genome editing components for treatment of eye diseases. *Advanced Drug Delivery Reviews, 168*, 181–195.

Zafar, K., et al. (2020). Genome editing technologies for rice improvement: Progress, prospects, and safety concerns. *Frontiers in Genome Editing, 2*, 1–16.

Zhang, B. (2021). CRISPR/Cas gene therapy. *Journal of Cellular Physiology, 236*, 2459–2481.

Zhang, J. P., et al. (2019). Curing hemophilia A by NHEJ-mediated ectopic F8 insertion in the mouse. *Genome Biology, 20*, 1–17.

Zhang, D., et al. (2020a). Genome editing with the CRISPR-Cas system: An art, ethics and global regulatory perspective. *Plant Biotechnology Journal, 18*, 1651–1669.

Zhang, S., Shen, J., Li, D., & Cheng, Y. (2020b). Strategies in the delivery of Cas9 ribonucleoprotein for CRISPR/Cas9 genome editing. *Theranostics, 11*, 614–648.

Zhang, F., Abudayyeh, O. O., & Gootenberg, J. S. (2020c). *A protocol for detection of COVID-19 using CRISPR diagnostics (v.20200321)* (pp. 1–8). Broad Institute of MIT and Harvard.

Zhang, D., Zhang, Z., Unver, T., & Zhang, B. (2021a). CRISPR/Cas: A powerful tool for gene function study and crop improvement. *Journal of Advanced Research, 29*, 207–221.

Zhang, H., et al. (2021b). Application of the CRISPR/Cas9-based gene editing technique in basic research, diagnosis, and therapy of cancer. *Molecular Cancer, 20*, 1–22.

Zhou, W., & Wang, X. (2021). Human gene therapy: A scientometric analysis. *Biomedicine & Pharmacotherapy, 138*, 111510.

Zhu, Z., Verma, N., González, F., Shi, Z. D., & Huangfu, D. (2015). A CRISPR/Cas-mediated selection-free Knockin strategy in human embryonic stem cells. *Stem Cell Reports, 4*, 1103–1111.

Zhu, H., Li, C., & Gao, C. (2020). Applications of CRISPR–Cas in agriculture and plant biotechnology. *Nature Reviews. Molecular Cell Biology, 21*, 661–677.

Zhu, L., et al. (2021). Selection and applications of functional nucleic acids for infectious disease detection and prevention. *Analytical and Bioanalytical Chemistry, 413*, 4563–4579.

Zirin, J., Bosch, J., Viswanatha, R., Mohr, S. E., & Perrimon, N. (2022). State-of-the-art CRISPR for in vivo and cell-based studies in Drosophila. *Trends in Genetics, 38*, 437–453.

Chapter 9
Plant System Biology

Imran Ul Haq, Maria Babar, Hayssam Mohamed Ali, Nasir Ahmad Khan, Amer Habib, Muhammad Waris, and Zobia Jabeen

Abstract Modern systems biology emerged owing to the necessity to translate the massive amounts of data produced by genome-scale investigations into physiologically relevant insights. Given the large datasets produced by several omics technologies, systems integration in plant biology is urgently needed. A smaller number of platforms have been developed that coalesce multidimensional information to create models for delineating biological interactions in plants, in contrast to the enormous recent efforts to generate large datasets. A comprehensive study covering multiple kinds of networks and strong standpoints is lacking yet for plant systems biology, and most prior reviews concentrated on the application in one specific area of plant biology. This chapter aims to offer a thorough overview of recent developments in incorporating multidimensional data into networks and review multidisciplinary subjects related to plant systems biology.

Keywords Systems biology · Networks · Plant systems biology · Systems integration

I. U. Haq (✉) · N. A. Khan · A. Habib
Department of Plant Pathology, University of Agriculture, Faisalabad, Pakistan

M. Babar
Centre of Agricultural Biochemistry and Biotechnology (CABB), University of Agriculture, Faisalabad, Pakistan

H. M. Ali
Department of Botany and Microbiology, College of Science, King Saud University, Riyadh, Saudi Arabia

M. Waris · Z. Jabeen
Balochistan Agriculture College, Quetta, Pakistan

S. Ijaz et al. (eds.), *Trends in Plant Biotechnology*,
https://doi.org/10.1007/978-981-97-0814-7_9

Introduction

Living organisms are highly organized collections of molecules with the capacity for self-replication. The astounding emergent qualities that set living things apart from the rest of the inanimate universe are produced by the physical and functional interactions between their constituent molecules. Therefore, biology ultimately revolves around deciphering a complex web of interactions and comprehending how the structure and characteristics of the web and its elements enable the living creature to develop, perceive, and respond to its environment and grow and reproduce. Systems biology studies interactions and associations between various biological components that emphasize integrating subcellular components, i.e., genomics, transcriptomics, interactome, and cellome levels (Kitano, 2002). Systems biology is a promising approach by which the whole system, including all its interacting units, is analyzed. Recently, the predominantly existing types of systems are top-down, bottom-up, and middle-out; top-down systems are composed of high-throughput technologies to acquire a holistic understanding of biological systems; bottom-up systems include the study of molecular properties and lead to model derivation and validation; and middle-out systems are based on the principle that starts in between top-down and bottom-up systems that constitute the study of the hierarchy of models (Hartwell et al., 1999; Ideker et al., 2001; Noble, 2002). Top-down and bottom-up systems have been extensively employed to describe biological systems (Sheth & Thaker, 2014). Robustness and complexity are significant characteristics of biological systems, in which robustness plays a crucial role in sustaining the complexity and proper functioning of the system, regardless of external or internal stimulating factors, such as various developmental signals, variation in environmental conditions, and genetic variation (Coruzzi et al., 2009). A systems-level perspective is emerging in biology due to the recent surge in high-throughput data. Practically, systems biology generates and combines several types of genome-wide information to create models of the system in living organisms. A deep understanding of biological systems is a primary requirement for precisely predicting the effects of these factors, as mentioned above, on the functionality of biological systems (Shannon et al., 2003; Sheth & Thaker, 2014).

Systems biology strives to connect the molecular disciplines of biology, such as genetics, biochemistry, and molecular biology, and the disciplines of biology that examine entire organisms (Qian et al., 2001; Wichert et al., 2004). It accomplishes this by quantitatively describing organisms' molecular and system-level molecular pathways. In other words, they show how molecular structure and organization play a role in the characteristics of organisms and how the system can be changed molecularly to behave in a particular way.

Nucleic acids, carbohydrates, proteins, lipids, and other tiny molecules are among the numerous chemically varied components of living systems. There is a close connection between these elements. For example, signal transduction cascades triggered by developmental or environmental stimuli can cause changes in the messenger RNA levels of hundreds of genes that encode thousands of distinct proteins. The robustness of biological systems is a result of their complexity. The ability of biological systems to sustain correct functions during various perturbations and uncertainties is a fundamental trait that unites all biological systems (Stelling et al., 2004; Kitano, 2007). For instance, robustness is a property that only occurs at the system level and cannot be comprehended at the level of an individual unit (Kitano, 2002). A system is robust until it sustains its functionality, even if this is accomplished by shifting to a new steady state. For example, yeast metabolism dramatically switches from anaerobic to aerobic during the diauxic shift. This transformation entails reprogramming metabolic pathways and gene expression alterations across the genome. The system's status is drastically changed in this example. The system is robust, nevertheless, as evidenced by its continuing generation of adenosine triphosphate. Robustness preserves certain functions while allowing for changes in the system's components and structure in reaction to disturbances (Kitano, 2002).

For biological system analysis, plants are highly suitable due to their ease of handling, the high production rate of offspring which is quite a requirement for genetic analysis, and the short generation/reproduction period (Keurentjes et al., 2008). Plant systems biology is the application of biochemical, physiological, molecular, morphological, and genetic knowledge to plant biology (Trewavas, 2006). The requirement to integrate and understand sizable datasets from high-throughput omics technologies led to the demand for the topic. Despite tremendous efforts to create vast experimental datasets, nothing has been done to establish computational platforms to evaluate and interpret the data as biological models (Yuan et al., 2008). It entails keeping an eye on gene, protein, and biochemical pathways, using omics techniques, integrating data, and developing mathematical models that describe how plants respond to specific perturbations in the form of physiologically significant perturbations.

One of the most popular methods for describing biological systems is the development and analysis of networks. Networks can be static or dynamic; the molecules that make them up can be any combination of genes, proteins, metabolites, or other substances.

Plant Gene–Metabolite Network

According to Oliver et al. (1998), the "metabolome" refers to the total collection of molecular metabolites any organism generates. Oliver Fiehn invented the term "metabolomics" in 2002 to describe the thorough examination of every metabolite in an organism under specific circumstances. Compared to genes and proteins, the collection of metabolites in an organism exhibits greater variability in terms of their physicochemical characteristics, differing greatly in size, stability, quantity, and polarity. This explains why the metabolome's dynamism is subject to temporal and geographical restrictions (Fiehn, 2002). Gene–metabolite networks describe the relationships between genes and metabolites. The distance between genes and metabolites based on their profile patterns is frequently used to illustrate the statistical analysis results. A gene shown to be "near" to a metabolite may be involved in the production or control of the latter. As plants produce a wider variety and more metabolites due to their sessile lifestyle, studying gene–metabolite networks in plants is more complicated than in mammals.

In earlier studies, gene–metabolite networks were employed to analyze the dynamic responses in *Arabidopsis* during sulfur and nitrogen shortages (Hirai et al., 2004; Nikiforova et al., 2005). For plant species such as *Catharanthus roseus* that have minimal genomic information, gene–metabolite networks have also been created (Rischer et al., 2006). In this instance, network analysis identified novel candidate genes for the biosynthesis of terpenoid indole alkaloids in *C. roseus*.

By applying *in silico* analysis of previously available data on gene expression, metabolite or gene–metabolite networks may be constructed. The premise of this research style is that the coregulation of genes in the same metabolic pathway occurs often. This meta-analysis of microarray data allows the identification of coregulated gene groupings, which may be linked to certain metabolic pathways. This method recently found 140 different transcriptional regulator pathways and genes involved in metabolite synthesis. MapMan is a program that enables statistical analysis of numerous microarray datasets to highlight the substantially altered genes in many linked metabolic pathways (Usadel et al., 2005). This method effectively pinpointed the genes and metabolic networks involved in the body's reaction to a lack of nitrogen during the day and night cycles (Blasing et al., 2005). Nuclear magnetic resonance (NMR), liquid chromatography-nuclear magnetic resonance, and mass spectrometry are a few examples of analytical techniques that can be used in metabolomics.

Over the last 4 years, many gene-to-metabolite networks have expanded plant science's horizons and provided novel insights. The gene–metabolite network, which first described the association between biological processes, led to significant advancements in the interpretation of omics data and improved system perturbation outcome prediction (Hesse & Hoefgen, 2006). Second, the systemic perspective improves gene function annotation by making it easier to identify key regulatory components in biological systems (Tohge et al., 2005). Third, establishing a link between genes and metabolites makes it possible to identify novel genes involved in synthesizing, regulating, and controlling metabolites (Hirai et al., 2004).

Although advances have been made, computational and technological constraints have made applying the gene-to-metabolite network strategy difficult. By using MS-based approaches, it is often difficult to resolve the structure of plant metabolites (Hesse & Hoefgen, 2006). A thorough understanding of the genome of the plant species under study is necessary for gene expression investigations employing microarrays or serial analysis of gene expression (Galbraith, 2006). Recent research on secondary metabolites has been performed on jasmonates, flavonoids, polyphenolics, alkaloids, and glucosinolates (De Geyter et al., 2012).

Additionally, most of the gene-to-metabolite network was created using linear correlation, which cannot accurately capture the system's dynamics. Improved MS and next-generation sequencing for transcriptome profiling can be used to overcome technological hurdles (Schauer & Fernie, 2006; Ahmadian et al., 2006). With rapid advancements in technology and algorithm development, more precise gene–metabolite networks will be able to be designed.

Gene Circuit Modeling

Gene circuit modeling includes describing the behavior of gene regulation networks using mathematical equations. These models can aid researchers in anticipating how variations in gene expression or genetic mutations may alter the behavior of the network as a whole. Gene circuit modeling uses mathematical equations to visualize the intricate connections between genes within a cell or organism. These models can aid researchers in better understanding how gene expression is controlled and how alterations within the gene network may impact the system's overall behavior. The representation of genes as nodes in a network with edges denoting regulatory interactions between genes is a fundamental component of gene circuit models. These edges might signify gene activation or repression, such as when a transcription factor binds to a gene and causes it to be active or inhibited. Gene circuit models include equations that explain how the expression of one gene is affected by the expression of other genes in the network (Brophy & Voigt, 2014).

Gene circuits can be modeled using various methods, such as deterministic and stochastic models (Gillespie simulations). In contrast to stochastic models, which consider the possibility of individual cells behaving differently due to random fluctuations in gene expression, deterministic models assume that all cells within a population of cells behave uniformly. Gene circuit modeling is a powerful tool that may be used to study various issues, from determining the mechanisms underlying certain genetic illnesses to forecasting how cells will react to various environmental cues (Weiss et al., 2003). Here are a few examples of gene circuit modeling:

A basic gene circuit known as a **toggle switch** has two stable states: on and off. It involves two genes, one of which acts as negative feedback on the other. The toggle switch may be switched between the two states by adding certain inputs, such as chemicals, to the system.

Oscillators: Gene circuits may be constructed to produce oscillations or cycles of gene expression. These can help comprehend biological rhythms or schedule gene expression in synthetic biology applications.

Logic gates: To enable complicated decision-making processes in biological systems, gene circuits can function as Boolean logic gates, such as AND or OR gates (Wang & Buck, 2012).

Gene Circuit Modeling in the Logical Domain

In the logical domain, gene circuit modeling entails simplification of the interconnections between genes using Boolean logic. The interconnections between genes are shown as edges, and each gene is shown as a node. The edges can be activated or inhibited depending on how one gene interacts. In the logical domain, gene circuit modeling entails utilizing Boolean logic to depict the interactions between genes. Boolean logic is a mathematical framework that represents and manipulates logical statements using variables and logical operations such as AND, OR, and NOT (Wang & Buck, 2012).

One must first identify the genes participating in a certain biological process to construct a gene circuit model. Take the lac operon in *E. coli* to illustrate how genes are regulated. Three genes, lacZ, lacY, and lacA, make up the lac operon and are involved in lactose degradation. The lac repressor attaches to the lac operon's operator region when lactose is absent, suppressing the transcription of the lacZ, lacY, and lacA genes. In the presence of lactose, the lac repressor attaches to it and undergoes a conformational shift that prevents it from binding to the operator region. It allows the transcription of the lacZ, lacY, and lacA genes (Wang & Buck, 2012).

Another illustration is the tumor suppressor protein p53, which controls the cell cycle and inhibits the development of malignant cells. The MDM2 protein, which functions as an E3 ubiquitin ligase and targets p53 for destruction, controls the protein levels of p53. p53 levels rise in response to DNA damage, binding to the MDM2 protein to limit its function and prevent p53 breakdown. As a result, a negative feedback loop is created, aiding in maintaining healthy p53 protein levels. This circuit may be used to explain the interactions between p53 and MDM2 in the logical domain utilizing Boolean variables and logical operations (Tu et al., 2007).

The biological clock, known as the circadian clock, controls the sleep-wake cycle in humans and other organisms. The genes that make up the clock interact in an intricate web of positive and negative feedback loops. For instance, the PER and CRY genes are activated by the CLOCK-BMAL1 complex, which in turn prevents CLOCK and BMAL1 from being transcribed. This results in a negative feedback loop that aids in controlling the expression of these genes throughout 24 h (Yan et al., 2021).

Gene Circuit Modeling in the Physical Domain

To understand the behavior of genetic circuits in living cells, scientists who work in the physical area of gene circuit modeling use mathematical equations and simulations. It can aid researchers in comprehending how various gene networks work and how modifications or mutations to these networks might impact cell behavior. Some important factors to consider when modeling gene circuits in the physical domain include the specific parameters of the system being studied, such as the transcription and translation rate, the kinetics of protein interactions, and the geometry of the cellular environment. Additionally, multiple modeling approaches, such as deterministic and stochastic models and continuous or discrete methods, can describe various gene regulation elements.

Understanding the intricate behavior of genes and their interconnections within live cells requires gene circuit modeling in the physical domain. Researchers may learn more about how genetic circuits behave and how adjustments or mutations might impact cellular behavior by utilizing mathematical simulations and models to reflect this behavior. The investigation of gene expression is a significant area of application for gene circuit modeling. The process by which the data contained in genes are translated into useful proteins is known as gene expression. Researchers can better comprehend the processes driving gene expression and the elements that regulate it by simulating the interactions between genes, mRNA, and proteins. For instance, recent research (Kung & Weber, 2022) examined how the association between two genes, ARF and MDM2, influences tumor cell proliferation.

The analysis of genetic networks is a significant application of gene circuit modeling. The intricate web of connections between genes in a cell is known as a genetic network. By modeling these networks, researchers may better understand how several genes interact and how changes to one gene might alter the entire network's behavior. For instance, recent research (Eliaš & Macnamara, 2021) examined the consequences of p53 gene mutations on the stability of genetic networks using gene circuit modeling.

Gene circuit modeling in the physical domain is a crucial tool for understanding the intricate relationships between genes and proteins in living cells and offers key insights into the molecular behavior of biological systems.

Genome-Level Cellular Metabolism

The intricate network of metabolic events that occur inside plant cells and are regulated by genetic and environmental variables is called genome-level cellular metabolism in plants. A plant's genome includes all the genetic material required for growth and development, including the genes that code for the enzymes involved in cellular metabolism. Plants undergo several activities in their cellular metabolism, including photosynthesis, respiration, and production of different organic substances, including lipids, nucleotides, and amino acids. An intricate network of enzymes, transporters, and signaling molecules controls and connects these activities. The control of metabolic pathways in response to environmental signals, including light, temperature, and food availability, is a crucial component of genome-level cellular metabolism in plants (Cultrera, 2023).

Genetic variables, such as gene expression and enzyme activity differences, impact plant genome-level cellular metabolism. The genetic components of cellular metabolism in plants have been better understood via developments in genome sequencing and other molecular methods, which have also made it possible to pinpoint important genes and metabolic pathways. Identifying important genes and pathways that control crop production, quality, and stress tolerance can be aided by a genome-level understanding of plant metabolism. Using this knowledge, new plant types that are more productive and more tolerant to environmental challenges such as pests, drought, and harsh temperatures can be created (Cultrera, 2023).

Designing and engineering plants that generate useful substances such as medications, biofuels, and industrial chemicals may be done using a genome-level understanding of plant metabolism. Metabolic engineering is a discipline that can completely alter how these goods are made. Additionally, knowing how plants react to environmental stress and how their metabolism is controlled might aid in developing ways to lessen the impact of climate change on agriculture (Kim & Buell, 2015).

Arabidopsis thaliana, a model plant for which a genome-scale metabolic network has been rebuilt, is one example of genome-level cellular metabolism in plants. It has been applied to research plant metabolism, how environmental conditions affect plant development, and how plant metabolic pathways have evolved. Rice's genome has been sequenced, and a metabolic network at the genome size has been created. This has been utilized to investigate the metabolism of rice and find fresh targets for crop enhancement. Additionally, the genome of maize has been sequenced, and a metabolic network at the genome size has been created. This has been utilized to investigate the metabolism of maize and find fresh targets for crop enhancement. The tomato genome has been sequenced, and a metabolic network at the genomic level has been recreated. This has been applied to research tomato metabolism and find fresh targets for crop enhancement. These are only a few examples, but many more plant species have genome-scale metabolic networks recreated and utilized to research plant cellular metabolism (Kim & Buell, 2015).

Principles Behind Constraint-Based Analysis (Flux Balance Analysis)

A computer method called flux-based analysis (FBA) makes it possible to model and examine plant metabolic networks. The fundamental idea underlying FBA is to identify a steady-state solution for a plant's metabolic network by maximizing a certain objective function, usually the growth rate. To do this, FBA adds constraints to the metabolic network, such as mass balance and thermodynamic constraints, to ensure that the network runs within the parameters of recognized biological limitations. According to the available nutrients and environmental parameters, FBA can forecast the metabolic fluxes of various pathways by solving this network with these restrictions and optimizing the objective function. It can also anticipate how the plant will react under various scenarios (Orth et al., 2010).

The foundation of FBA is the mass balance principle, which stipulates that the overall concentration of each chemical species inside a system must stay constant throughout time. When discussing plant metabolism, FBA entails modeling the metabolic network of the plant cell as a collection of connected biochemical reactions, followed by the solution of a mathematical optimization problem to predict the metabolic fluxes (i.e., the rates at which metabolites are produced and consumed) that are consistent with the recorded growth phenotype of the plant. In plant biology research, FBA has been used extensively to understand plant metabolism, pinpoint metabolic engineering targets for enhancing crop output and quality, and foresee the impacts of genetic alterations on plant metabolism (Orth et al., 2010).

Flux balance analysis (FBA) models a plant cell's metabolic network as a collection of interrelated biochemical events. Building a metabolic network model—a mathematical depiction of all the biochemical processes within the plant cell—is the first stage in FBA. Once the metabolic network model has been built, a set of mass balance equations that characterize the steady-state behavior of the plant's metabolism are solved by FBA using linear programming techniques. The stoichiometry of the biological processes is related to each metabolite's production and consumption rates in the network using these mass balance equations. FBA aims to identify a set of metabolic flux distributions consistent with the experimental data on the growth and metabolism of plants that are currently available. These flux distributions indicate the rates at which the plant produces and consumes metabolites, and they may be used to forecast how genetic or environmental changes would affect the plant's growth and metabolism. The yields of significant cofactors such as ATP, NADH, or NADPH can be predicted using FBA (Orth et al., 2010; Varma & Palsson, 1993).

FBA can be used to foresee how genetic alterations affect plant metabolism. FBA has been used, for instance, to forecast the results of overexpressing the Rubisco enzyme in tobacco plants. Experimental verification of the FBA analysis predicts that Rubisco overexpression would result in higher plant growth and biomass output. Identifying prospective targets for metabolic engineering is another use. FBA has been used, for instance, to pinpoint the genes and enzymes necessary for plant development and metabolism and foretell the consequences of perturbing these targets on plant metabolism. FBA also examines how plants react to environmental challenges such as drought and excessive salt. FBA has been used, for instance, to forecast the metabolic changes that occur in plants under drought stress and to pinpoint possible targets for enhancing crop resistance to drought. The interactions between plants and microorganisms can also be studied using FBA. In general, FBA is an effective tool for comprehending how plants use their energy, and it has significant implications for engineering plants to be more productive and resilient to stress (Orth et al., 2010).

Rab ESCORT Proteins

These proteins are present in plants and control the movement of cargo-carrying vesicles across various cell regions. They play a part in regulating the fusion of vesicles with the target membrane and ensuring that they are transported to the proper location within the cell. These proteins are called after the tiny G proteins known as Rab GTPases, which control vesicle transit. The ESCORT proteins are one of several varieties of Rab proteins found in plants. Plants have RabE1c, RabE1d, RabE2b, and RabE3f, which are Rab ESCORT proteins. Numerous features of intracellular trafficking, including vesicle formation, cargo choice, and membrane fusion, have been linked to these proteins (Speth et al., 2009). Transport between the ER and Golgi complex is mediated by Rab1 and Rab2, two GTPases that coordinate cytoskeletal connections and selectivity in fusing with membranes at the cis-Golgi complex (Moyer et al., 2001). Various novel paths begin at the trans-Golgi face, some of which travel to the lysosome and others of which sort endosomes. Retrograde transport via the Golgi complex, which Rab6 likewise regulates, makes it easier to recover and retrieve proteins (Barr & Short, 2003).

Research on plant Rab ESCORT proteins in systems biology can provide information on the molecular processes that regulate cell vesicle transit. Researchers can better understand how cells work and the processes involved in maintaining cellular homeostasis by knowing the involvement of these proteins in the transfer of cargo between various sections of the cell. This information can also be used to create fresh medications and cures for various illnesses, including those caused by flaws in vesicle trafficking and transport. Rab ESCORT proteins are a crucial component of the plant's cellular infrastructure and are essential for intracellular transport. They aid in the movement of membrane-bound vesicles between various cell compartments. They are crucial in the context of systems biology for comprehending how various plant cell components communicate and interact with one another (Brighouse et al., 2010).

The study of plant growth and development is one possible real-time use of Rab ESCORT proteins in systems biology. One can better understand the underlying mechanisms regulating plant growth and development by comprehending how these proteins transport significant chemicals throughout various cell compartments. Furthermore, Rab ESCORT proteins can be utilized to create novel techniques for enhancing plant development and boosting crop yields. Among other things, one might be able to improve stress tolerance, boost nutrient absorption, and boost photosynthetic efficiency by modifying the activity of these proteins. In general, research on Rab ESCORT proteins in plants has the potential to provide significant molecular level insights into the behavior of plant cells and may open the door to new advancements in plant biology and agriculture (Brighouse et al., 2010).

Protein Prenylation

Prenyl groups (such as farnesyl or geranylgeranyl) are added to certain amino acid residues of proteins as part of the posttranslational modification process known as plant protein prenylation. This posttranslational alteration is essential for controlling several biological functions, including membrane targeting, protein–protein interactions, and signal transduction. Additionally, it is essential for controlling the positioning and function of many different plant proteins, including signaling proteins involved in stress response and developmental processes. Prenyltransferase enzymes catalyze the addition of the prenyl group to the protein, and inhibitors of these enzymes have been investigated as possible herbicides and anticancer medications. In particular, prenylated proteins play a crucial role in signal transduction pathways, where they serve as intermediaries between receptors and downstream signaling elements. Prenylation can also alter protein–protein interactions, membrane association, and subcellular localization, which can all affect the protein's overall function and the biological processes in which it is involved (Hála & Žárský, 2019).

By studying how plant proteins are prenylated, systems biologists can learn more about the signaling pathways involved in critical activities, including growth, development, and stress response. One can improve crop yields, increase plant resistance to environmental stress, and create new plant-based products with potential health benefits by better understanding how plant protein prenylation works. This increased understanding of how plants function at the molecular level can also help develop more effective strategies.

The protein Ras, commonly mutated in cancer, is one example of protein prenylation. Ras is prenylated in healthy cells and assists in transmitting signals from growth factors to various downstream signaling pathways. However, Ras can develop mutations that make it permanently active, resulting in uncontrolled cell growth and division. Another example is the prenylated protein lamin, which secures the nuclear envelope to the cytoskeleton. Laminopathies are diseases caused by mutations in lamin and are characterized by problems in nuclear structure and function (Konstantinopoulos et al., 2007).

Prenylation is a step in the production of numerous significant plant substances, including carotenoids, chlorophylls, and gibberellins. For instance, the production of carotenoids, significant pigments involved in photosynthesis and plant growth, depends on the prenylation of the protein phytoene synthase. Prenylation has also been shown to have a role in plant defense mechanisms. Prenylation directs the immune system to target and degrade the plant immunity-regulating protein RIN4 for destruction. The plant's immune system is triggered by this degradation to combat infections (Zhou et al., 2022). Researchers can better understand the complex processes that take place inside plant cells by gaining an understanding of the role of plant protein prenylation in systems biology. Researchers may also be able to create new tools and strategies for manipulating these processes to increase crop yields, create new medicines, and address other significant issues in agriculture and human health (Zhou et al., 2022).

Protein Geranylgeranyl Transferases in Plants

Enzymes known as protein geranylgeranyl transferases add a geranylgeranyl moiety to proteins. Protein geranylgeranyl transferases are crucial for protein prenylation, a posttranslational modification that changes target proteins' subcellular distribution and function. Numerous physiological functions of plants, such as photosynthesis, hormone signaling, and pathogen defense, depend on this mechanism.

In plants, protein geranylgeranyl transferases have several crucial activities. Modifying small GTPases, which are essential for controlling various cellular activities, is one of the most crucial. Geranylgeranyl transferases aid in targeting these proteins to certain membranes where they may more efficiently carry out their roles by adding a geranylgeranyl group. Protein geranylgeranyl transferases modify GTPases and manufacture and signal plant hormones, i.e., abscisic acid and gibberellins. They are crucial for controlling hormone levels and responses, which are required for the growth and development of plants. Additionally, GGTs play roles in plant pathogen defense mechanisms. They make it easier for proteins involved in plant defense responses to undergo posttranslational modification, which enables them to localize to the proper cell membrane and carry out their intended roles (Hála & Žárský, 2019).

Their function in controlling signaling pathways in *Arabidopsis thaliana* provides a real-time illustration of the function of the plant protein geranylgeranyl transferase in systems biology. Studies have revealed the significance of geranylgeranyl alterations for the correct localization and operation of Rho family GTPases, which are involved in several signaling pathways. These signaling networks rely primarily on Rho-family GTPases, the majority of which behave as molecular switches that transition between a GTP-bound (active) and GDP-bound (inactive) state (Jaffe & Hall, 2005). According to Poraty-Gavra et al. (2013), the geranylgeranyl modification of the ROP6 protein is necessary for correctly localizing the protein to the plasma membrane and its role in controlling stomatal opening in response to light and CO2. Additionally, the ROP2 protein's role in controlling pollen tube formation requires geranylgeranylation (Feiguelman et al., 2018). Protein geranylgeranyl transferases are essential in plants and participate in various cellular activities, including growth, development, and defense against biotic stresses.

Mitophagy Pathways in Plants

Mitophagy is a mechanism that helps maintain cellular homeostasis by directing the cell's lysosomes to specifically target and destroy damaged or defective mitochondria. Although animal cells have been the focus of most recent studies on mitophagy, new studies have revealed that plants contain mitophagy pathways comparable to those found in animals. Plants use mitophagy to respond to environmental challenges and maintain mitochondrial function. The most extensively researched mitophagy route in mammals, the PINK1/Parkin pathway, has been found in plants and demonstrated to be engaged in mitophagy under specific circumstances. Furthermore, recent research has linked several other proteins and signaling pathways, including ATG32, ATG11, and ATG5, essential elements of the autophagy process, to mitophagy in plants. Plant mitophagy, the process of removing damaged mitochondria from plant cells, involves the proteins ATG32, ATG11, and ATG5. An outer membrane protein of the mitochondria called ATG32 interacts with ATG11 and ATG8 to bring the autophagy machinery to the mitochondria, which is necessary for the selective autophagic destruction of mitochondria. ATG5 is a crucial part of the autophagy machinery necessary for autophagosome membrane formation, while ATG11 is an adapter protein that connects ATG32 to the rest of the autophagy machinery. Together, these proteins operate to guarantee that damaged mitochondria are specifically targeted for destruction and that plant cells are operating as intended (Ashrafi & Schwarz, 2013).

In eukaryotes, the maintenance of mitochondrial quality control is carried out through a conserved system called the PINK1/Parkin pathway. Recent research has revealed that this route also exists in plants, although it was first discovered in mammalian cells. This route is essential for controlling mitophagy, mitochondrial dynamics, and response of plants to varied environmental stimuli. Insights into the basic biology of plant mitochondria may be gained through understanding the molecular processes governing the PINK1/Parkin pathway in plants, which may also help create innovative methods for enhancing plant stress resistance and productivity. The PINK1/Parkin pathway in plants uses molecular processes comparable to those of mammals. When mitochondria are harmed, PINK1 builds up on the outer membrane, attracting Parkin, an E3 ubiquitin ligase, to the mitochondria. The subsequent ubiquitination of several mitochondrial proteins by Parkin results in the activation of autophagy receptors. Through mitophagy, damaged mitochondria are eventually consumed and destroyed (Ashrafi & Schwarz, 2013).

Other proteins, such as the mitochondrial kinase *Arabidopsis* mitochondrial kinase (AtMKS1) and the mitochondrial ubiquitin ligase mitogen-activated protein kinase kinase 3 (AtMAPKK3), modulate the molecular processes of the PINK1/Parkin pathway in plants. In response to diverse conditions, including oxidative stress, food restriction, and pathogen infection, these proteins regulate the activity of PINK1 and Parkin (Jagodzik et al., 2018). Furthermore, it has been demonstrated that the PINK1/Parkin pathway controls mitochondrial dynamics and biogenesis in plants. For instance, the route is required to maintain the appropriate balance between mitochondrial fission and fusion, which is essential for maintaining the optimal size, shape, and distribution of mitochondria inside the cell. This is significant because mitochondria are the cell's powerhouses, producing ATP as an energy source. Cellular signaling, apoptosis, and stress reaction are further areas where fusion and fission are involved (Ashrafi & Schwarz, 2013).

In systems biology, plant mitophagy is a crucial activity because it maintains the healthy operation of plant cells. As the cell's powerhouse, mitochondria employ oxidative phosphorylation to produce energy. They can, however, also generate damaging reactive oxygen species (ROS) that can injure the cell. Mitophagy aids in eliminating damaged mitochondria and reducing ROS buildup, both of which can result in oxidative stress and cell death. Additionally, nutrient recycling and carbon metabolism depend on plant mitophagy. The cell can use mitochondrial breakdown products to promote other metabolic pathways, such as amino acid synthesis. This guarantees that nutrients are utilized effectively and are not lost. Plant mitophagy is essential to systems biology studies because it is vital in preserving cellular homeostasis and enhancing plant growth and development (Broda et al., 2018).

Autophagy Pathways in Plants

Cellular homeostasis and nutrition recycling depend on the highly preserved mechanism of autophagy. Despite substantial research on autophagy in mammals and yeast, the function of autophagy in plants is still unknown. In response to environmental challenges such as nutrient deprivation, oxidative stress, and pathogen assault, autophagy is crucial in plants for remobilizing nutrients during senescence. Recent research has illuminated the molecular processes underpinning the development and function of autophagosomes in plants. Numerous autophagy-related genes have been discovered in plants. The ATG8 pathway, which is involved in the production of autophagosomes, and the ATG18 pathway, which is involved in the docking and fusing of autophagosomes with lysosomes, are two of the important autophagy processes in plants. The ESCRT-mediated autophagy pathway, which destroys endosomal vesicles, is one of the distinctive autophagy mechanisms discovered in plants (Kalinowska & Isono, 2018; Wang et al., 2021).

The ATG8 pathway in plants is a convoluted network of molecular events involving a variety of proteins and cellular structures. ATG8 proteins are recruited to the site of autophagy by several protein complexes, notably the ATG12-ATG5-ATG16 complex, which is at the center of the ATG8 pathway. Once enlisted, ATG8 proteins can attach to autophagosomal membranes because they have been conjugated to phosphatidylethanolamine (PE) lipids. ATG8 proteins extend and close the membrane as the autophagosome develops around the targeted cellular component, confining it inside the autophagosome. A lysosome or vacuole joins the autophagosome toward the end, where the contents are broken down, and the resultant molecules are recycled for future use. Plants' ability to react to stress and maintain cellular homeostasis depends on the ATG8 pathway (Wang et al., 2021).

The ATG18 pathway plays a significant role in plant autophagy. In this process, the ATG18 protein interacts with other proteins to create a complex that produces autophagosomes, membrane-bound structures that swallow cellular material to be broken down. Certain cellular cues that trigger the autophagy process start the ATG18 pathway. Once in the location where the autophagosome will develop, the ATG18 protein interacts with other proteins, such as ATG2, ATG9, and ATG11, to create a complex. Then, this complex aids in attracting other proteins, such as ATG8, that assist in developing autophagosomes. Overall, the ATG18 pathway is essential for controlling plant autophagy, and comprehension of the processes behind this activity might shed light on how plants react to environmental stimuli and other parameters (Wang et al., 2021).

A collection of protein complexes known as ESCRT (endosomal sorting complexes needed for transport) are involved in several cellular activities, including autophagy. By recycling damaged organelles and proteins, autophagy is a crucial mechanism that enables cells to maintain their integrity and functionality. The cellular waste-engulfing and degradation structures known as autophagosomes are regulated by the ESCRT-mediated autophagy process. Endosomes are cellular vesicles that move materials from the plasma membrane to the lysosome for destruction. The ESCRT-0 subcomplex recognizes and binds to ubiquitinated cargo proteins on the surface of endosomes. After that, the ESCRT-I and ESCRT-II subcomplexes combine to create a larger complex that recognizes and arranges the ubiquitinated cargo proteins for degradation. The ESCRT-III subcomplex, which is responsible for creating autophagosomes, is subsequently recruited by this complex. According to Lefebvre et al. (2018) and Kalinowska and Isono (2018), the ESCRT-III complex creates the membrane curvature necessary for the formation of the autophagosome, which subsequently engulfs the cargo proteins that have been ubiquitinated.

As it is directly related to plant growth, development, and response to various stresses, plant autophagy plays a significant role in systems biology. This mechanism is crucial for both the growth of specialized cells such as pollen and seeds and the survival of plants in nutrient-limited environments. Additionally, autophagy controls cell death and eliminates diseased cells in the plant's response to biotic stresses such as pathogens (Kalinowska & Isono, 2018).

Plant Interactomics

Interactomics is the study of genome-wide interactions between different proteins. The dynamic and coordinated interactions of a cell's macromolecular components with various biochemical properties allow a system to work. The most frequently reported among these are protein–protein interactions in plants. The histone proteins attached to DNA to create the chromatin structure and play a role in the epigenetic regulation of many physiological processes are examples of DNA–protein interactions. Therefore, interactomics, the thorough examination of interactions between various macromolecules in an organism, primarily protein–protein interactions, is essential to comprehend biological systems (Morsy et al., 2008). Studies of protein–protein interactions (PPIs) help clarify the biophysical underpinnings of all biological functions and create a framework for functionalizing specific proteins. In recent years, interactomics has mainly focused on profiling protein–protein interactions (Charbonnier et al., 2008).

Physical and genetic interactomes have recently been identified as distinct forms of the interactome. A network of genes known as a genetic interactome is defined by genetic interactions that reveal how certain genes participate in physiological processes (Boone et al., 2007). Systematic genetic analysis (SGA) for yeast illustrates this strategy (Tong et al., 2001). Genes with significant biological functions frequently form interaction hubs in the yeast genetic interactome, serving as the focal point of numerous interactions with various other genes (Tong et al., 2004). *Caenorhabditis elegans*, a nematode, has similarly benefited from applying the SGA-based genetic interactome (Lehner et al., 2006). Despite its potential, polyploidy and a relatively long plant life cycle that reduces precision and lengthens the time required for genetic manipulation hamper the frequent use of SGA in plants. Due to redundant and noisy data, these factors make it difficult for phenotypes to emerge. The local genetic interaction network of proteins can help comprehend how genes interact and function, even though creating a map of the genome's genetic interactome may be challenging. Considering the comparatively simple transformation of *Arabidopsis*, RNA interference-based methods of downregulating target genes can also aid in creating genetic interactome maps in addition to the SGA interactome (To, 2004; Riefler et al., 2006). Physical interactions between proteins are generally simpler to characterize in plants than genetic connections. The Y2H system and anti-tag immunoprecipitation coupled with tandem MS characterization are the two primary methods (Fields & Song, 1989). In biological systems with a lower level of complexity than plants, genome-wide studies of the physical interactome are more advanced (Uetz et al., 2000). Yeast two-hybrid (Y2H) and fluorescence resonance energy transfer are a few of the in vivo technologies used for plant interactome mapping studies (Morsy et al., 2008). Affinity purification mass spectrometry, surface plasmon resonance, and protein microarrays are a few of the several in vitro analytics. *In silico* prediction methods frequently incorporate various data types from different analyses, such as coexpression, colocalization, coevolution, and interologs (Sharan et al., 2007).

For yeast and *C. elegans*, comprehensive physical interactomes based on the Y2H method have been reported (Li et al., 2004). The yeast interactome has also been described at the genomic level using MS techniques (Ho et al., 2002). Both methods have not yet been used in analogous interactome studies in plants, but preliminary efforts are being made in this direction, such as the discovery of recombinant clones containing full-length cDNAs for *Arabidopsis* (Bevan & Walsh, 2005). In silico methods on a broad scale or network investigations of protein–protein interactions at the gene and family levels have been used in published physical interactome work in plants. The maps of the in silico technique are known as "interologs" (interacting orthologs). The anticipated interacting protein pairs known as "interologs" are recognized by their sequence similarity to proteins from yeast and humans (Matthews et al., 2001). Various protein–protein interaction web resources in plants include rice kinase-protein interaction maps, auxin signaling networks, InAct, TAIR protein interaction data, and STRING (Yu et al., 2004; Geisler-Lee et al., 2007; Aranda et al., 2010; Szklarczyk et al., 2011; Mukhtar et al., 2011; Vernoux et al., 2011; Cowley et al., 2012).

Protein–protein interaction networks are important in various biological events, including signal transduction, response to abiotic stressors, cell cycle, protein ubiquitination, and RNA processing (Van Leene et al., 2007; Hunt et al., 2008). To better understand plants at the systems level, network dynamics and comprehensiveness of protein–protein interaction networks are challenges demanding significant attention.

Plant Gene Regulatory Networks

A gene regulatory network illustrates how genes associate with one another to perform a function during the biological process (Ueda, 2006; Alvarez-Buylla et al., 2007). These networks are frequently created using computer models based on data from mutant studies and previously published gene expression studies. Gene knockouts and other biotic or abiotic treatments that perturb the system are used to verify and enhance the model (Ueda, 2006; Alvarez-Buylla et al., 2007). The ability of the gene regulation network to represent the dynamics of a biological system is a significant advantage. Gene regulatory network models frequently exhibit dynamic behavior in contrast to the static behavior of other networks (Li et al., 2004). Plants' diverse physiological and developmental processes have been studied using gene regulatory networks (Ueda, 2006; Locke et al., 2006). For instance, a discrete dynamic model has been created that incorporates genetic information and previously reported gene expression data during the growth of an *Arabidopsis* flower. The findings suggested that the gene network dynamics, rather than a specific signaling channel, influence cell destiny in flower development (Espinosa-soto et al., 2004).

In another instance, gene regulation network analysis similarly modeled the crucial elements regulating guard cell size (Li et al., 2004). Earlier research data on network organization were used to build the basic network structure. When the network system is affected by some stimuli, a dynamic model may predict the chance of guard cell closure. An accurate prediction of the likelihood that guard cells would shut when the network system was disrupted by external stimulation supported the validity of the resultant model, which provided a detailed description of the genes that interact to regulate stomata (Li et al., 2004). Consequently, a gene regulatory network is a practical method for dynamically describing gene functioning in biological processes. The advantages of networks emerge from their capacity to understand the inherent biological complexity by identifying key components and links for system regulation. Even though a network may have been constructed flawlessly, it is crucial to understand that it frequently contains data predictions for modeling and cannot accurately represent a biological system. Despite their shortcomings, networks have opened up fresh avenues for analyzing omics data and have been used to investigate various plant biological issues.

Plant Transcriptional Regulatory Networks

For building transcriptional regulatory networks, it is crucial to understand how transcription factors and cis-elements interact (Palaniswamy et al., 2006). This method has recently been applied to research the transcription factors forming glucosinolates. By creating a network of gene-to-gene interactions, over 1000 microarray trials were first meta-analyzed to investigate the genes accountable for glucosinolate biosynthesis. It is essential to comprehend how transcription factors and cis-elements interact to construct transcriptional regulatory networks (Hirai et al., 2004).

Gene-Specific Genome-Wide ChIP

Recognizing certain cis-regulatory elements (CREs) through transcription factors (TFs) and correct chromatin assembly are two examples of how protein:DNA interactions are essential for life. It is possible to find protein:DNA interactions using experimental, computational, or a combination of these methods (Grotewold & Springer, 2009). The creation of protein:DNA complexes can be studied experimentally, for instance, using electrophoretic mobility shift assays (EMSAs). Due to the in vitro protein–DNA interactions involved, implementing these approaches is contingent (Nowack et al., 2005). For ChIP, various cross-linking agents with various spacer lengths are available. However, imino group reactions are probably more prevalent in single-stranded DNA areas. Lys, Arg, Trp, and His are some amino acids that formaldehyde targets in proteins. The reaction results in the formation of a methylene bridge that joins the two amino groups. Being reversible in aqueous solutions makes this reaction special, and incubating it at 65 °C can considerably speed up the process. The reversibility of the formaldehyde-mediated reaction has made it the preferred reagent for cross-linking. However, it must be emphasized that if formaldehyde does not produce the required effects, in some cases, it may be worthwhile to try alternate cross-linking agents (Nowack et al., 2005).

Whole-Genome Microarrays

The increased accessibility of genome sequences has enabled a wide range of experimental and computational research at the whole-genome level. The development of whole-genome tiling microarrays has made it possible to query an entire genome through arrays, with recent advancements in sequence data accessibility. These DNA microarrays provide a potent platform for various experimental methods to explore the breadth of the structural and functional information inside a whole genome in a single experiment. The most popular high-density microarray platform types are two general categories. Short (up to 100 mer) oligonucleotide probes are the primary component of the first class of microarrays. These arrays can be further categorized according to the types of probes they include. Some so-called semiwhole-genome expression arrays only show annotated portions of a genome, such as exons. However, true whole-genome tiling arrays (referred to as WGAs from here on) are made to impartially examine a full genome (Kakutani, 2002; Kankel et al., 2003; Penterman et al., 2007). These microarrays cover the entire genome from beginning to end by tiling or spacing nonoverlapping or partially overlapping probes at regular intervals. More than 6,000,000 features per array of the WGAs are already being produced, with each feature including millions of copies of the particular probe sequence. For example, the whole nonrepetitive region in the genome of *A. thaliana* is covered by the Affymetrix1 GeneChip1 *Arabidopsis* tiling array. However, spotted arrays typically have a substantially lower feature density than in situ-produced oligonucleotide arrays, with 10,000–40,000 spots per chip. WGAs have become the go-to platform for whole-genome analysis because of their great repeatability and enhanced feature density. Furthermore, WGAs are the best option for detecting the widest variety of genomic features (Gregory & Belostotsky, 2009).

Although gene prediction methods using computers have improved substantially over the past 10 years, they cannot precisely identify every transcription unit in an organism or establish the structure of a gene. Furthermore, the conventional method for finding coding areas overlooks transcripts with very low abundance. Therefore, WGAs encompassing the whole sequence of the relevant genome constitute a desirable alternative that mainly avoids such problems. Surprisingly, the first datasets using WGAs to examine the transcriptional process in the multiple tissues of *Arabidopsis* found significant unique expression sites that computational gene identification methods and cDNA collections had overlooked (Kakutani, 2002; Zhang et al., 2006). This method provides an objective count of novel genic units entirely dependent on the probe intensity values for the regions outside identified gene structures.

Contrary to previously annotated transcripts (Kakutani, 2002), an interesting number of these newly discovered transcripts express themselves from the antisense strand, and several have intriguing regulatory potential. For instance, FLOWERING LOCUS C, a crucial repressor that controls flowering time, contains an antisense transcript that overlaps the 3′ end of its mRNA (FLC). This antisense transcript might be a biochemical building block for small interfering RNAs (siRNAs) that undergo heterochromatinization and then do gene silencing.

RNA–Protein Interactions in Plants

Regulated gene expression is necessary for homeostasis, environmental adaptability, and organismal development. The output of each gene is determined by superimposing the transcription rate with the rates of a variety of posttranscriptional events; in some situations, posttranscriptional processes take the limelight. Consequently, RNA-binding proteins that affect RNA subset translation, stability, or processing are probably responsible for the extensive coordination of gene expression (Hieronymus & Silver, 2004; Keene, 2007). In addition, RNA-binding proteins are at the center of the mechanisms that localize particular mRNAs within cells, which affects how plants, animals, and fungi localize protein production to particular subcellular domains (Okita & Choi, 2002; Jambhekar & Derisi, 2007).

The importance of posttranscriptional processes in regulating the levels of gene expression in plant mitochondria and chloroplasts has long been recognized (Knoop & Brennicke, 2002; Bollenbach et al., 2004; Bonen, 2004; Barkan, 2004; Zerges, 2004). The discovery of genes encoding nuclear/cytosolic RNA-binding proteins affecting various processes, including flowering and hormone responses, has recently brought attention to the significance of posttranscriptional events in determining various plant traits (Fedoroff, 2002; Carrington & Ambros, 2003; Kuhn & Schroeder, 2003; Belostotsky & Rose, 2005). Posttranscriptional regulation systems in plants have recently been the subject of genome-wide experiments, but the data thus far point to a significant impact. For instance, circadian control over mRNA subset stabilities exists (Lidder et al., 2005), and stress has been shown to alter the translation of large plant mRNA sets (Kawaguchi et al., 2004; Branco-Price et al., 2005). There is proof that biotic and abiotic pressures affect the alternative splicing of plant pre-mRNAs (Reddy, 2007).

Biosynthesis of mRNA, localization, translation, turnover, and production of noncoding RNAs are all regulated by association among RNAs and RNA-binding proteins in ribonucleoprotein particles (Dreyfuss et al., 2002). However, the most anticipated RNA-binding protein–RNA ligands and functions are still unknown, even in the most well-researched fungal and mammalian systems (Hieronymus & Silver, 2004). The difficulty is considerably greater in plants because their anticipated RNA-binding protein repertoire is far larger than that of metazoa (Wang & Brendel, 2004; Barkan, 2004; Belostotsky & Rose, 2005; Barkan et al., 2007; Walker et al., 2007).

Furthermore, RNA metabolism is substantially more advanced in plant mitochondria than in metazoa, and it resembles chloroplast metabolism in many ways (Bonen, 2004). The CRM and PPR families, two RNA-binding protein families uniquely grown in the plant lineage, seem connected to the significant functions of posttranscriptional aspects of gene expression (Barkan et al., 2007). Several techniques have been used for affinity purification of RNA-binding proteins and identification of various linked RNAs, but approaches for studying native ribonucleoproteins are still under process.

Chromatin Charting and Epigenetic Effect

Modeling epigenetic regulatory mechanisms requires a thorough understanding of the relationship between chromatin structure and gene expression. It has been demonstrated that eukaryotic interphase chromatin exists in distinctive subnuclear compartments known as chromatin territories (CTs), with only a small level of intermixing among nearby chromosomes (Kosak & Groudine, 2004; Cremer et al., 2006). The structure of euchromatin in the PR can be modified using epigenetic mechanisms, including DNA methylation and histone deacetylation. Thus, this regional model of chromatin arrangement offers a structural justification for regulating gene expression by epigenetic means (Luo & Lam, 2009).

Without affecting the gene sequence in question, epigenetic mechanisms cause heritable changes in gene expression. Paramutation is a well-known instance of this in plants, and it was initially identified in maize more than 50 years ago (Brink, 1956). Brink discovered that even after an allele (R-r) segregates from R-stippled at the R locus, which controls seed color, it is still possible for it to be changed by another (R-stippled) allele in a heritable (but reversible) manner. Since then, both animal and plant systems, including *Arabidopsis*, have been shown to undergo paramutation (Chandler & Stam, 2004). Various epigenetic control mechanisms in eukaryotes have been reported, such as gene silencing at the centromeres of many organisms and X-inactivation (Chandler & Stam, 2004; Hsieh & Fischer, 2005). The common mediator of the numerous epigenetic silencing mechanisms is the development of highly condensed heterochromatin and its distribution throughout the chromosome. The regulation of meristem size and identity, vernalization, and genomic imprinting during gametogenesis are examples of how epigenetic mechanisms can significantly impact plant development (Hsieh & Fischer, 2005; Reyes, 2006). DNA methylation at cytosine residues and particular forms of covalent modification, such as methylation, are two primary regulatory mechanisms frequently implicated in epigenetic regulation (Bender, 2004; Chan et al., 2005). Gene silencing is typically tied with histone hypermethylation of cytosines and decreased acetylation, while specific methylation states of histone H3's amino terminus are linked to either active (H3K4m) or inhibited (H3K9m) gene expression (Fuchs et al., 2006). Recently, it has been demonstrated that elements in the RNA interference (RNAi) and small interfering RNA (siRNA) pathways are necessary for DNA methylation (Lippman & Martienssen, 2004; Matzke & Birchler, 2005; Pontes et al., 2006). In each of these situations, reactivation of the transgenes is connected with a change in the DNA methylation status at the transgene locus. However, in two of the mutants, the transgene continued to function even after the wild-type allele was introduced, suggesting that the chromatin state at the locus for the transgene was altered in a heritable way. These findings and investigations into the impact of RNAi mutants on transposon and transgene silencing (Elmayan et al., 2005; Cremer et al., 2006) imply the existence of various distinctively characterized epigenetic silencing pathways. Additionally, these genetic studies demonstrate that transgenes can serve as helpful markers for observing plant epigenetic processes in maize and *Arabidopsis* (Chandler & Stam, 2004).

Discrete Dynamic Modeling

The methods for performing extensive analyses of the relationships between the elements of these datasets across different biological levels are developing quickly. For instance, techniques such as yeast two-hybrid analysis (Legrain & Selig, 2000; Walhout & Vidal, 2001; Fields, 2005) and its derivatives (Obrdlik et al., 2004), as well as mass spectrometric identification of proteins after coimmunoprecipitation (Figeys et al., 2001; Berggard et al., 2007), have made it feasible for researchers to build networks that depict the protein interactomes of model species (Walhout & Vidal, 2001; Rain et al., 2001). The "ChIP-chip method" (Buck & Lieb, 2004; Haring et al., 2007) and protein microarrays (Hudson & Snyder, 2006) have both made it possible to identify the global targets of important transcription factors, allowing links to be made between the proteome and the transcriptome (de Folter et al., 2007; Lee et al., 2007). One of the most challenging types of networks is identifying and modeling fast cell signaling pathways. These pathways frequently involve the metabolome and proteome, which comprise at least two biological levels. Furthermore, these networks frequently contain posttranslational processes that modify the activity but not always the quantity of important signaling proteins. The ability to assess the effects of such alterations on protein behavior is frequently lacking despite advances in global approaches for identifying posttranslational modifications such as protein phosphorylation (Peck, 2006).

Here, a technique is briefed that, despite the restrictions above, allows cellular signaling networks to be put together and modeled in a predictive manner. These networks include an input (the original signal), an output (the final result of the signaling cascade), and a variable, frequently a high number of internal nodes. The method overcomes the lack of comprehensive kinetic and temporal information on signal propagation by taking into account all biologically significant initial states of all internal nodes, with each node only being allowed to exist in one of the two potential states: "active" (ON) or "not active" (OFF). Random updates are made to the internal nodes. After updating each internal node in this manner, the result is evaluated. The model is called dynamic since it tracks the nodes' state changes over time. The notion of two distinct activity levels rather than a continuum of levels makes it discrete. "Asynchronous update" describes the nonsynchronous, one-by-one updating of the internal nodes' status (Assmann & Albert, 2009).

Plant Metabolic Pathways and Their Coexpression Analysis

Through coexpression analysis, researchers can use archived expression microarray data to discover previously hidden functional links among various genes. The observed coexpression of a group of genes in various experimental settings suggests that they might have comparable cell functions. Using the data from many hundred expression microarray tests conducted on *Arabidopsis thaliana* samples to study plant metabolic networks is now possible. DNA microarrays have sparked a molecular biology revolution. The enormous amount of data that even a single well-designed microarray experiment may generate has encouraged molecular biology labs to acquire data management and statistical analysis expertise that they can redeploy in many situations (Loraine, 2009).

Approximately 500,000 25 bp oligonucleotide probes, organized into probe sets with 11 perfect matches and 11 mismatches or control probes in each set, make up the Affymetrix ATH1 array. Each probe in a set is chosen from a 200–300 base region adjacent to a target transcript's three prime ends; depending on the transcript, the probes may overlap. The 22,814 probe sets of the ATH1 array include control probe sets that help determine the quality of the samples. Technical challenges in the design of probe sets are frequently brought on by quality problems with EST data, such as the inability to identify the transcribed strand consistently (Loraine, 2009).

The Gene Expression Omnibus had approximately 3000 ATH1 sample records at the beginning of 2008, each matching one array hybridization (Wheeler et al., 2006). It is alluring to try meta-analysis techniques that compare samples from several trials because such a large amount of data is readily available in one place. However, this requires some caution. Users should use caution when comparing expressions between samples from various origins, for instance, as any detected variances may result from sample origin and not necessarily an interesting biological variation (Loraine, 2009).

Gene Regulation and Metabolic Reaction Integration

Gene regulation and metabolic processes are vital and closely related aspects of life. On the one hand, metabolism supports the chemical processes that allow living things to produce energy, synthesize basic substances, and eliminate harmful waste. On the other hand, through modifying RNA and protein synthesis, gene regulation performs a controlling role. The relative importance of these two functions to all living things has made the order in which they occur a contested topic in evolutionary biology (Maynard Smith & Szathmary, 1999). The two processes are now interconnected, regardless of where they came from. Cells frequently modify their regulatory plans and vary their metabolic states in response to environmental changes (Yeang, 2011).

The connection between gene regulatory and metabolic systems is far more complex in plants and other multicellular organisms than it is in unicellular systems. Fast sequencing technology and comparative research will allow the exploration of species' genetic and molecular relationships. Metabolomics, the study of how the metabolic networks in a particular area interact, may help us understand how carbon, nitrogen, hazardous substances, and energy are transported throughout the ecosystem (Yeang, 2011).

There is a vast quantity of information and data about metabolism, gene regulation, and how these three processes interact as a result of decades of research. Although research on the interaction between metabolism and gene expression in plant systems is still in its infancy, this chapter primarily concentrates on unicellular organisms such as *E. coli* and *S. cerevisiae* due to the vastly greater data availability and simpler processes in these systems. However, plants should also be able to use the same fundamental ideas (Yeang, 2011).

All species devote many genetic resources to metabolism since it is a necessary component of life. In *E. coli* and *S. cerevisiae*, approximately half of the genes are involved in metabolism. Naturally, the metabolic responses are modulated by the regulation of those genes. Specific genes regulate some metabolic process enzymes and transporters. These genes and models offer details on how gene regulation influences metabolic processes at the local and global levels (Yeang, 2011).

Numerous crucial metabolic enzyme regulators have been discovered in various species. Several prominent instances of regulators of the primary metabolic pathways are briefly listed here. A two-component regulatory system for respiratory regulation in *E. coli* comprises ArcA (response regulator) and ArcB (membrane-bound sensor kinase) (Iuchi, 1993). Under oxygen deprivation state, ArcB does phosphorylation of ArcA that obstructs the production of several enzymes essential for aerobic respiration. The TCA cycle and the glyoxylate shunt are two of the enzymes that are catalyzed by this regulation. These enzymes include acnAB, sucABCD, aceB, and sdhCDAB (Iuchi, 1993; Perrenoud & Sauer, 2005).

In *E. coli*, CRP acts as a dual transcriptional regulator (Saier et al., 1996). The promoters of operons involved in the metabolism of glucose, lactose, electron transport, and various substances are bound by CRP (Zheng et al., 2004; Grainger et al., 2005). The CRP-cAMP complex is the best characterized bacterial catabolite suppression system (Saier et al., 1996). By attaching to promoters, the dimeric CRP-cAMP complex promotes transcription. Exogenous glucose decreases the levels of the CRP-cAMP complex and induces glucose repression by inhibiting cAMP production and stimulating cAMP efflux from the cytoplasm (Makman & Sutherland, 1965).

Cra (FurR) is another universal regulator of bacterial carbon metabolism (Saier et al., 1996). The core carbon pathway enzymes pfkA, pykF, and zwf are repressed by Cra, while ppsA, pckA, icd, and aceB are activated (Saier & Ramseier, 1997). Research on Cra-deficient variants of *E. coli* shows that Cra promotes glucogenesis while suppressing the catabolic routes for glucose, such as glycolysis (Saier et al., 1996; Perrenoud & Sauer, 2005). The Met*J* transcriptional repressor controls several genes involved in the synthesis of methionine (Su & Greene, 1971). The TyrR transcriptional dual regulator controls genes producing and transporting aromatic amino acids (Pittard et al., 2005).

Gcn4p is a chief transcription factor that regulates the genes involved in amino acid production in yeast. Additionally, large-scale chromatin immunoprecipitation (ChIPChip) tests reveal that the promoters of the genes involved in amino acid synthesis are enriched (Lee et al., 2002). Different transcription factors also activate the enzymes of various amino acid production pathways in addition to Gcn4p. Leu3 for the synthesis of leucine (Xiao & Rank, 1990), Bas1 for the synthesis of histidine and arginine (Denis et al., 1998), and Cbf1 for the synthesis of methionine are some examples (O'Connel et al., 1995).

Specialized analyses of the overall characteristics and interactions of biological system components have grown significantly and effectively in the postgenomic age due to the availability of a vast quantity of data. Systems biology is being tested on integrating metabolic processes and gene regulation. Whole metabolic networks (or a sizable fraction) in various animals have been mapped through decades of biochemical research. There are already databases that combine various metabolic and regulatory data and some integrated models that use these databases. The outcomes of a few integrated research projects of this type show considerable promise for applications in environmental and energy sciences, as well as in agriculture and medicine (Yeang, 2011).

Numerous applications can result from understanding the connections between gene regulation and metabolic systems. Genetic and metabolic engineering has already significantly impacted the pharmaceutical and agricultural industries. Breakthroughs in systems biology and synthetic biology enable the reprogramming or creation of organisms to produce or consume the proper quantities of molecules at the proper times in response to environmental inputs. These novel organisms have the potential to advance fundamental scientific understanding as well as agricultural productivity, biofuel production, antibiotic and protein manufacturing, and environmental waste remediation (Yeang, 2011).

Word-Based Algorithm

The idea of a word is extremely helpful in language and sequence analysis. In writing, words rather than letters are used to express information. In biology, strings with recognized functional activities include the amino acids RGD (extracellular matrix attachment) and GAATTC (EcoRI). Languages lacking delimiters, such as ancient Roman, transform characters into words using context. Introns were discovered relatively recently, and it was soon realized that removing introns reduced the expression of multiple genes. In contrast, adding introns to reporter genes lacking them, such as bacterial genes lacZ or GUS, frequently boosted their expression in transgenic species (Korf & Rose, 2009).

Attempts to pinpoint the intron sequences responsible for boosting expression have led to a conundrum. It is obvious that some introns that are efficiently spliced increase expression far more than others do, which implies that some introns include boosting sequences. Conventional homology searches, however, could not identify features shared by stimulatory introns because of the significant variation in intron size and sequence and the relatively small number of introns known to increase expression. Deletion analysis attempts to identify boosting regions have also been unsuccessful since the intron does not require specific sequences to promote mRNA accumulation (Rose & Beliakoff, 2000; Korf & Rose, 2009).

Introns close to the gene's promoter are more likely to trigger expression than other introns, according to three lines of evidence. First, although not all first introns have this property, almost all first introns are known to increase expression. Second, when introns are relocated to the 3′-UTR from the 5′-UTR, they no longer have the same influence on expression. Third, an enhancing intron positioned further downstream in a gene begins to lose its ability to enhance at approximately 500 bp from the promoter and has no impact at 1 kb or more from the 50 ends (Rose, 2004). Defining the sequences necessary for IME would be advantageous since this would provide biochemical separation of transacting components that bind to those sequences. This might be a key step toward comprehending the unique processes by which introns regulate expression. Finding the enhancing sequences would also make it possible to forecast whether an untested intron will increase expression. In the future, to enhance gene expression in transgenic applications, artificial introns that are more effective than any naturally occurring intron might be engineered (Korf & Rose, 2009).

To comprehend the unique method by which introns regulate expression, it would be helpful to define the sequences necessary for intron-mediated enhancement (IME) (Mascarenhas et al., 1990). This would provide biochemists with a foothold to isolate transacting factors that bind to those sequences. The idea that IME signals are enhanced in introns close to the promoter (proximal) instead of introns further down the transcript is crucial for modeling IMEs (distal). The sequence makeup of proximal and distal introns can be examined by beginning with a straightforward letter count. Since the components are different, proximal and distal introns may differ significantly. However, even if the compositions were similar, there might still be variances at higher orders that are not apparent at the 0th order (Korf & Rose, 2009).

The IMEter program (Rose et al., 2008), which forecasts IMEs, is remarkably similar to algorithms that forecast coding areas. An intron is more likely to include components that cause IMEs if it has a high score because it resembles proximal introns more than distal introns. The IMEter excludes these areas because one does not believe that the splice donor and acceptor sequences were produced using the same model as the intron body. One must decide on several parameters before training the IMEter: word size, cutoff for proximal as well as distal introns, and length of the splice acceptor site are among the factors that must be taken into consideration (Korf & Rose, 2009).

The IMEter demonstrates some of the benefits and drawbacks of word-based algorithms. Positively, the IMEter found previously unknown variations in the makeup of introns, a sizable collection of extremely different components. No prior knowledge of the length or locations of the pertinent sequences was necessary. Not all sequence parts, meanwhile, lend themselves to word frequency analysis. The biological and statistical models work well together because IME signals are distributed and redundant (Korf & Rose, 2009). One would not anticipate word frequency analysis to be very helpful if someone had been seeking an isolated signal where the position was a key component, such as the TATA box. The difficulty in locating the functional components recognized among the complete dictionary of terms is perhaps the most significant flaw in word-based analysis. The biological significance of any candidates must finally be assessed experimentally. A variety of statistical techniques might be used. Despite their inherent drawbacks, word-based studies can be very helpful tools for systems biologists since they allow for detecting previously undetected patterns in complicated data collections, leading to new linkages (Korf & Rose, 2009).

Challenges of Plant Systems Biology

To fully utilize plant systems biology, three areas must be addressed: (i) omics technology development, (ii) data integration into accessible formats, and (iii) data analysis within the bioinformatics domain. Given the necessity of normalizing, standardizing, and visualizing biological data to construct integrated models, bioinformatics likely requires the greatest attention (Tokimatsu et al., 2005; Bornholdt, 2005; Di Ventura et al., 2006). Systems biology networks may never fully capture the nature of the biological system. Reductionist approaches can complement systems biology methodologies, and systems biology interpretation of data must always be cautious.

However, plant systems biology has advanced due to its ability to illuminate regulatory networks that control the physiological, developmental, and pathological processes that occur in plants. To fully benefit from the expanding array of omics technologies, advancements in plant systems biology are needed. Improvements are needed in fundamental plant biology and in identifying the critical regulatory genes for agriculture (Yuan et al., 2008).

References

Ahmadian, A., et al. (2006). Pyrosequencing: History, biochemistry and future. *Clinica Chimica Acta, 363*, 83–94.

Alvarez-Buylla, E. R., et al. (2007). Gene regulatory network models for plant development. *Current Opinion in Plant Biology, 10*, 83–91.

Aranda, B., Achuthan, P., Alam-Faruque, Y., et al. (2010). The IntAct molecular interaction database in 2010. *Nucleic Acids Research, 38*, D525–D531.

Ashrafi, G., & Schwarz, T. L. (2013). The pathways of mitophagy for quality control and clearance of mitochondria. *Cell Death and Differentiation, 20*(1), 31–42.

Assmann, S. M., & Albert, R. (2009). Discrete dynamic modeling with asynchronous update, or how to model complex systems in the absence of quantitative information. *Plant Systems Biology*, 207–225.

Barkan, A. (2004). Intron splicing in plant organelles. In H. Daniell & C. Chase (Eds.), *Molecular biology and biotechnology of plant organelles* (pp. 281–308). Kluwer Academic Publishers.

Barkan, A., Klipcan, L., Ostersetzer, O., Kawamura, T., Asakura, Y., & Watkins, K. (2007). The CRM domain: An RNA binding module derived from an ancient ribosome-associated protein. *RNA, 13*, 55–64.

Barr, F. A., & Short, B. (2003). Golgins in the structure and dynamics of the Golgi apparatus. *Current Opinion in Cell Biology, 15*(4), 405–413.

Belostotsky, D. A., & Rose, A. B. (2005). Plant gene expression in the age of systems biology: Integrating transcriptional and posttranscriptional events. *Trends in Plant Science, 10*, 347–353.

Bender, J. (2004). DNA methylation and epigenetics. *Annual Review of Plant Biology, 55*, 41–68.

Berggard, T., Linse, S., & James, P. (2007). Methods for the detection and analysis of protein–protein interactions. *Proteomics, 7*(16), 2833–2842.

Bevan, M., & Walsh, S. (2005). The Arabidopsis genome: A foundation for plant research. *Genome Research, 15*, 1632–1642.

Blasing, O. E., et al. (2005). Sugars and circadian regulation make major contributions to the global regulation of diurnal gene expression in Arabidopsis. *Plant Cell, 17*, 3257–3281.

Bollenbach, T. J., Schuster, G., & Stern, D. B. (2004). Cooperation of endo- and exoribonucleases in chloroplast mRNA turnover. *Progress in Nucleic Acid Research and Molecular Biology, 78*, 305–337.

Bonen, L. (2004). Translational machinery in plant organelles. In H. Daniell & C. Chase (Eds.), *Molecular biology and biotechnology of plant organelles* (pp. 323–345). Springer.

Boone, C., et al. (2007). Exploring genetic interactions and networks with yeast. *Nature Reviews Genetics, 8*, 437–449.

Bornholdt, S. (2005). Less is more in modeling large genetic networks. *Science, 310*, 449–451.

Branco-Price, C., Kawaguchi, R., Ferreira, R. B., & Bailey-Serres, J. (2005). Genome-wide analysis of transcript abundance and translation in Arabidopsis seedlings subjected to oxygen deprivation. *Annals of Botany, 96*, 647–660.

Brighouse, A., Dacks, J. B., & Field, M. C. (2010). Rab protein evolution and the history of the eukaryotic endomembrane system. *Cellular and Molecular Life Sciences, 67*, 3449–3465.

Brink, R. A. (1956). A genetic change associated with the R locus in maize which is directed and potentially reversible. *Genetics, 41*, 872–889.

Broda, M., Millar, A. H., & Van Aken, O. (2018). Mitophagy: A mechanism for plant growth and survival. *Trends in Plant Science, 23*(5), 434–450.

Brophy, J. A., & Voigt, C. A. (2014). Principles of genetic circuit design. *Nature Methods, 11*(5), 508–520.

Buck, M. J., & Lieb, J. D. (2004). ChIP-chip: Considerations for the design, analysis, and application of genome-wide chromatin immunoprecipitation experiments. *Genomics, 83*(3), 349–360.

Carrington, J., & Ambros, V. (2003). Role of MicroRNAs in plant and animal development. *Science, 301*, 336–338.

Chan, S. W. L., Henderson, I. R., & Jacobsen, S. E. (2005). Gardening the genome: DNA methylation in Arabidopsis thaliana. *Nature Reviews. Genetics, 6*, 351–360.

Chandler, V., & Stam, M. (2004). Chromatin conversations: Mechanisms and implications of paramutation. *Nature Reviews Genetics, 5*, 532–544.

Charbonnier, S., Gallego, O., & Gavin, A. C. (2008). The social network of a cell: Recent advances in interactome mapping. *Biotechnology Annual Review, 14*, 1–28.

Coruzzi, G. M., Burga, A. R., Katari, M. S., & Gutiérrez, R. A. (2009). Systems biology: Principles and applications in plant research. *Annual Plant Reviews, 35*, 3–40.

Cowley, M. J., Pinese, M., Kassahn, K. S., et al. (2012). PINA v2. 0: Mining interactome modules. *Nucleic Acids Research, 40*(D1), D862–D865.

Cremer, T., Cremer, M., Dietzel, S., Muller, S., Solovei, I., & Fakan, S. (2006). Chromosome territories—A functional nuclear landscape. *Current Opinion in Cell Biology, 18*, 307–316.
Cultrera, N. G. (2023). Genetics of plant metabolism. *International Journal of Molecular Sciences, 24*(8), 6890.
de Folter, S., Urbanus, S. L., van Zuijlen, L. G., Kaufmann, K., & Angenent, G. C. (2007). Tagging of MADS domain proteins for chromatin immunoprecipitation. *BMC Plant Biology, 7*, 47.
De Geyter, N., Gholami, A., Goormachtig, S., & Goossens, A. (2012). Transcriptional machineries in jasmonate-elicited plant secondary metabolism. *Trends in Plant Science, 17*(6), 349–359.
Denis, V., Boucherie, H., Monribot, C., & Daignan-Fornier, B. (1998). Role of the Myb-like protein Bas1p in Saccharomyces cerevisiae: A proteome analysis. *Molecular Microbiology, 30*(3), 557–566.
Di Ventura, B., et al. (2006). From in vivo to in silico biology and back. *Nature, 443*, 527–533.
Dreyfuss, G., Kim, V. N., & Kataoka, N. (2002). Messenger-RNA-binding proteins and the messages they carry. *Nature Reviews. Molecular Cell Biology, 3*, 195–205.
Eliaš, J., & Macnamara, C. K. (2021). Mathematical modelling of p53 signalling during DNA damage response: A survey. *International Journal of Molecular Sciences, 22*(19), 10590.
Elmayan, T., Proux, F., & Vaucheret, H. (2005). Arabidopsis RPA2: A genetic link among transcriptional gene silencing, DNA repair, and DNA replication. *Current Biology, 15*, 1919–1925.
Espinosa-soto, C., et al. (2004). A gene regulatory network model for cell-fate determination during Arabidopsis thaliana flower development that is robust and recovers experimental gene expression profiles. *Plant Cell, 16*, 2923–2939.
Fedoroff, N. (2002). RNA-binding proteins in plants: The tip of an iceberg? *The Plant Journal, 5*, 452–459.
Feiguelman, G., Fu, Y., & Yalovsky, S. (2018). ROP GTPases structure-function and signaling pathways. *Plant Physiology, 176*(1), 57–79.
Fiehn, O. (2002). Metabolomics—The link between genotypes and phenotypes. *Plant Molecular Biology, 48*(1–2), 155–171.
Fields, S. (2005). High-throughput two hybrid analysis. The promise and the peril. *The FEBS Journal, 272*(21), 5391–5399.
Fields, S., & Song, O. K. (1989). A novel genetic system to detect protein-protein interactions. *Nature, 340*, 245–246.
Figeys, D., McBroom, L. D., & Moran, M. F. (2001). Mass spectrometry for the study of protein–protein interactions. *Methods, 24*(3), 230–239.
Fuchs, J., Demidov, D., Houben, A., & Schubert, I. (2006). Chromosomal histone modification patterns—From conservation to diversity. *Trends in Plant Science, 11*, 199–208.
Galbraith, D. W. (2006). DNA microarray analyses in higher plants. *OMICS, 10*, 455–473.
Geisler-Lee, J., et al. (2007). A predicted interactome for Arabidopsis. *Plant Physiology, 145*, 317–329.
Grainger, D. C., Hurd, D., Harrison, M., Holdstock, J., & Busby, S. J. (2005). Studies of the distribution of Escherichia coli cAMP-receptor protein and RNA polymerase along the E. coli chromosome. *Proceedings of the National Academy of Sciences of the United States of America, 102*(49), 17693–17698.
Gregory, B. D., & Belostotsky, D. A. (2009). Whole-genome microarrays: Applications and technical issues. *Plant Systems Biology*, 39–56.
Grotewold, E., & Springer, N. (2009). Decoding the transcriptional hardwiring of the plant genome. In G. Coruzzi & R. Gutierrez (Eds.), *In systems biology*. Blackwell Publishing.
Hála, M., & Žárský, V. (2019). Protein prenylation in plant stress responses. *Molecules, 24*(21), 3906.
Haring, M., Offermann, S., Danker, T., Horst, I., Peterhansel, C., & Stam, M. (2007). Chromatin immunoprecipitation: Optimization, quantitative analysis and data normalization. *Plant Methods, 3*, 11.
Hartwell, L. H., Hopfield, J. J., Leibler, S., et al. (1999). From molecular to modular cell biology. *Nature, 402*, 47–51.

Hesse, H., & Hoefgen, R. (2006). On the way to understand biological complexity in plants: S-nutrition as a case study for systems biology. *Cellular & Molecular Biology Letters, 11*, 37–56.
Hieronymus, H., & Silver, P. A. (2004). A systems view of mRNP biology. *Genes & Development, 18*, 2845–2860.
Hirai, M. Y., et al. (2004). Integration of transcriptomics and metabolomics for understanding of global responses to nutritional stresses in Arabidopsis thaliana. *Proceedings of the National Academy of Sciences of the United States of America, 101*, 10205–10210.
Ho, Y., et al. (2002). Systematic identification of protein complexes in Saccharomyces cerevisiae by mass spectrometry. *Nature, 415*, 180–183.
Hsieh, T.-F., & Fischer, R. L. (2005). Biology of chromatin dynamics. *Annual Review of Plant Biology, 56*, 327–351.
Hudson, M. E., & Snyder, M. (2006). High-throughput methods of regulatory element discovery. *BioTechniques, 41*(6), 673., 675, 677 passim. https://doi.org/10.2144/000112322
Hunt, A. G., Xu, R., Addepalli, B., et al. (2008). Arabidopsis mRNA polyadenylation machinery: Comprehensive analysis of protein–protein interactions and gene expression profiling. *BMC Genomics, 9*(1), 220.
Ideker, T., Galitski, T., & Hood, L. (2001). A new approach to decoding life: Systems biology. *Annual Review of Genomics and Human Genetics, 2*(1), 343–372.
Iuchi, S. (1993). Phosphorylation/dephosphorylation of the receiver module at the conserved aspartate residue controls transphosphorylation activity of histidine kinase in sensor protein ArcB of Escherichia coli. *The Journal of Biological Chemistry, 268*(32), 23972–23980.
Jaffe, A. B., & Hall, A. (2005). Rho GTPases: Biochemistry and biology. *Annual Review of Cell and Developmental Biology, 21*, 247–269.
Jagodzik, P., Tajdel-Zielinska, M., Ciesla, A., Marczak, M., & Ludwikow, A. (2018). Mitogen-activated protein kinase cascades in plant hormone signaling. *Frontiers in Plant Science, 9*, 1387.
Jambhekar, A., & Derisi, J. L. (2007). Cis-acting determinants of asymmetric, cytoplasmic RNA transport. *RNA, 13*, 625–642.
Kakutani, T. (2002). Epi-alleles in plants: Inheritance of epigenetic information over generations. *Plant & Cell Physiology, 43*, 1106–1111.
Kalinowska, K., & Isono, E. (2018). All roads lead to the vacuole—Autophagic transport as part of the endomembrane trafficking network in plants. *Journal of Experimental Botany, 69*(6), 1313–1324.
Kankel, M. W., Ramsey, D. E., Stokes, T. L., Flowers, S. K., Haag, J. R., Jeddeloh, J. A., Riddle, N. C., Verbsky, M. L., & Richards, E. J. (2003). Arabidopsis MET1 cytosine methyltransferase mutants. *Genetics, 163*, 1109–1122.
Kawaguchi, R., Girke, T., Bray, E. A., & Bailey-Serres, J. (2004). Differential mRNA translation contributes to gene regulation under non-stress and dehydration stress conditions in Arabidopsis thaliana. *The Plant Journal, 38*, 823–839.
Keene, J. D. (2007). RNA regulons: Coordination of post-transcriptional events. *Nature Reviews. Genetics, 8*, 533–543.
Keurentjes, J. J., Koornneef, M., & Vreugdenhil, D. (2008). Quantitative genetics in the age of omics. *Current Opinion in Plant Biology, 11*(2), 123–128.
Kim, J., & Buell, C. R. (2015). A revolution in plant metabolism: Genome-enabled pathway discovery. *Plant Physiology, 169*(3), 1532–1539.
Kitano, H. (2002). Systems biology: A brief overview. *Science, 295*(5560), 1662–1664.
Kitano, H. (2007). Towards a theory of biological robustness. *Molecular Systems Biology, 3*(1), 137.
Knoop, V., & Brennicke, A. (2002). Molecular biology of the plant mitochondrion. *Critical Reviews in Plant Sciences, 21*, 111–126.
Konstantinopoulos, P. A., Karamouzis, M. V., & Papavassiliou, A. G. (2007). Post-translational modifications and regulation of the RAS superfamily of GTPases as anticancer targets. *Nature Reviews Drug Discovery, 6*(7), 541–555.

Korf, I. F., & Rose, A. B. (2009). Applying word-based algorithms: The IMEter. *Plant Systems Biology*, 287–301.
Kosak, S. T., & Groudine, M. (2004). Gene order and dynamic domains. *Science, 306*, 644–647.
Kuhn, J., & Schroeder, J. (2003). Impacts of altered RNA metabolism on abscisic acid signalling. *Current Opinion in Plant Biology, 6*, 463–469.
Kung, C. P., & Weber, J. D. (2022). It's getting complicated—A fresh look at p53-MDM2-ARF triangle in tumorigenesis and cancer therapy. *Frontiers cell Developmental Biology, 10*, 818744.
Lee, T., Rinaldi, N., Robert, F., Odom, D., Bar-Joseph, Z., et al. (2002). A transcriptional regulatory network map for Saccharomyces cerevisiae. *Science, 298*, 799–804.
Lee, J., He, K., Stolc, V., et al. (2007). Analysis of transcription factor HY5 genomic binding sites revealed its hierarchical role in light regulation of development. *Plant Cell, 19*(3), 731–749.
Lefebvre, C., Legouis, R., & Culetto, E. (2018). ESCRT and autophagies: Endosomal functions and beyond. In *Seminars in cell & developmental biology* (Vol. 74, pp. 21–28). Academic Press.
Legrain, P., & Selig, L. (2000). Genome-wide protein interaction maps using two-hybrid systems. *FEBS Letters, 480*(1), 32–36.
Lehner, B., et al. (2006). Systematic mapping of genetic interactions in Caenorhabditis elegans identifies common modifiers of diverse signaling pathways. *Nature Genetics, 38*, 896–903.
Li, S., et al. (2004). A map of the interactome network of the metazoan C. elegans. *Science, 303*, 540–543.
Lidder, P., Gutierrez, R. A., Salome, P. A., McClung, C. R., & Green, P. J. (2005). Circadian control of messenger RNA stability. Association with a sequence-specific messenger RNA decay pathway. *Plant Physiology, 138*, 2374–2385.
Lippman, Z., & Martienssen, R. (2004). The role of RNA interference in heterochromatic silencing. *Nature, 431*, 364–370.
Locke, J. C. W., et al. (2006). Experimental validation of a predicted feedback loop in the multi-oscillator clock of Arabidopsis thaliana. *Molecular Systems Biology, 2*, 59.
Loraine, A. (2009). Co-expression analysis of metabolic pathways in plants. *Plant Systems Biology*, 247–264.
Luo, C., & Lam, E. (2009). Chromatin charting: Global mapping of epigenetic effects. *Plant Systems Biology*, 127–139.
Makman, R. S., & Sutherland, E. W. (1965). Adenosine 3′,5′-phosphate in Escherichia coli. *The Journal of Biological Chemistry, 240*, 1309–1314.
Mascarenhas, D., Mettler, I. J., Pierce, D. A., & Lowe, H. W. (1990). Intron-mediated enhancement of heterologous gene expression in maize. *Plant Molecular Biology, 15*, 913–920.
Matthews, L. R., et al. (2001). Identification of potential interaction networks using sequence-based searches for conserved protein-protein interactions or 'interologs'. *Genome Research, 11*, 2120–2126.
Matzke, M. A., & Birchler, J. A. (2005). RNAi-mediated pathways in the nucleus. *Nature Reviews. Genetics, 6*, 24–35.
Maynard Smith, J., & Szathmary, E. (1999). *The origins of life: From the birth of life to the origin of language*. Oxford University Press.
Morsy, M., Gouthu, S., Orchard, S., et al. (2008). Charting plant interactomes: Possibilities and challenges. *Trends in Plant Science, 13*(4), 183–191.
Moyer, B. D., Allan, B. B., & Balch, W. E. (2001). Rab1 interaction with a GM130 effector complex regulates COPII vesicle cis-Golgi tethering. *Traffic, 2*(4), 268–276.
Mukhtar, M. S., Carvunis, A. R., Dreze, M., et al. (2011). Independently evolved virulence effectors converge onto hubs in a plant immune system network. *Science, 333*, 596–601.
Nikiforova, V. J., et al. (2005). Integrative gene-metabolite network with implemented causality deciphers informational fluxes of sulphur stress response. *Journal of Experimental Botany, 56*, 1887–1896.
Noble, D. (2002). Modeling the heart—From genes to cells to the whole organ. *Science, 295*(5560), 1678–1682.

Nowack, D. E., Tian, B., & Brasier, A. R. (2005). Two-step cross-linking method for identification of NF-KB gene network by chromatin immunoprecipitation. *BioTechniques, 39*, 715–725.

O'Connel, K., Surdin-Kerjan, Y., & Baker, R. (1995). Role of the Saccharomyces cerevisiae general regulatory factor cp1 in methionine biosynthetic gene transcription. *Molecular and Cellular Biology, 15*, 1879–1888.

Obrdlik, P., El-Bakkoury, M., Hamacher, T., et al. (2004). Kþ channel interactions detected by a genetic system optimized for systematic studies of membrane protein interactions. *Proceedings of the National Academy of Sciences of the United States of America, 101*(33), 12242–12247.

Okita, T. W., & Choi, S. B. (2002). mRNA localization in plants: Targeting to the cell's cortical region and beyond. *Current Opinion in Plant Biology, 5*, 553–559.

Oliver, S. G., Winson, M. K., Kell, D. B., & Baganz, F. (1998). Systematic functional analysis of the yeast genome. *TIPTECH, 16*, 373–373.

Orth, J. D., Thiele, I., & Palsson, B. Ø. (2010). What is flux balance analysis? *Nature Biotechnology, 28*(3), 245–248.

Palaniswamy, S. K., et al. (2006). AGRIS and AtRegNet. A platform to link cis-regulatory elements and transcription factors into regulatory networks. *Plant Physiology, 140*, 818–829.

Peck, S. C. (2006). Phosphoproteomics in Arabidopsis: Moving from empirical to predictive science. *Journal of Experimental Botany, 57*(7), 1523–1527.

Penterman, J., Zilberman, D., Huh, J. H., Ballinger, T., Henikoff, S., & Fischer, R. L. (2007). DNA demethylation in the Arabidopsis genome. *Proceedings of the National Academy of Sciences of the United States of America, 104*, 6752–6757.

Perrenoud, A., & Sauer, U. (2005). Impact of global transcriptional regulation by ArcA, ArcB, Cra, Crp, Cya, Fnr, and Mlc on glucose catabolism in Escherichia coli. *Journal of Bacteriology, 187*(9), 3171–3179.

Pittard, J., Camakaris, H., & Yang, J. (2005). The TyrR regulon. *Molecular Microbiology, 55*(1), 16–26.

Pontes, O., Li, C. F., Nunes, P. C., Haag, J., Ream, T., Vitins, A., Jacobsen, S. E., & Pikaard, C. S. (2006). The Arabidopsis chromatin-modifying nuclear siRNA pathway involves a nucleolar RNA processing center. *Cell, 126*, 79–92.

Poraty-Gavra, L., Zimmermann, P., Haigis, S., Bednarek, P., Hazak, O., Stelmakh, O. R., Sadot, E., Schulze-Lefert, P., Gruissem, W., & Yalovsky, S. (2013). The Arabidopsis Rho of plants GTPase AtROP6 functions in developmental and pathogen response pathways. *Plant Physiology, 161*(3), 1172–1188.

Qian, J., Dolled-Filhart, M., Lin, J., Yu, H., & Gerstein, M. (2001). Beyond synexpression relationships: Local clustering of time-shifted and inverted gene expression profiles identifies new, biologically relevant interactions. *Journal of Molecular Biology, 314*, 1053–1066.

Rain, J. C., Selig, L., De Reuse, H., et al. (2001). The protein–protein interaction map of Helicobacter pylori. *Nature, 409*(6817), 211–215.

Reddy, A. S. (2007). Alternative splicing of pre-messenger RNAs in plants in the genomic era. *Annual Review of Plant Biology, 58*, 267–294.

Reyes, J. C. (2006). Chromatin modifiers that control plant development. *Current Opinion in Plant Biology, 9*, 21–27.

Riefler, M., et al. (2006). Arabidopsis cytokinin receptor mutants reveal functions in shoot growth, leaf senescence, seed size, germination, root development, and cytokinin metabolism. *The Plant Cell, 18*, 40–54.

Rischer, H., et al. (2006). Gene-to-metabolite networks for terpenoid indole alkaloid biosynthesis in Catharanthus roseus cells. *Proceedings of the National Academy of Sciences of the United States of America, 103*, 5614–5619.

Rose, A. B. (2004). The effect of intron location on intron-mediated enhancement of gene expression in Arabidopsis. *The Plant Journal, 40*, 744–751.

Rose, A. B., & Beliakoff, J. A. (2000). Intron-mediated enhancement of gene expression independent of unique intron sequences and splicing. *Plant Physiology, 122*, 535–542.

Rose, A. B., Elfersi, T., Parra, G., & Korf, I. (2008). Promoter-proximal introns in Arabidopsis thaliana are enriched in dispersed signals that elevate gene expression. *The Plant Cell, 20*, 543–551.

Saier, M. H., & Ramseier, T. M. (1997). The catabolite repressor/activator (Cra) protein of enteric bacteria. *Journal of Bacteriology, 178*, 3411–3417.

Saier, M. H., Ramseier, T. M., & Reizer, J. (1996). Regulation of carbon utilization. In F. C. Neidhardt et al. (Eds.), *Escherichia coli and Salmonella: Cellular and molecular biology*. ASM Press.

Schauer, N., & Fernie, A. R. (2006). Plant metabolomics: Towards biological function and mechanism. *Trends in Plant Science, 11*, 508–516.

Shannon, P., Markiel, A., Ozier, O., et al. (2003). Cytoscape: A software environment for integrated models of biomolecular interaction networks. *Genome Research, 13*(11), 2498–2504.

Sharan, R., Ulitsky, I., & Shamir, R. (2007). Network-based prediction of protein function. *Mol Sys Biol, 3*(1), 88.

Sheth, B. P., & Thaker, V. S. (2014). Plant systems biology: Insights, advances and challenges. *Planta, 240*(1), 33–54.

Speth, E. B., Imboden, L., Hauck, P., & He, S. Y. (2009). Subcellular localization and functional analysis of the Arabidopsis GTPase RabE. *Plant Physiology, 149*(4), 1824–1837.

Stelling, J., Sauer, U., Szallasi, Z., et al. (2004). Robustness of cellular functions. *Cell, 118*(6), 675–685.

Su, C. H., & Greene, R. C. (1971). Regulation of methionine biosynthesis in Escherichia coli: Mapping of the metJ locus and properties of a metJ plus-metJ minus diploid. *Proceedings of the National Academy of Sciences of the United States of America, 68*(2), 367–371.

Szklarczyk, D., Franceschini, A., Kuhn, M., et al. (2011). The STRING database in 2011: Functional interaction networks of proteins, globally integrated and scored. *Nucleic Acids Research, 39*(1), D561–D568.

To, J. P. C., et al. (2004). Type-A Arabidopsis response regulators are partially redundant negative regulators of cytokinin signaling. *Plant Cell, 16*, 658–671.

Tohge, T., et al. (2005). Functional genomics by integrated analysis of metabolome and transcriptome of Arabidopsis plants over-expressing an MYB transcription factor. *The Plant Journal, 42*, 218–235.

Tokimatsu, T., et al. (2005). KaPPA-view: A web-based analysis tool for integration of transcript and metabolite data on plant metabolic pathway maps. *Plant Physiology, 138*, 1289–1300.

Tong, A. H. Y., et al. (2001). Systematic genetic analysis with ordered arrays of yeast deletion mutants. *Science, 294*, 2364–2368.

Tong, A. H. Y., et al. (2004). Global mapping of the yeast genetic interaction network. *Science, 303*, 808–813.

Trewavas, A. (2006). A brief history of systems biology "Every object that biology studies is a system of systems". Francois Jacob (1974). *Plant Cell, 18*(10), 2420–2430.

Tu, D., Lee, J., Ozdere, T., Lee, T., & You, L. (2007). Engineering gene circuits: Foundations and applications. In T. Vo-Dinh (Ed.), *Nanotechnology in biology and medicine: Methods, devices and applications*.

Ueda, H. R. (2006). Systems biology flowering in the plant clock field. *Molecular Systems Biology, 2*, 60.

Uetz, P., et al. (2000). A comprehensive analysis of protein-protein interactions in Saccharomyces cerevisiae. *Nature, 403*, 623–627.

Usadel, B., et al. (2005). Extension of the visualization tool MapMan to allow statistical analysis of arrays, display of corresponding genes, and comparison with known responses. *Plant Physiology, 138*, 1195–1204.

Van Leene, J., Stals, H., Eeckhout, D., et al. (2007). Tandem affinity purification-based technology platform to study the cell cycle interactome in Arabidopsis thaliana. *Molecular & Cellular Proteomics, 6*, 1226–1238.

Varma, A., & Palsson, B. O. (1993). Metabolic capabilities of Escherichia coli: I. Synthesis of biosynthetic precursors and cofactors. *Journal of Theoretical Biology, 165*(4), 477–502.

Vernoux, T., Brunoud, G., Farcot, E., et al. (2011). The auxin signalling network translates dynamic input into robust patterning at the shoot apex. *Molecular Systems Biology, 7*(508).

Walhout, A. J., & Vidal, M. (2001). High-throughput yeast two-hybrid assays for large-scale protein interaction mapping. *Methods, 24*(3), 297–306.

Walker, N. S., Stiffler, N., & Barkan, A. (2007). POGs/PlantRBP: A resource for comparative genomics in plants. *Nucleic Acids Research, 35*, D852–D856.

Wang, B. B., & Brendel, V. (2004). The ASRG database: Identification and survey of Arabidopsis thaliana genes involved in pre-mRNA splicing. *Genome Biology, 5*, R102.

Wang, B., & Buck, M. (2012). Customizing cell signaling using engineered genetic logic circuits. *Trends in Microbiology, 20*(8), 376–384.

Wang, P., Wang, T., Han, J., Li, M., Zhao, Y., Su, T., & Ma, C. (2021). Pl autophagy: An intricate process controlled by various signaling pathways. *Frontiers in Plant Science, 12*, 754982.

Weiss, R., Basu, S., Hooshangi, S., Kalmbach, A., Karig, D., Mehreja, R., & Netravali, I. (2003). Genetic circuit building blocks for cellular computation, communications, and signal processing. *Natural Computing, 2*, 47–84.

Wheeler, D. L., Barrett, T., Benson, D. A., Bryant, S.H., Canese, K., Chetvernin, V., Church, D.M., DiCuccio, M., Edgar, R., Federhen, S., Geer, L.Y., Helmberg, W., Kapustin, Y., Kenton, D.L., Khovayko, O., Lipman, D.J., Madden, T.L., Maglott, T.R., Ostell, J., … Yaschenko, E. (2006). Database resources of the National Center for Biotechnology Information. *Nucleic Acids Research, 34*, D173–D180.

Wichert, S., Fokianos, K., & Strimmer, K. (2004). Identifying periodically expressed transcripts in microarray time series data. *Bioinformatics, 20*, 5–20.

Xiao, W., & Rank, G. (1990). Branched chain amino acid regulation of the ilv2 locus in Saccharomyces cerevisiae. *Genome, 33*(4), 596–603.

Yan, J., Kim, Y. J., & Somers, D. E. (2021). Post-translational mechanisms of plant circadian regulation. *Genes, 12*(3), 325.

Yeang, C. H. (2011). Integration of metabolic reactions and gene regulation. *Molecular Biotechnology, 47*(1), 70–82.

Yu, H., et al. (2004). Annotation transfer between genomes: Protein-protein interologs and protein-DNA regulogs. *Genome Research, 14*, 1107–1118.

Yuan, J. S., Galbraith, D. W., Dai, S. Y., Griffin, P., & Stewart, C. N., Jr. (2008). Plant systems biology comes of age. *Trends in Plant Science, 13*(4), 165–171.

Zerges, W. (2004). Regulation of translation in chloroplasts. In H. Daniell & C. Chase (Eds.), *Molecular biology and biotechnology of plant organelles* (pp. 347–383). Springer.

Zhang, X., Yazaki, J., Sundaresan, A., Cokus, S., Chan, S. W., Chen, H., Henderson, I. R., Shinn, P., Pellegrini, M., Jacobsen, S. E., & Ecker, J. R. (2006). Genome-wide high-resolution mapping and functional analysis of DNA methylation in Arabidopsis. *Cell, 126*, 1189–1201.

Zheng, D., Constantinidou, C., Hobman, J. L., & Minchin, S. D. (2004). Identification of the CRP regulon using in vitro and in vivo transcriptional profiling. *Nucleic Acids Research, 32*(19), 5874–5893.

Zhou, X., Rao, S., Wrightstone, E., Sun, T., Lui, A. C. W., Welsch, R., & Li, L. (2022). Phytoene synthase: The key rate-limiting enzyme of carotenoid biosynthesis in plants. *Frontiers in Plant Science, 13*, 977.

Chapter 10
Bioinformatics and Computational Biology

Muhammad Amjad Ali, Adil Zahoor, Zeenat Niaz, Muhammad Jabran, Muhammad Anas, Ikhlas Shafique, Hafiz Muhammad Ahmad, Muhammad Usama, and Amjad Abbas

Abstract Bioinformatics and computational biology integrate computer science, molecular biology, statistics, mathematics, and engineering to collect, manage, and analyze biological data. This interdisciplinary synergy has led to innovative advances in our understanding of plant diseases and the development of effective strategies for disease management. Bioinformatics plays a diverse and influential role in addressing the fundamental challenges faced by plant pathologists. Bioinformatics in plant pathology is vital for analyzing pathogen genomes, achieved through DNA sequencing and annotation. Comparative genomics involves comparing different pathogen genomes to identify commonalities and differences, shedding light on the genetic basis of pathogenicity, virulence, drug resistance, and evolutionary relationships. Bioinformatics also aids in studying pathogen evolution, reconstructing their history, and tracking new variants, essential for predicting disease outbreaks and developing mitigation strategies. Additionally, bioinformatics contributes to advanced diagnostic tools, detecting specific genetic markers for rapid and accurate disease identification, crucial for minimizing crop losses and ensuring food security. This chapter introduces and uses different bioinformatics tools and resources used to study plant pathogens. Moreover, applications and methods of the tools involved in BLAST, multiple sequence alignment, phylogenet-

M. A. Ali · M. Anas · I. Shafique · M. Usama · A. Abbas (✉)
Department of Plant Pathology, University of Agriculture, Faisalabad, Pakistan
e-mail: amjad.ali@uaf.edu.pk; amjad.abbas@uaf.edu.pk

A. Zahoor
Department of Biotechnology, Chonnam National University, Yeosu, South Korea

Z. Niaz
Graduate School of Agricultural Science, Tohoku University, Sendai, Japan

M. Jabran
State Key Laboratory for Biology of Plant Diseases, Institute of Plant Protection, Beijing, China

H. M. Ahmad
Department of Bioinformatics and Biotechnology, Government College University, Faisalabad, Pakistan

S. Ijaz et al. (eds.), *Trends in Plant Biotechnology*,
https://doi.org/10.1007/978-981-97-0814-7_10

ics, gene structure analysis, protein domain and motif, gene localization on the chromosome, protein expression analysis, and molecular docking and interaction networks are also discussed.

Keywords Computational biology · Databases · Bioinformatics · Software · Programs

Introduction

In a broader sense, "bioinformatics" is "using computers for acquiring, managing, and analyzing biological information" (Bayat, 2002). It incorporates elements of many fields together. Bioinformatics involves conceptualizing biological processes in molecular terms, such as the principles of physical chemistry. It includes applying techniques from fields such as applied mathematics, computer science, and statistics to comprehend and systematize the vast information linked to these molecules. In principle, bioinformatics is an information management system for molecular biology and boasts numerous practical applications (Anonymous, 1999). It is a multidisciplinary field, including computer system management, networking, database design, computer programming, and molecular biology. It is a multifaceted discipline combining multiple scientific fields, including computer science and biology, emphasizing molecular biology and genetics, statistics, engineering, mathematics, chemistry, and biochemistry of biological molecules (Lesk, 2019).

Bioinformatics and computational biology play a crucial role in plant pathology by integrating computer science, molecular biology, statistics, mathematics, and engineering to acquire, manage, and analyze biological data (Bayat, 2002). The marriage of these disciplines has led to groundbreaking advancements in understanding plant diseases and developing effective strategies for disease management (Jabran et al., 2023). The applications of bioinformatics in plant pathology are diverse and impactful, addressing key challenges plant pathologists face. For instance, bioinformatics and computational biology have enabled us to perform comparative analyses of plant pathogen genomes, transcriptomes, and proteomes. The genome analysis of pathogens provides us with information regarding many important aspects, and it has been improved with the help of better computational resources (Zahoor et al., 2023). As sequencing has become more accessible, computer-based genetic analysis has played a significant role in infectious disease research. Sequencing of pathogens can confirm suspected cases of infectious disease, discriminate between different strains, and classify novel pathogens (Carlton, 2003). Recent advances in bioinformatics have made comparative genomics of pathogens easy. It provides information for identifying disease-causing agents, pathogen genotyping, and predicting virulence and antibiotic resistance (Gabaldón et al., 2013).

This chapter comprises the importance and application of bioinformatics and computational biology in plant pathology. Moreover, it also discusses the different bioinformatics tools used to study plant pathogens, host genetics, and their interactions at the molecular level.

Applications of Bioinformatics in Plant Pathology

One of the fundamental applications of bioinformatics in plant pathology is the analysis of pathogen genomes (Abad et al., 2008). Through DNA sequencing and annotation, researchers can study the genetic makeup of various plant pathogens, including fungi, bacteria, viruses, and nematodes. Comparative genomics involves comparing the genomes of different pathogen strains or species to identify common features and differences (Sivashankari & Shanmughavel, 2007). This information aids in understanding the genetic basis of pathogenicity, virulence, and drug/pesticide resistance and provides insights into the evolutionary relationships between different pathogens. Likewise, bioinformatics facilitates the study of pathogen evolution by analyzing genetic data (Zahoor et al., 2023). Researchers can reconstruct the evolutionary history of pathogens and track the emergence and spread of new pathogen variants or strains. Therefore, understanding pathogen evolution is crucial for predicting disease outbreaks, identifying potential sources of infection, and developing strategies to mitigate the impact of rapidly evolving pathogens on crops (Gilbert & Parker, 2010).

Similarly, bioinformatics is pivotal in developing new and advanced diagnostic tools for identifying and characterizing plant pathogens. These tools identify genetic information and molecular markers specific to certain pathogens to detect diseases rapidly and accurately in crops (Toktay et al., 2020). Early and precise diagnosis is essential for implementing timely and targeted control measures, reducing crop losses, and ensuring food security (Jabbar et al., 2021).

Pathogens employ specific molecules known as effectors to manipulate the host plant's cellular processes and establish infection (Ali et al., 2017b, 2018). Bioinformatics aids in the identification and prediction of these effector molecules based on genetic data. Understanding pathogenicity factors and effector molecules provides crucial insights into the molecular mechanisms of plant-pathogen interactions, enabling the development of resistant crop varieties and targeted disease management strategies (Ali et al., 2017a, 2017b). Moreover, bioinformatics tools are utilized to analyze vast amounts of genetic data generated through DNA sequencing and RNA sequencing. Additionally, bioinformatics allows for studying gene expression patterns in infected plants, revealing how pathogens manipulate host gene regulation to facilitate infection and disease development (Zahoor et al., 2023; Jabran et al., 2023).

In plant pathology, bioinformatics is instrumental in organizing and managing genomic data of pathogens. It primarily includes annotating genes, identifying functional elements, and understanding the structure and organization of pathogen genomes (Tan et al., 2022). By identifying essential genes and functional elements, researchers gain insights into potential targets for disease control and control interventions. Furthermore, phylogenetic analysis is essential for understanding the evolutionary relationships between different plant pathogen species (Zahoor et al., 2023). By aligning and comparing genetic sequences from various organisms, bioinformatics generates phylogenetic trees that depict their evolutionary history. These models help categorize pathogens into related families, predict their potential behavior and characteristics, and identify common ancestors (Aono et al., 2020).

Bioinformatics uses information science and technology principles to analyze complex life science data. The primary focus is now on interpreting various data types, such as nucleotide and amino acid sequences, protein domains, and protein structures (Azeem et al., 2021). The mapping of entire genomes of numerous organisms has been made possible by bioinformatics advancements, enabling a better understanding of the underlying causes of plant diseases and the discovery of potential remedies (Kushwaha et al., 2017). In current plant disease management, bioinformatics has practical applications in studying host-pathogen interactions, understanding disease genetics, and identifying pathogenicity factors (Ali et al., 2018). This knowledge assists in designing effective management strategies. Many aspects of biology, including nucleic acid sequences, protein structure/function relationships, genome organization, and gene expression regulation, benefit from a bioinformatics approach, which has become indispensable in modern research (Alemu, 2015; Rao et al., 2008).

Bioinformatics Used for Molecular Characterization

BLAST at NCBI

In the field of bioinformatics, basic local alignment search tool (BLAST) is an algorithm and software program used to compare primary biological sequence data, such as amino acid sequences in proteins or nucleotide sequences in DNA and RNA (Altschul et al., 1990). It is connected to DNA and protein databases linked to the National Center of Biotechnology Information (NCBI), New York, USA. BLAST search allows researchers to compare a specific protein or nucleotide sequence (a query) with a vast library or database of sequences, identifying library sequences that closely resemble the query sequence beyond a predefined threshold. The summary of nucleotide BLAST is shown in Fig. 10.1.

Often called the "Google of Biological Sequences," BLAST is one of the most widely utilized bioinformatics tools for sequence searching. It addresses a fundamental challenge in bioinformatics research, where speed is vital in processing the extensive genome databases available today. BLAST achieves this speed through a heuristic algorithm that, while not as precise as calculating an optimal alignment, is significantly faster.

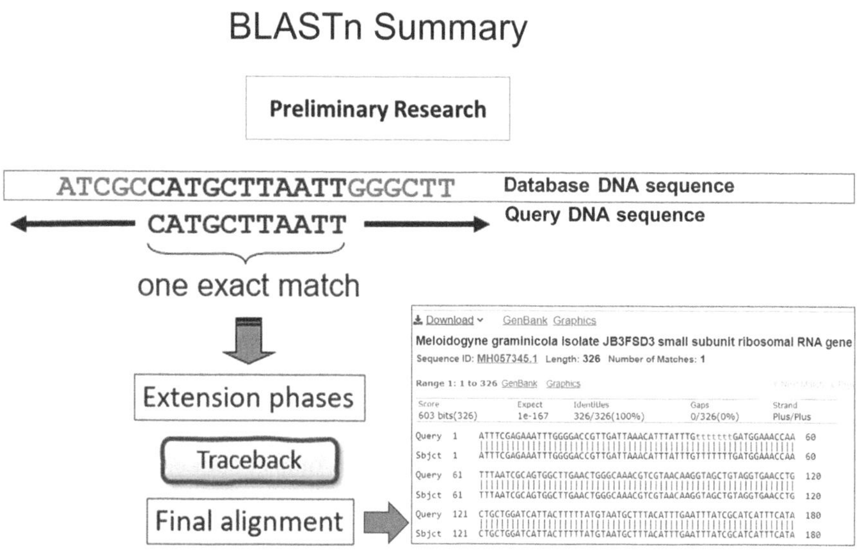

Fig. 10.1 How BLAST works. A simple showcasing of the BLAST working mechanism. (Courtesy: NCBI)

How Does BLAST Work?

To run BLAST, a researcher provides a query sequence to search for a target sequence (in the sequence database). Typically, the query sequence is much smaller than the database, with the database containing billions of nucleotides. BLAST seeks high-scoring sequence alignments based on similarity between the query and existing sequences in the database using a heuristic approach that approximates the Smith-Waterman algorithm (Smith & Waterman, 1981). This empirical approach sacrifices some accuracy with high speed, making BLAST a practical tool for searching large genomic databases such as GenBank. BLAST's speed and reasonably good accuracy represent key technical innovations in bioinformatics (Schaffer et al., 2001).

The BLAST Program

The BLAST program is a downloadable command-line utility called "blast" or can be accessed for free via the NCBI-hosted BLAST web server. BLAST's open-source nature allows access and modification of its code, creating several BLAST "spin-offs." There are various BLAST programs tailored to specific needs and types of sequences. For instance, **nucleotide-nucleotide BLAST (blastn)** (https://blast.ncbi.nlm.nih.gov/Blast.cgi?PAGE_TYPE=BlastSearch) compares a DNA query with similar DNA sequences in a specified DNA database, while **protein-protein BLAST (BLASTP)** (https://blast.ncbi.nlm.nih.gov/Blast.cgi?PAGE=Proteins) compares a protein query with similar protein sequences in a specified protein database. Similarly, **Position-Specific Iterative BLAST (PSI-BLAST) (blastpgp)** (https://www.ebi.ac.uk/Tools/sss/psiblast/) is used to find distant protein relatives by iteratively refining the search based on profile sequences (Schaffer et al., 2001). **The nucleotide 6-frame translation-protein (blastx)** (https://blast.ncbi.nlm.nih.gov/Blast.cgi?PROGRAM=blastx&PAGE_TYPE=BlastSearch&LINK_LOC=blasthome) compares the translation products of a DNA query with protein sequences in a database.

Nucleotide 6-frame translation-nucleotide 6-frame translation (tblastx) (https://blast.ncbi.nlm.nih.gov/Blast.cgi?PROGRAM=tblastx&PAGE_TYPE=BlastSearch&BLAST_SPEC=&LINK_LOC=blasttab) finds the translation products of a DNA query with translations of DNA sequences in a database, whereas **protein-nucleotide 6-frame translation (tblastn)** (https://blast.ncbi.nlm.nih.gov/Blast.cgi?PROGRAM=tblastn&PAGE_TYPE=BlastSearch&LINK_LOC=blasthome) compares a protein query with translations of DNA sequences in a database. However, **large numbers of query sequences (megablast)** (https://www.ultimateears.com/en-us/wireless-speakers/megablast.html) efficiently compare multiple input sequences to a database, saving time compared to running BLAST multiple times (Morgulis et al., 2008). Likewise, Primer-BLAST (https://www.ncbi.nlm.nih.gov/tools/primerblast/index.cgi?ORGANISM=1235996&INPUT_SEQUENCE=JX869059.1&LINK_LOC=nuccore) helps in designing primers for specific sequences (Ye et al., 2012).

Multiple Sequence Alignment

Multiple sequence alignment (MSA) is a fundamental technique in bioinformatics used to compare and align multiple biological sequences, such as DNA, RNA, or proteins (Chatzou et al., 2016). Its significance lies in elucidating conserved regions, insertions, deletions, and other evolutionary events within sequences, aiding in inferring structural, functional, and evolutionary relationships. MSA is crucial in diverse biological applications, including phylogenetic analysis, protein structure prediction, and functional annotation (Chatzou et al., 2016).

The MSA holds considerable importance in the field of plant pathology. It enables the identification of conserved virulence factors responsible for infection and disease development in plant pathogens (Saeed et al., 2012). By comparing the genetic sequences of different strains, researchers can elucidate the evolutionary relationships among pathogens, which is essential for epidemiological studies and disease management strategies. Additionally, MSA assists in deciphering mutations linked to pathogen resistance and susceptibility, informing the design of resilient crops and targeted treatments (Saeed et al., 2012).

Analyzing the genetic diversity of pathogens using MSA involves aligning sequences such as ITS regions in fungi or coat proteins in viruses from different isolates (Alemu, 2015). Conserved and variable regions in the alignment indicate evolutionary relationships and allow molecular typing. MSA also aids the development of molecular diagnostics by identifying unique pathogen sequences. Conserved PCR primer binding sites flanking variable sequences can be determined by aligning genomes. These variable regions serve as molecular biomarkers for pathogen detection (Satpathy, 2022).

The relevance of MSA in plant pathology is multifaceted. It aids in identifying conserved motifs within pathogenic genes, thereby enhancing the understanding of vital pathogenicity mechanisms. Furthermore, MSA facilitates the differentiation of pathogenic strains based on genetic variations, enhancing the precision of diagnostics (Alemu, 2015). The impact of mutations on protein structure and function can be predicted by analyzing MSA outcomes, shedding light on the influence of genetic changes on pathogen-host interactions (Alemu, 2015).

Multiple Sequence Alignment Techniques

MSA arranges homologous sequences to identify conserved regions and maintain the positional homology of residues across sequences (Thompson et al., 1994). Before constructing a multiple alignment, pairwise alignments must be generated between sequence pairs. Two foundational pairwise alignment methods are Needleman–Wunsch (Needleman & Wunsch, 1970) and Smith-Waterman (Smith & Waterman, 1981). These employ dynamic programming to systematically fill out a scoring matrix, assigning scores for matches, mismatches, and gaps. The optimal alignment is traced back, allowing for the aligned introduction of gaps.

Progressive methods such as ClustalW extend pairwise alignments to construct multiple sequence alignments (Thompson et al., 1994). ClustalW first calculates a distance matrix comparing all sequence pairs. A guide tree is constructed to reflect evolutionary relationships. The most closely related sequences are aligned first using pairwise dynamic programming. Progressively, more distant ones are aligned to the emerging multiple alignments. Gap penalties are applied to introduce gaps and maximize aligned identities. ClustalW produces a global alignment, optimizing the overall alignment.

Other programs have been introduced to improve on early limitations. MUSCLE uses a distance measure for more accurate guide trees and refines alignments with iterative steps (Edgar, 2004). MAFFT rapidly identifies homologous regions using fast Fourier transform approaches (Katoh et al., 2002). T-coffee combines information from a library of local and global alignments (Notredame et al., 2000). Profile alignment methods such as Clustal Omega incorporate knowledge of residue frequencies from alignments of homologous sequence families (Sievers et al., 2011).

Several factors guide the selection of appropriate MSA tools. Sequence divergence and number impact algorithm accuracy and speed. Local aligners such as

MAFFT-L-INS-i (Katoh & Standley, 2013) better handle highly divergent sequences than global aligners such as ClustalW. Longer sequences require more computer-intensive approaches. Characteristics of sequences, such as protein domains, also influence the choice of penalties and scoring matrices.

Phylogenetic Analysis

Phylogenetic analysis is the study of the evolutionary relationships between different organisms. It is based on the idea that all organisms share a common ancestor and that their differences can be explained by changes that have occurred over time (Khan et al., 2021). Phylogenetic trees are a graphical representation of these evolutionary relationships (Fig. 10.2). They are branching diagrams that show how different organisms are related to each other. The tree branches represent the evolutionary events that have led to the diversification of life (Satpathy, 2022).

There are four main types of phylogenetic trees: unrooted, rooted, ultrametric, and additive trees. Unrooted trees do not have a designated ancestor. They represent the relationships between a group of organisms without specifying which one is the ancestor. Rooted trees have a designated ancestor. They represent the evolutionary history of a group of organisms, starting from the common ancestor. Ultrametric trees have all the branch lengths equal to each other (Satpathy, 2022). It means that

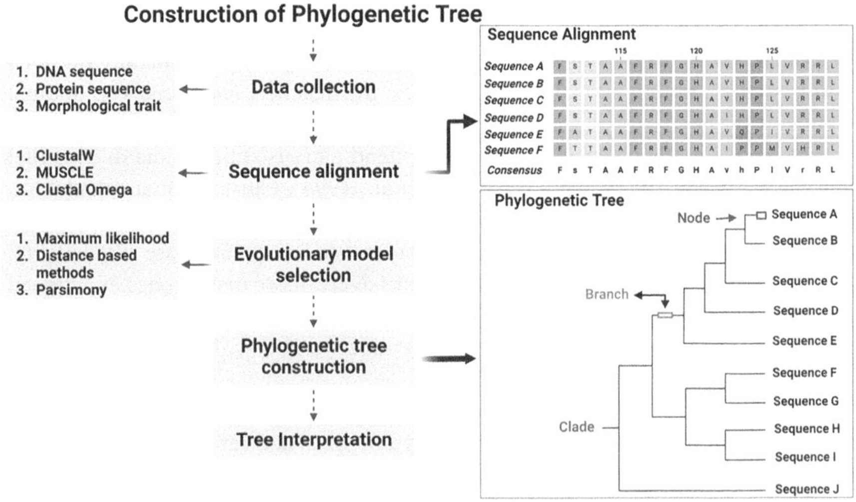

Fig. 10.2 Overview of the methodology for constructing the phylogenetic tree. The construction of a phylogenetic tree involves multiple key steps. Homologous sequences are aligned to identify common regions. An appropriate evolutionary model accounting for substitution patterns is selected. A tree topology is derived using neighbor-joining and likelihood optimization under the model parameters. The resulting tree visually depicts the evolutionary relationships hypothesized among taxa based on genetic/molecular data, offering insights into diversification

they assume that all of the evolutionary events that have occurred since the common ancestor have occurred at the same rate. Additive trees have branch lengths that represent the amount of evolutionary change that has occurred along each branch.

The construction of phylogenetic trees typically involves several steps. The first step is to collect data that can be used to infer the evolutionary relationships between the organisms being studied (Alemu, 2015). These data can include DNA sequences, protein sequences, morphological characteristics, or other data types. The next step is to align the data. It means arranging the sequences or characters to align the homologous (corresponding) positions. The final step is constructing the phylogenetic tree (Alemu, 2015). It can be done using various methods, including distance-based, character-based, and model-based methods.

Distance-based methods infer the evolutionary relationships between organisms by calculating the distances between their sequences or characters (Bogusz & Whelan, 2017). The most common distance-based method is the neighbor-joining algorithm. The neighbor-joining algorithm works by iteratively joining the two sequences or characters that are closest together. This process is repeated until all sequences or characters are joined into a single tree (Bogusz & Whelan, 2017). Character-based methods infer the evolutionary relationships between organisms by considering the shared and derived characteristics of the organisms (Munjal et al., 2019). The most common character-based methods are maximum parsimony and maximum likelihood, which find the tree most likely to have produced the observed data. Maximum parsimony methods find the tree that requires the fewest evolutionary changes (Munjal et al., 2019). Model-based methods combine the principles of distance-based and character-based methods. They use a statistical model to estimate the evolutionary relationships between organisms (Munjal et al., 2019). The most common model-based method is Bayesian inference. Bayesian inference uses Bayes' theorem to calculate the probability of each tree given the observed data. This method can account for uncertainty in the data and estimate the confidence in the results.

Phylogenetic trees can be interpreted in a variety of ways. Some of the most common interpretations include several terms, such as clade, branch, and node. A clade is a group of organisms with a common ancestor (Bawono & Heringa, 2014). A branch is a segment of a phylogenetic tree that connects two nodes. A node is a point in a phylogenetic tree where two or more branches meet.

Phylogenetic trees can infer various aspects of the evolutionary relationships between organisms (Choudhuri, 2014). For example, they can be used to determine the order in which different species evolved, to identify the common ancestor of a group of organisms, and to infer the functions of genes and proteins.

Applications of Phylogenetic Analysis

Phylogenetic analysis has emerged as a transformative tool in plant pathology, revolutionizing our understanding of plant diseases and offering critical insights into their management (Alemu, 2015). This powerful technique harnesses evolutionary relationships to unravel the genetic intricacies of plant pathogens, ultimately shaping strategies for disease control, monitoring, and crop improvement.

One of the foremost applications of phylogenetic analysis in plant pathology is its ability to unveil the evolutionary relationships among different strains and species of plant pathogens (Satpathy, 2022). By constructing phylogenetic trees, researchers can trace the evolutionary history of these pathogens, identify the emergence of new strains, track the transmission pathways, and decipher the genetic factors contributing to virulence (Satpathy, 2022). Likewise, phylogenetic studies play a crucial role in understanding the spread of plant diseases across geographical regions. By comparing the genetic sequences of pathogens collected from different locations, researchers can reconstruct the paths of disease transmission (Khan et al., 2021). This information is vital for understanding the movement of diseases and predicting potential outbreaks.

Predicting virulence factors and host adaptation is another significant application of phylogenetics (Khan et al., 2021). The application of phylogenetic analysis extends to predicting the virulence factors of plant pathogens. Researchers can identify specific genes and mutations associated with increased virulence or host adaptation by examining the genetic variations within and between strains. In pursuing disease-resistant crops, phylogenomics has illuminated a new avenue for crop improvement. Comparative genomics and phylogenetic insights enable researchers to identify resistance genes in wild or less susceptible plant relatives (Soltis & Soltis, 2003). This information guides breeding efforts, facilitating the transfer of resistance traits to cultivated crops through conventional breeding or genetic engineering. By incorporating phylogenetic knowledge, breeders can accelerate the development of crops that are resilient to diseases, contributing to sustainable agriculture and global food security (Soltis & Soltis, 2003).

Gene Structure Analysis

Functions of Gene Structure

The collection of specific sequence components that make up a gene is known as its structure. Most of the essential knowledge for live cells to survive and proliferate is found in their genes (Hartwell et al., 1999). DNA makes up genes in many organisms, and a gene's function is determined by its specific DNA sequence. In another way, gene structure refers to the arrangement and structuring of genetic data within a DNA molecule (Carroll et al., 2013). The basic building blocks of heredity, the genes, contain the blueprints for developing and operating living things. Different components of a gene structure have functions in the regulation and expression of genes (Longabaugh et al., 2005). The main elements of gene structure are listed below:

- *Promoter*

- This area near the start of a gene functions as a recognition site for the enzyme RNA polymerase, which starts transcription (Glover-Cutter et al., 2008). Specific sequences in the promoter help control when and how frequently a gene is expressed.
- *Transcription start site*
- Transcription, the process of creating an RNA molecule using the DNA template as a template, starts at this moment.
- *Exons*
- The DNA sections known as exons are responsible for encoding a portion of the final RNA molecule, which will ultimately be translated into a protein. They contain the knowledge required to make the proteins' functional parts (Brown, 2023). Exons are the coding sections of a gene that hold the blueprints needed to create useful proteins. They are essential to gene expression, in which the genetic material encoded in DNA is transformed into proteins that perform different cellular activities (Feitelson & Treinin, 2002).
- *Introns*
- Noncoding DNA regions called introns can be discovered within genes. They are eliminated through splicing before the RNA is translated into a protein (Diederichs et al., 2016). Formerly considered to be "junk DNA," introns are now understood to play regulatory roles in the production of genes. New genes can be derived from introns. Noncoding areas called introns are found in the genes of eukaryotic organisms. Introns perform several significant functions in the expression of genes, evolution, and biological activities, even though they do not directly encode functional proteins (Doolittle, 2013).
- *3′ and 5′ Untranslated Regions (UTRs)*
- Both upstream (5′) and downstream (3′) of the coding sequence, these areas are found at the ends of genes. They participate in posttranscriptional control, influencing localization, translation effectiveness, and mRNA stability (Matoulkova et al., 2012).
- *Terminator*

- A segment known as a terminator indicates that transcription has finished. It designates when the freshly created RNA molecule is released, and the RNA polymerase separates from the DNA template (Kessler & Maraia, 2021).

Role of Gene Structure in Plant-Pathogen Interactions at the Molecular Level

The intron-exon structure of genes is important in plant-pathogen interactions at the molecular level, the development of plant diseases, and their effects on plant health, agriculture, and ecosystems (Yao et al., 2021). Understanding how the intron-exon organization of genes relates to plant-pathogen interactions can provide insights into disease resistance, susceptibility, and development of plant defenses (Kaur et al., 2017). "Gene structure" in plant pathology refers to the organization and arrangement of genes within plant DNA, as well as the structure of genes in pathogens (such as bacteria, fungi, and viruses) that cause plant diseases (Osuna-Cruz et al., 2018). Understanding the structure of genes in plants and pathogens is critical for understanding disease mechanisms, host-pathogen interactions, and evolution of resistance and virulence (Neik et al., 2020). Plant pathology, or the study of plant diseases and their causes, relies heavily on gene structure. Understanding the gene structure of plants and pathogens (such as bacteria, fungi, and viruses) is critical for understanding host-pathogen interactions and devising strategies to manage and mitigate plant diseases (Hunjan & Lore, 2020). Here are some examples of how gene structure influences plant pathology:

By studying pathogenicity-related genes' gene structure and function, researchers can gain insights into how pathogens interact with host plants and develop strategies to disrupt these interactions (Guttman et al., 2014). Likewise, plant defense mechanisms rely on specific pathogen compounds/patterns being recognized (Amil-Ruiz et al., 2011). Therefore, knowledge of the gene structure of plant immunity genes can aid in developing resistant plant types using genetic engineering or traditional breeding procedures. Similarly, pathogen gene structures can reveal information about their evolutionary history and methods of adaptability to new settings or hosts (Antonovics et al., 2013; Zahoor et al., 2023). This information on gene structure can be utilized to create focused control tactics, such as building RNA interference (RNAi) molecules to silence specific pathogen genes that disrupt crucial pathogen activities (Bocos-Asenjo et al., 2022).

Pathogens can be accurately identified and classified by studying their gene composition. It is critical for tracking disease spread and comprehending disease epidemiology (Martinelli et al., 2015). Introns and exons can contain regulatory elements that control gene expression. In plant-pathogen interactions, these elements can modulate the expression of genes involved in defense pathways, such as pathogen recognition, signaling, and defense response. When the intron-exon architectures of orthologous genes from different plant species are compared, they can show both conserved and divergent traits. This investigation can reveal shared defense mechanisms as well as pathogen-specific modifications.

Gene Structure Analysis

Assessment of the properties, distribution, conservation, and potential functional relevance of introns and exons within the genes is an important attribute of gene structures that group similar genes (Sturgeon & Gardiner, 2011). This type of analysis can reveal information on gene regulation, evolutionary patterns, and diversity of gene products. Here is a step-by-step process for analyzing intron and exon structure:

Data Retrieval and Preprocessing:

- Genomic sequences of interest were obtained from reliable databases.
- Gene annotations and exon-intron boundaries were identified.

Intron Analysis:

- Determine intron lengths: Calculate the lengths of individual introns and generate statistical summaries, such as the mean, median, and distribution histograms (Lee et al., 2011).
- Conservation analysis: Intron sequences across related species were compared to identify conserved regions that may have functional significance.

Exon Analysis:

- Exon length analysis: The exon lengths were calculated, and their distribution was analyzed. Comparison of exon lengths among different genes or species.
- Alternative splicing analysis: Identify and categorize different isoforms produced by alternative splicing (Vitulo et al., 2014). Analyze how different combinations of exons affect protein structure and function.

Splice Site Analysis:

- Study donor and acceptor splice site motifs: Analyze the consensus sequences of 5′ and 3′ splice sites to understand their sequence preferences and variations. Comparison of splice site motifs across different introns and exons.

Conservation Analysis:

- For comparing exon and intron sequences across species, sequence conservation was investigated to identify functionally important regions preserved during evolution (Frazer et al., 2003).
- Study differences in intron-exon structure between paralogous genes (genes derived from a common ancestral gene) or orthologous genes (genes in different species with a common ancestor).

Functional and Expression Correlation:

- Correlate intron-exon features with gene expression patterns, tissue specificity, and developmental stages using experimental data such as RNA sequencing or microarray data.

Phylogenetic Analysis:

- Phylogenetic trees were constructed using intron and exon sequences to study evolutionary relationships and divergence patterns.

Bioinformatics Tools and Software

Several bioinformatics tools and software packages are available for analyzing gene structure, including intron-exon structure, splice sites, alternative splicing, and more. Here are some commonly used tools for various aspects of gene structure analysis:

- GSDS (gene structure display server): A tool used to show the length of introns and exons in gene structure using RNA-seq (Hu et al., 2015). The flow sheet and real-time structure prediction diagrams for gene structure analysis using GSDS are given in Figs. 10.3 and 10.4.
- SpliceGrapher: A tool for visualizing and analyzing alternative splicing events using RNA-seq data.
- MEME Suite: A collection of motif discovery and analysis tools that can identify regulatory elements within exons or introns (Bailey et al., 2009).
- BLAST: Basic local alignment search tool can compare gene sequences and identify homologous genes with similar structures (Altschul et al., 1990).

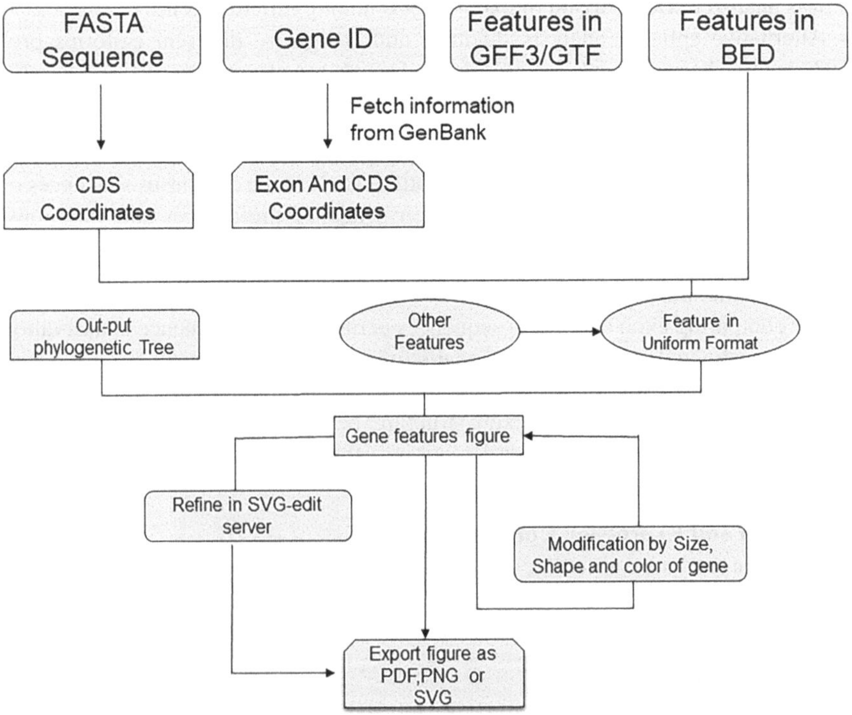

Fig. 10.3 Schematic workflow diagram for gene structure analysis

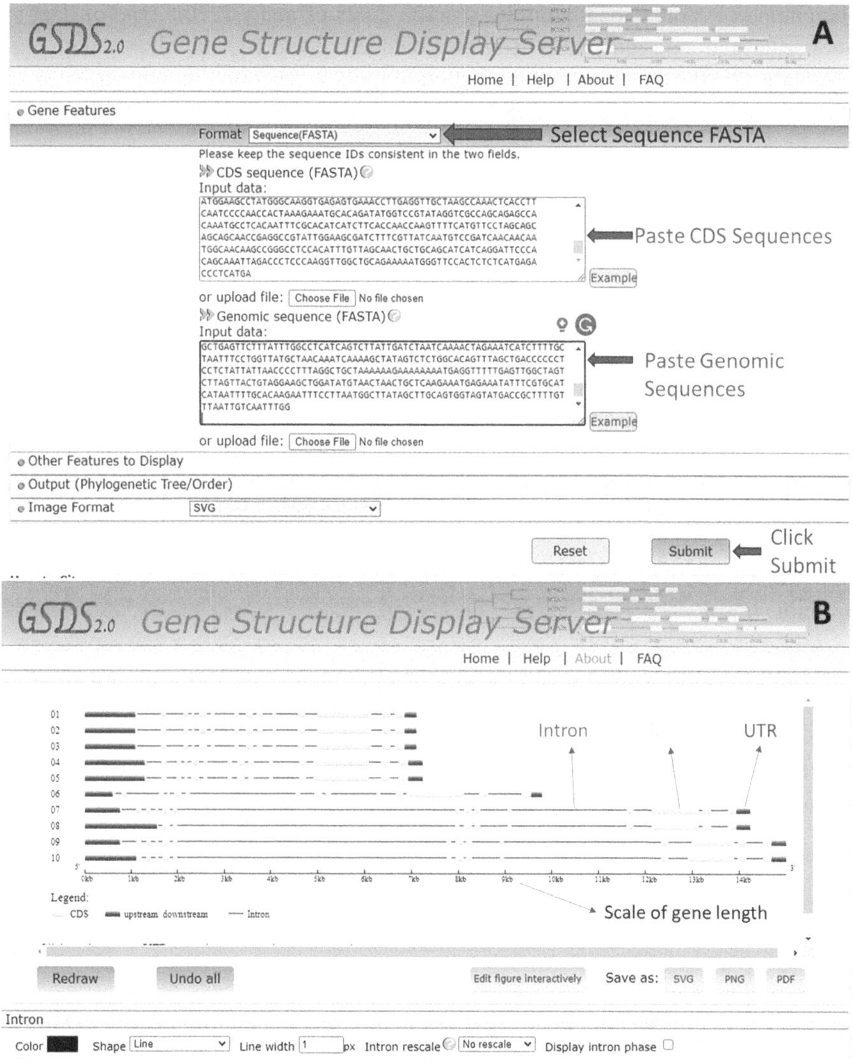

Fig. 10.4 The interface of GSDS for intron-exon structure analysis. A represents the input files and data of CDS and genomic DNA of genes, and B represents the output as an intron-exon structure

- BEDTools: A suite of tools for manipulating genomic features, which can be useful for extracting and analyzing exon-intron boundaries and other gene-related features.
- Exon-intron database (EID): A database that provides information about exon-intron structures and alternative splicing events.
- JASPAR: A database of transcription factor binding site motifs that can aid in identifying regulatory elements within gene structures (Tompa et al., 2005).

The choice of tools depends on the specific goals of analysis, the type of data to work with (genomic sequences, RNA-seq data, etc.), and your familiarity with bioinformatics software. Using a combination of tools to perform comprehensive gene structure analysis is often beneficial.

Motif and Domain Analyses

Motif

"Motif" refers to a brief, repeating sequence pattern or structural element in biological sequences such as DNA, RNA, or proteins (Macke et al., 2001). These patterns frequently have structural or functional importance and can be utilized to locate common elements in a collection of connected sequences. Motifs in proteins can designate certain functional or structural regions, while motifs in DNA and RNA can serve as protein binding sites.

Domain

A domain is a unique, useful unit inside a larger biological entity, usually a protein. It describes a distinct area of a protein sequence that functions in a particular way, frequently independently of the remainder of the protein (Vogel et al., 2004). Enzymatic processes, molecule binding sites, and areas involved in protein-protein interactions are examples of domains.

Motif and Domain Analysis

Understanding many facets of genetic regulation, evolution, and function requires motif analysis, a critical approach in molecular biology and bioinformatics. Motifs can be found in DNA, RNA, or protein sequences (Rehm, 2001). Researchers can learn more about how genes are regulated by recognizing and researching these motifs. Identification of possible transcription factor binding sites is aided by motif analysis. These binding sites control how and where genes are activated or inactivated. The regulatory networks that regulate different biological processes can be better understood by understanding the binding locations. The annotation of the roles of genes and gene products can benefit from discovering motifs (Brazma & Vilo, 2000). The roles of genes in cellular pathways and activities can be inferred from motif associations with biological processes or protein domains. DNA sequence enhancer and promoter potential can be predicted using motif analysis. Promoters and enhancers are crucial regulatory components that regulate gene expression (Hernandez & Finer, 2014). The identification of conserved and lineage-specific motifs can be performed by comparing motifs across different genomes.

Role of Motif Structure

Motif analysis provides a variety of useful uses in the study of plant pathology, including assisting researchers in understanding different facets of plant-microbe interactions, disease resistance, and pathogen virulence (Jackson & Taylor, 1996). The promoters or regulatory regions of genes activated in response to pathogen infection may contain overrepresented DNA or RNA motifs, which can be found via motif analysis. By recognizing these patterns, researchers can discover putative binding sites for transcription factors that control the plant's defense response (Ali et al., 2017a, 2017b). Similarly, effector proteins secreted by pathogenic microorganisms frequently alter the host defenses of plants. By providing information about their roles and potential interactions with plant host proteins, conserved patterns within these effector proteins can be predicted using motif analysis. Researchers can also utilize motif analysis to find possible plant proteins that function as immune receptors and recognize specific effector motifs (Zahoor et al., 2023).

Additionally, conserved motifs are frequently found in the leucine-rich repeat (LRR) and nucleotide-binding (NB) domains of plant resistance genes, among other domains. Based on similar motif or domain composition, phylogenetic analysis of the motifs conserved in different genes of the defense-related cystine-rich protein (CRP) family demonstrated that proteins with similar motifs clustered together in the same clade (Almaghrabi et al., 2019) (Fig. 10.5). For predicting the function of resistance genes and their probable interaction partners, motif analysis can help find these domains and motifs. Moreover, scientists working on plant-pathogen interactions can find possible transcriptional regulatory networks that control plant defense responses by looking at the regulatory motifs in the promoters of coregulated defense-related genes. It leads to a better understanding of how plants coordinate their defense mechanisms.

Comparative motif analysis between several pathogen species or strains might discover divergent or conserved motifs connected to virulence or pathogenicity (Baltrus et al., 2011). Motif analysis can be used to predict the secondary structures of RNAs implicated in regulatory processes. These secondary structures can influence the stability and performance of regulatory RNAs that support interactions between plants and microbes. Likewise, the plant microbiome contains a variety of diseases, and motif analysis can help discover conserved DNA or RNA motifs within their genomes (Röttjers & Faust, 2018). In plant communities, this can be used to characterize the diversity and potential interactions of harmful species.

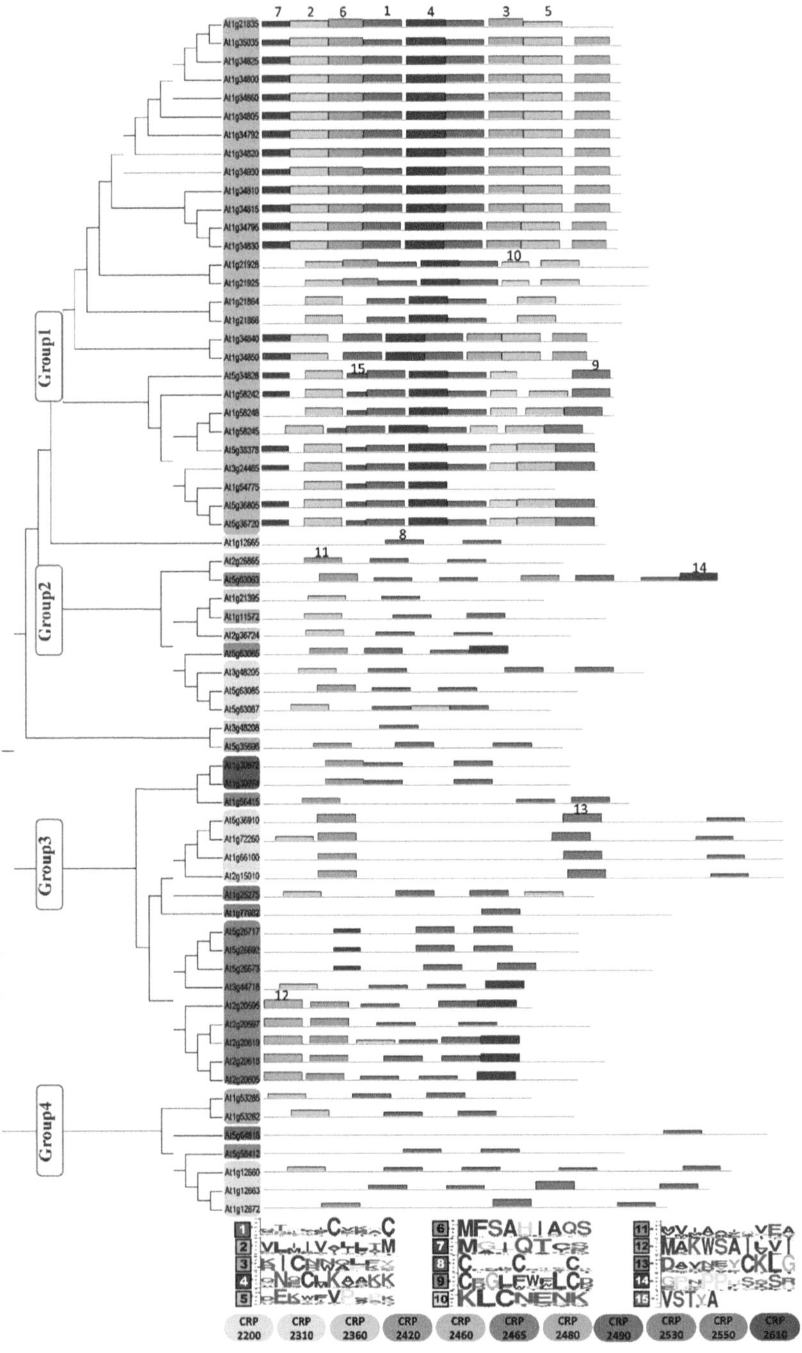

Fig. 10.5 Phylogenetic analysis representing the motifs conserved in different defense-related cystine-rich protein (CRP) family genes. The motifs and domains are the output of the MEME Suite analysis. The position and length of the mature peptide and acidic domain of thionins are labeled with blue and red boxes, respectively. Members of different CRP families are highlighted along with the name of their CRP family. Proteins with similar motifs are clustered together in the same clade

Tools Used for Motif Analysis

Several techniques are available for motif analysis, each intended to find and examine certain sequence motifs in DNA, RNA, or protein sequences. The following are a few frequently used tools for motif analysis:

- Multiple EM for Motif Elicitation (MEME): MEME is a popular technique for locating new motifs among a group of connected sequences. The expectation-maximization approach is used to find conserved motifs and the probability matrices associated with each one of their positions (Bailey et al., 2009). The interface of MEME Suite used for motif and domain analysis with the instructions to use it is given in Fig. 10.6.

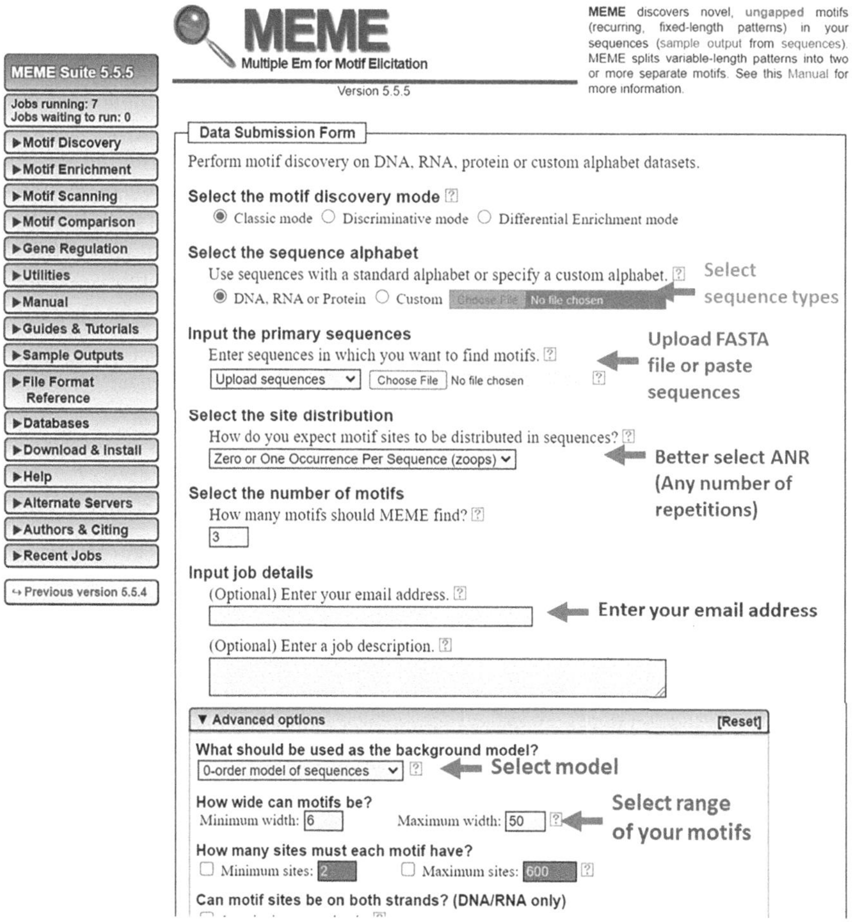

Fig. 10.6 The interface of MEME Suite used for motif and domain analysis with its instructions

- Hypergeometric Optimization of Motif EnRichment (HOMER): HOMER offers enrichment analysis, annotation, and motif discovery (Meng et al., 2014).
- Regulatory Sequence Analysis Tools (RSAT): RSAT provides a user-friendly interface for examining regulatory regions in DNA sequences.
- MEME-ChIP: Specifically created for motif analysis in ChIP-seq data, MEME-ChIP is an extension of MEME.
- Discriminative Regular Expression Motif Elicitation (DREME): The DREME program, which is a component of the MEME Suite, focuses on discriminatively locating brief, ungapped patterns in DNA sequences.
- ChIPMunk: ChIPMunk is designed to examine ChIP-seq data to find and examine transcription factor binding motifs. The strength and locations of the binding sites are also disclosed (Bailey et al., 2013).
- BioProspector: A technique for finding overrepresented motifs in sequences is called BioProspector. Gibbs sampling is used to find conserved motifs.

Promoter Analysis

A promoter, also known as a cis-regulatory element, is a DNA region where the transcription of a specific gene is initiated. A promoter is a section of the genome where the transcription apparatus can attach to initiate the transcription of the adjacent gene into an RNA molecule. Promoter regions frequently exhibit elevated CG content (Bosserhoff & Kappelmann-Fenzi, 2021). Typically, it comprises three segments:

- The core section holds the transcription start site and RNA polymerase binding site, which are crucial for proper transcription initiation.
- The proximal promoter contains primary regulatory elements.
- The distal region carries additional elements.

Figure 10.7 represents the structure of a typical gene and promoter, showing these elements of the promoter.

Promoters are vital for transgene performance. Various promoters, such as 35S, actin, ubiquitin, 2A11, Lhcb3, Opaque-2, and APase, have been used to manage gene expression based on whether it needs to be constant or tissue specific (Sangwan et al., 2018).

Plants encounter diverse challenges and stresses throughout their growth. Certain transcription factors (TFs) in plants are targeted in regulating gene expression, bolstering plants' capacity to adjust to stressful environments (Tan et al., 2022). The

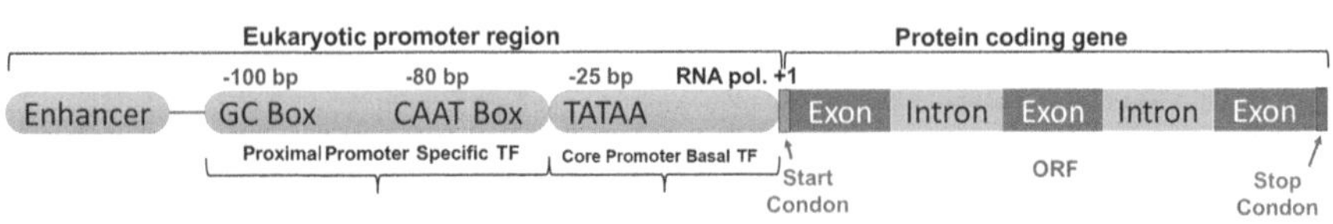

Fig. 10.7 Structure of a typical gene and promoter

involvement of TFs in the gene regulatory network of plants is vital for mediating stress response pathways (Li et al., 2019). While the coding regions of genes that produce a specific protein tend to exhibit conservation across different species, the similarity in their promoter sequences can vary (Luo & Wei, 2019; An et al., 2019). Studying plant promoters extensively, including identifying core activating sequences and analyzing the quantity, types, and distribution of cis-regulatory elements, not only enhances our comprehension of gene regulatory mechanisms but also furnishes diverse functional promoters for genetic engineering applications (Li et al., 2020).

The typical gene promoter region is approximately 1000 base pairs upstream of the gene's coding sequence transcription start site (Eun et al., 2017). Cis-regulatory elements within the promoter region enable plants to respond to biotic or abiotic shifts by controlling associated downstream genes during transcription (Li et al., 2020; Sazegari et al., 2015). Typically, cis-elements consist of short DNA motifs spanning 5–25 base pairs (Rani, 2007). Detecting these cis-elements aids in forecasting their regular function in transcriptional processes (Tang et al., 2023). Upstream of the transcription start site (TSS) exists a promoter, a cis-acting element facilitating transcription initiation (Grienberg & Benayahu, 2005). Cis-elements in plant gene promoters responsive to pathogens, controlled by effectors or regulatory proteins, modulate gene expression postinvasion (Yuan & Wang, 2011). Discovering internal cis-regulatory DNA elements (CREs) that respond to intrinsic and environmental signals is vital for understanding gene regulation (Roccaro et al., 2013) and plant-microbe interactions.

Applications

Two primary strategies in cis-regulatory research involve analyzing element positioning, orientation, and interactions, resulting in the discovery of enhanced and innovative applications (Table 10.1; Fig. 10.8) (Biłas et al., 2016). Concerning their application, the promoters are grouped as follows:

- **Constitutive promoters** exhibit activity across various tissues and developmental phases.
- **Spatiotemporally specific promoters** furnish expression data specific to particular tissues or developmental stages.
- **Inducible promoters** are controlled by the introduction of an external chemical or physical stimulus.
- **Synthetic promoters** comprise specific regulatory elements positioned near or within promoter sequences (Hernandez & Finer, 2014).

Table 10.1 Databases

URL	Resource	Reference
https://agris-knowledgebase.org/	AGRIS	Palaniswamy et al. (2006)
http://bioinformatics.psb.ugent.be/ATCOECIS/	AtCOECiS	Vandepoele et al. (2009)
http://www.athamap.de/	AthaMap	Steffens et al. (2004)
http://bar.utoronto.ca/ntools/cgi-bin/BAR_Promomer.cgi	BAR Promomer	Toufighi et al. (2005)
https://users.soe.ucsc.edu/~kent/improbizer/improbizer.html	Improbizer	Priest et al. (2009)
https://www.dna.affrc.go.jp/PLACE/?action=newplace	New PLACE	Higo et al. (1999)
http://bioinformatics.psb.ugent.be/webtools/plantcare/html/	PlantCARE	Rombauts et al. (1999), Lescot et al. (2002)
http://planttfdb.cbi.pku.edu.cn/	Plant TF DB	Guo et al. (2007)
https://www.arabidopsis.org/cgi-bin/patmatch/nph-patmatch.pl	Tair	Swarbreck et al. (2007)
http://gene-regulation.com/pub/databases.html	Transfac	Matys et al. (2003)
https://jaspar.genereg.net/	JASPAR	Mathelier et al. (2016)
http://cisbp.ccbr.utoronto.ca/	Cis-BP	Weirauch et al. (2014)
https://regulondb.ccg.unam.mx/	RegulonDB	Gama-Castro et al. (2016)
http://plantregmap.gao-lab.org/	Plant Reg Map	Tian et al. (2020)
https://grassius.org/	Grassius	Yilmaz et al. (2009)
http://ted.bti.cornell.edu/	Tomato Functional Genomic Database	Fei et al. (2010)

Methods

1. **Data retrieval:** Genomic sequences up to 1 kb or 1000 bp from the upstream region were obtained from databases, i.e., NCBI and Phytozome (Fig. 10.9).
2. **Database or web tool:**
   - Click on Search for CARE, put email in the first box and the third box, paste 1 kb genome sequence, and search.
   - An email with an attachment will be sent after 2–3 min. Open it in the Chrome web (Fig. 10.10).
3. **Prepare an Excel sheet:**
   - Get every motif function count on an Excel sheet with their Gene-ID, e.g., in the above picture, the AT TATA box contains no function, while Box 4 contains three functions, and so on.
   - Prepare an Excel sheet with the counts for every gene family member (Fig. 10.11).
4. **Visualize:**

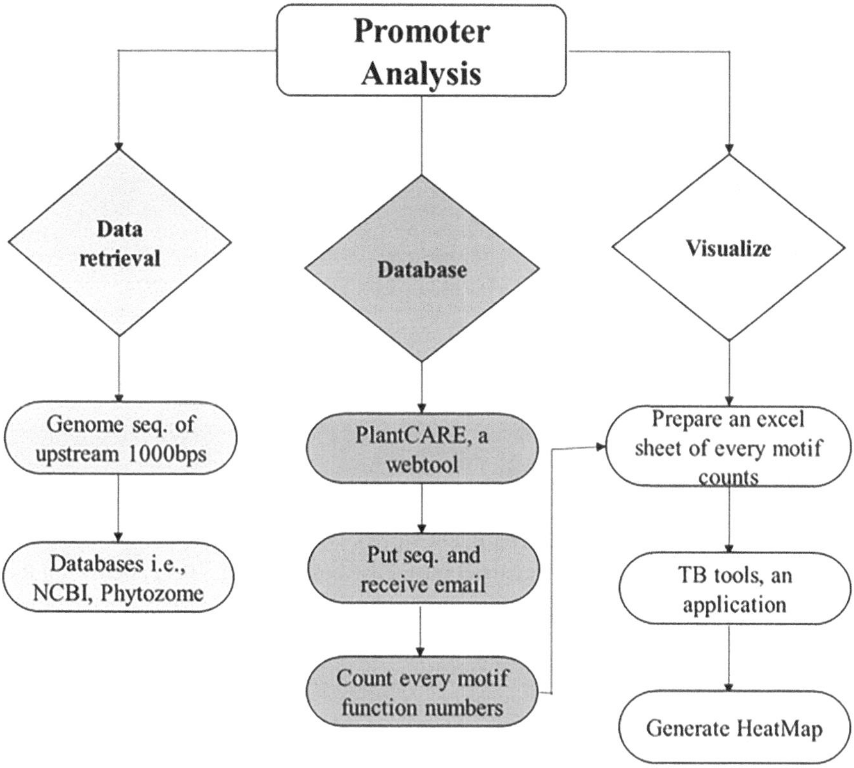

Fig. 10.8 Flow sheet for the promoter analysis to predict cis-regulatory elements

Fig. 10.9 Data retrieval from databases

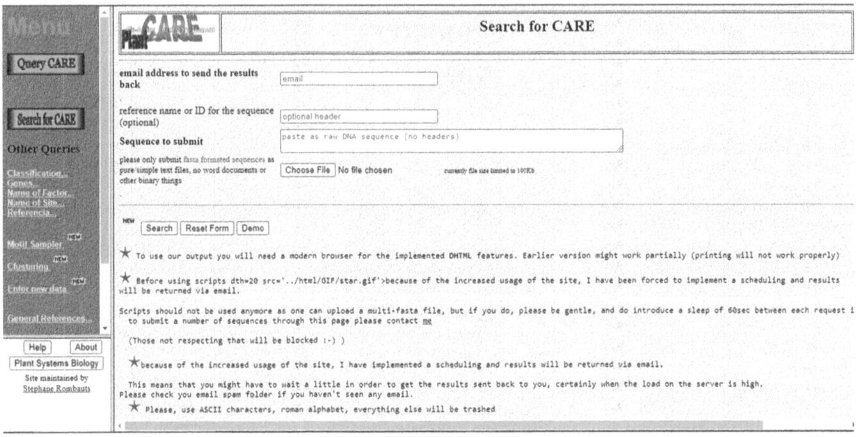

Fig. 10.10 Database for searching CARE

- Obtain TB tools, and click on the heatmap.
- All the data are in an Excel sheet paste, as shown in Fig. 10.12, below on the left side, and click on start.
- Similar to the right side, a heatmap will be generated.

Different Bioinformatics Approaches for Gene Expression Analysis

RNA Sequencing

RNA sequencing (RNA-seq) has emerged as a transformative technology in molecular biology, offering unparalleled insights into the transcriptome of an organism at an unprecedented resolution (Wang et al., 2009). The procedure involves several key steps, including extracting total RNA from the biological sample of interest, such as cells or tissues. The extracted RNA underwent treatment with enzymes to remove contaminating DNA and ribosomal RNA. Subsequently, the remaining RNA molecules are reverse-transcribed into complementary DNA (cDNA) to amplify and preserve the RNA sequences. Specific adapters were added to the fragmented cDNA to prepare the cDNA library for sequencing. These adapters contain unique sequences that facilitate identifying and sorting individual RNA molecules during sequencing. The prepared cDNA library was then subjected to high-throughput sequencing using platforms like Illumina. The resulting short reads are aligned to a reference genome or transcriptome, allowing for the quantification of gene expression levels and the detection of alternative splicing events (Ozsolak & Milos, 2011; Kukurba & Montgomery, 2015). RNA-seq allows the study of RNA expression differences in various tissues and facilitates the analysis of diverse

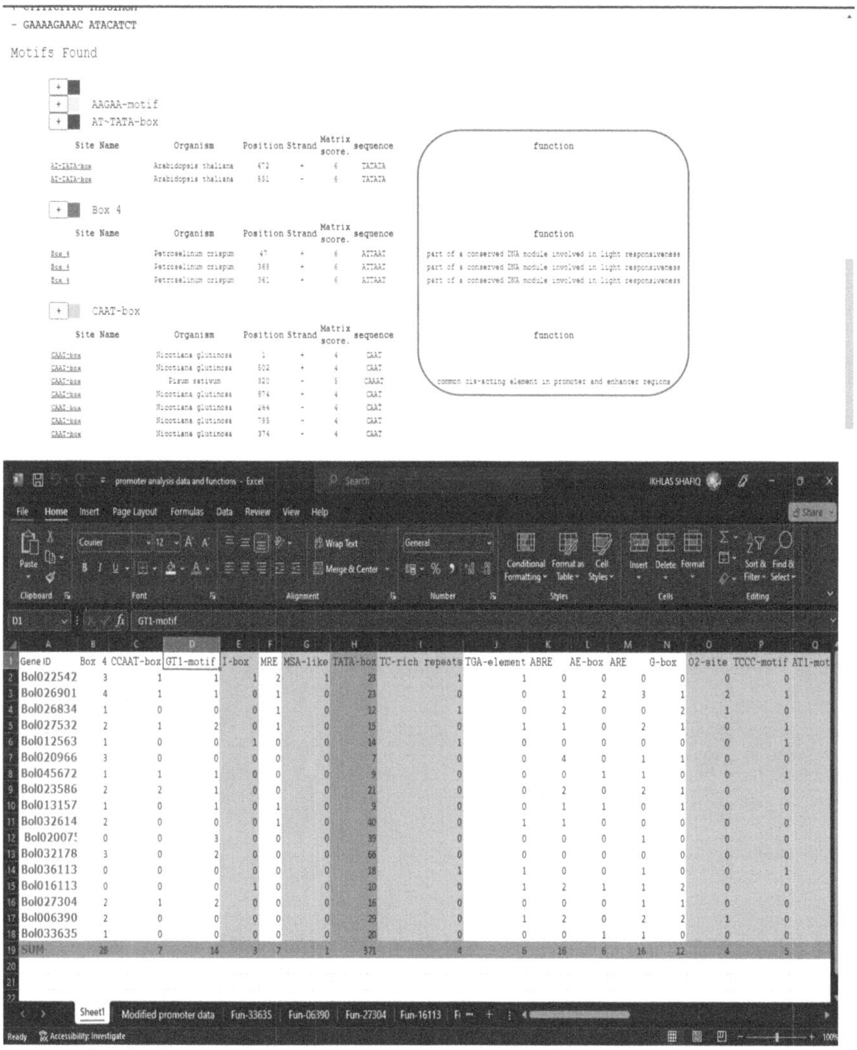

Fig. 10.11 Preparation on input file

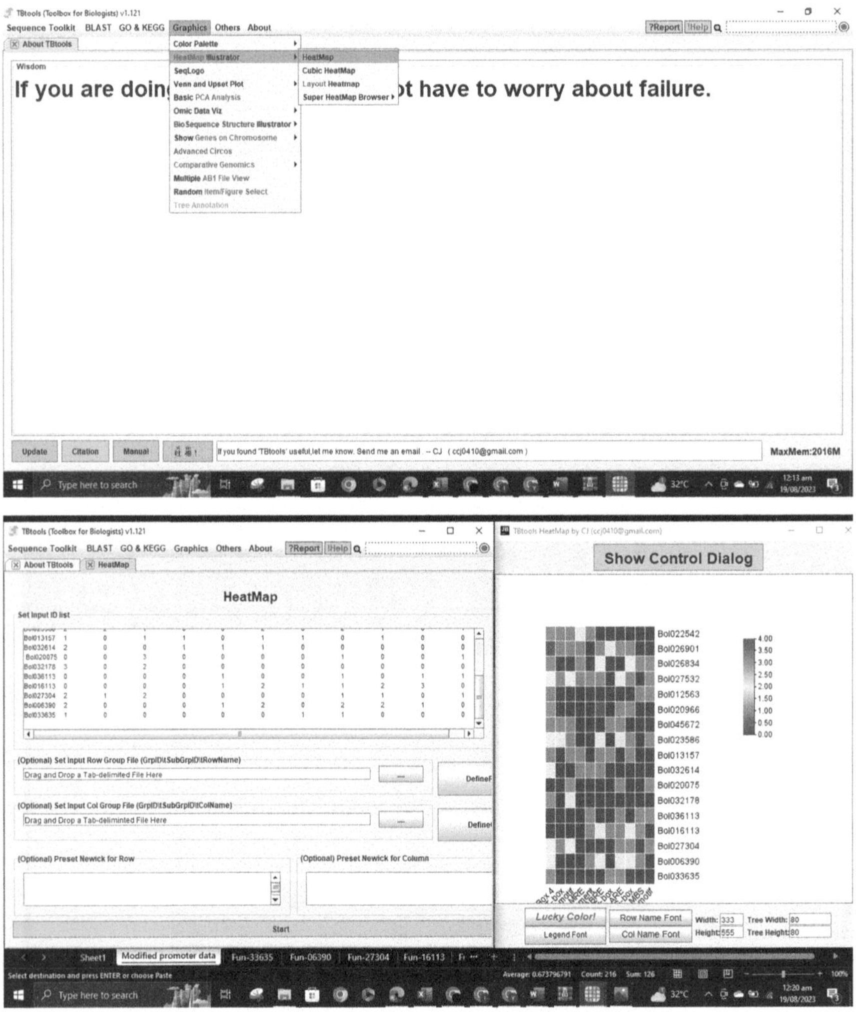

Fig. 10.12 Visualization of result

aspects, such as altered axon usage, gene fusions, and genetic variations such as SNPs, indels, and mutations (Bosserhoff & Kappelmann-Fenzi, 2021).

Importance of In Silico RNA-Seq Analysis

This revolutionary technique allows researchers to study gene expression alternative splicing events and even discover novel transcripts, providing a deeper understanding of cellular processes (Hong et al., 2020). By leveraging powerful bioinformatics tools and algorithms, researchers can obtain and align raw sequencing reads (obtained from RNA-seq databases such as NCBI-SRA and GEO), quantify gene expression levels, and perform differential expression analysis, enabling the identification of key genes and pathways underlying diverse biological phenomena. This analytical approach reduces experimental costs and time and enhances the reproducibility and rigor of scientific findings. Furthermore, in silico analysis opens doors to integrating multiomics data, enabling a more holistic view of complex biological systems (Conesa et al., 2016).

RNA-Seq Analysis Using the Galaxy Server

In silico RNA-seq analysis involves several key steps, including data preprocessing, alignment, quantification, and further downstream analyses. To analyze the expression response of a particular set of genes against abiotic and biotic stresses, first, RNA-seq data were retrieved from the NCBI-Sequence Read Archive (SRA) database (Sayers et al., 2022). A step-by-step guide to performing RNA-seq analysis is provided in Fig. 10.13.

1. Data Retrieval:
   - The SRA accession numbers of the RNA-seq datasets were retrieved from the NCBI-SRA database (https://www.ncbi.nlm.nih.gov/sra) using filters such as organism, tissue, and experimental condition.
   - Relevant details of each experiment are recorded along with the SRA run accession numbers.
2. Data Preprocessing:
   - Genomic sequence files and the organism's general feature format (GFF) files were extracted from the NCBI assembly database (https://www.ncbi.nlm.nih.gov/assembly).
   - These files are uploaded to the Galaxy Europe server (https://usegalaxy.eu/).
3. Read Mapping:
   - The "Faster Download and Extract Reads in FASTQ" tool was run on Galaxy to retrieve the raw sequencing reads in FASTQ format from SRA using the accession numbers.
   - The Bowtie 2 read aligner was used to map the Illumina sequencing reads to the reference genome (Langmead & Salzberg, 2012). SAMtools was used to convert the alignment outputs to BAM format.
4. Transcript Quantification:
   - Cufflinks were run on the mapped reads in BAM format to assemble transcripts and calculate their abundance in fragments per kilobase of exon per million fragments mapped (FPKM) (Trapnell et al., 2010).

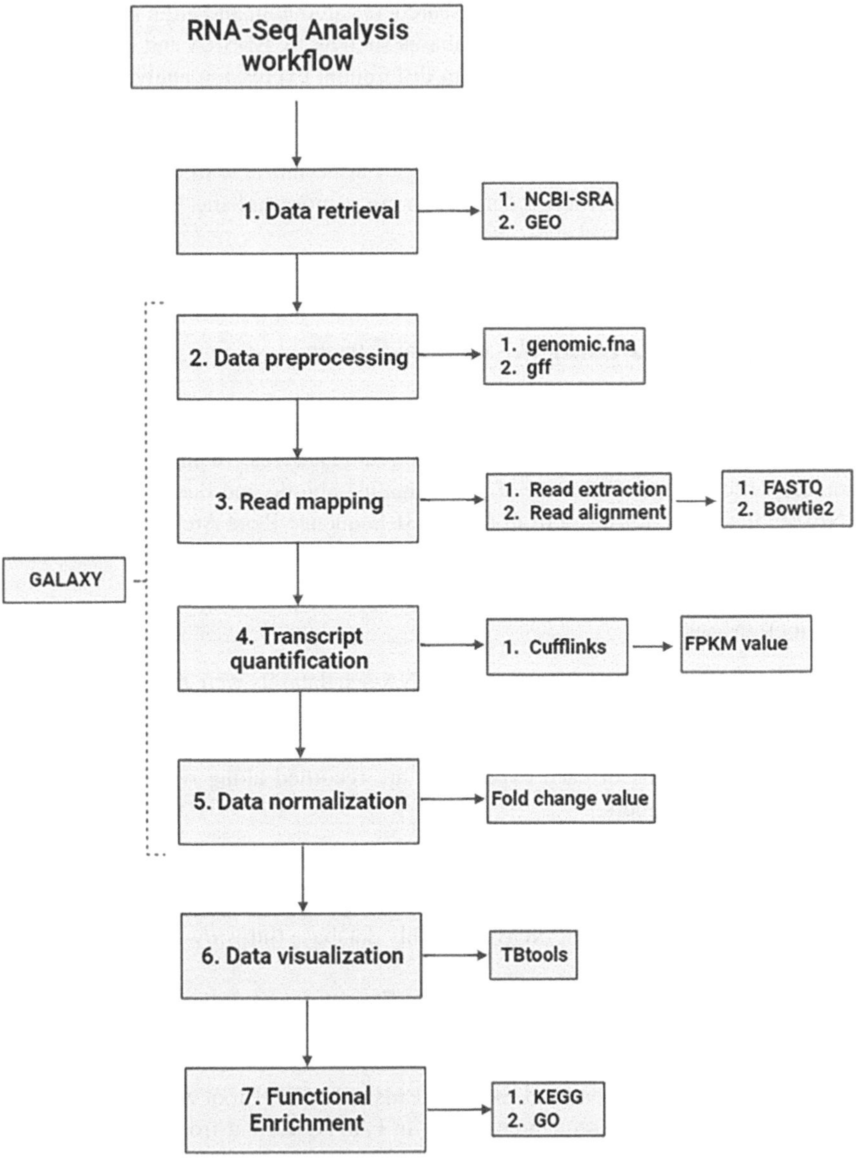

Fig. 10.13 Schematic representation of the RNA-seq analysis workflow using the Galaxy server. The analysis involves seven major steps: (1) data retrieval from the NCBI-SRA database, (2) data preprocessing involving genome extraction from the assembly database and uploading files to Galaxy, (3) read mapping of FASTQ files using Bowtie 2, (4) transcript quantification with Cufflinks, (5) data normalization by calculating mean expression across replicates and fold changes between conditions, (6) data visualization using heatmaps, and (7) downstream functional analysis and validation

5. Data Normalization:
   - Samples are classified into different groups based on treatment conditions.
   - The mean FPKM value was calculated for each gene across replicates within each group.
   - Fold change in expression was determined by dividing the mean FPKM of the treatment group by the control group.
6. Data Visualization:
   - Heatmaps showing differential gene expression between treatment groups can be generated using several software programs, such as TBtools (Toolbox for biologists) (Chen et al., 2020).
7. Downstream Analysis:
   - Functional enrichment analysis of differentially expressed genes was performed using appropriate tools/databases.
   - Qualitative and quantitative PCR techniques were used to validate the results.

DNA Microarray/Gene Chip Analysis

Microarrays are a high-throughput and versatile technique used to simultaneously detect and study the expression profiles of thousands of genes with known and unknown functions. Similarly, this technique could also be used to detect polymorphisms and mutations in genomic DNA. More specifically, DNA microarrays or GeneChips are microscope slides, silicon chips, or nylon membranes with several thousands of microscopic DNA, cDNA, or oligonucleotide spots with defined positions on the chip. Each spot contains a known DNA sequence of the genes under study, known as probes or reporters. These probes are primarily used to detect gene expression for a particular set of genes for which the probes were designed. However, genome-wide expression or transcriptome analysis is used to study gene expression at the whole plant or tissue level.

Recently, comparative transcriptomes have become more frequent, in which we mostly perform a microarray analysis on contrasting samples, such as mRNA molecules collected from contrasting pairs of samples, i.e., an experimental sample and a control sample. For instance, the control sample may be collected from a healthy individual, while an experimental or treated sample may be taken from a diseased individual, i.e., healthy versus cancer-affected cells.

Nucleic acid hybridization is exploited by microarray technology based on the complementarity of nucleotide base pairs, i.e., the complementarity rule of the DNA double-helix model. Therefore, the fundamental principle behind microarrays is hybridization between two DNA strands from the probe and the target DNA. The target sequences are labeled with fluorescent dyes that bind to a specific complementary probe sequence, leading to the generation of a fluorescent signal primarily based on the hybridization strength determined by the number of paired bases hybridized between the target and probe sequences.

The microarray analysis determining comparative transcriptomes mainly involves quantifying mRNA molecules in the contrasting samples. These two contrasting mRNA samples are then converted into complementary DNA (cDNA) through reverse transcriptase-polymerase chain reaction (RT-PCR), and each sample is labeled with a fluorescent probe of a different color. For instance, the experimental or treated cDNA sample could be labeled with a red fluorescent dye, while the reference or control cDNA sample could be labeled with a green fluorescent dye. It is followed by mixing two contrasting samples and allowed to hybridize to the probes on the surface of the microarray gene chip. Hybridization is a process that is characterized by the binding of cDNA molecules to DNA probes on a microarray gene chip. Following hybridization, the cDNA is washed away from the microarray slide, and a particular wavelength of the fluorescent dye is detected and scanned to quantify the expression of each probe (for each gene) present on the microarray slide.

If the expression of a particular gene is higher in the diseased sample than in the control sample, then the spot corresponding to that particular gene will appear red after scanning on the microarray. However, the spot appears green if that gene is downregulated in an experimental sample compared to the control sample. If the expression of the gene is not affected by the treatment, the expression will be the same in both samples, and the spot will appear yellow. The data gathered through microarrays can be used to create gene expression profiles, which show simultaneous changes in the expression of many genes in response to a particular condition or treatment. For instance, see the comparative microarray analysis for normal and tumor tissues in Fig. 10.14.

The principle behind microarrays is that complementary sequences will bind (hybridize) to each other. Some of the important properties of the hybridizing sequences that determine the level of sensitivity include accuracy, precision, specificity, variations in hybridization temperature, purity and degradation rate of genetic material, and amplification process. Similarly, probe length, GC content, hairpin, and self-dimerization of the hybridizing probes are other important features of hybridization that may lead to changes in the sensitivity of the microarray. Figure 10.14 demonstrates the overview of the microarray data analysis workflow.

Preprocessing of Microarray Data

In microarray analysis, preprocessing of the raw data provides us with the transformation of the raw fluorescence signal detected at the scanner into a signal normalized for various experimental errors. There are three main steps in the data preprocessing for microarray analysis: background correction, experimental normalization, and summarization of the transcript intensity. Removal of

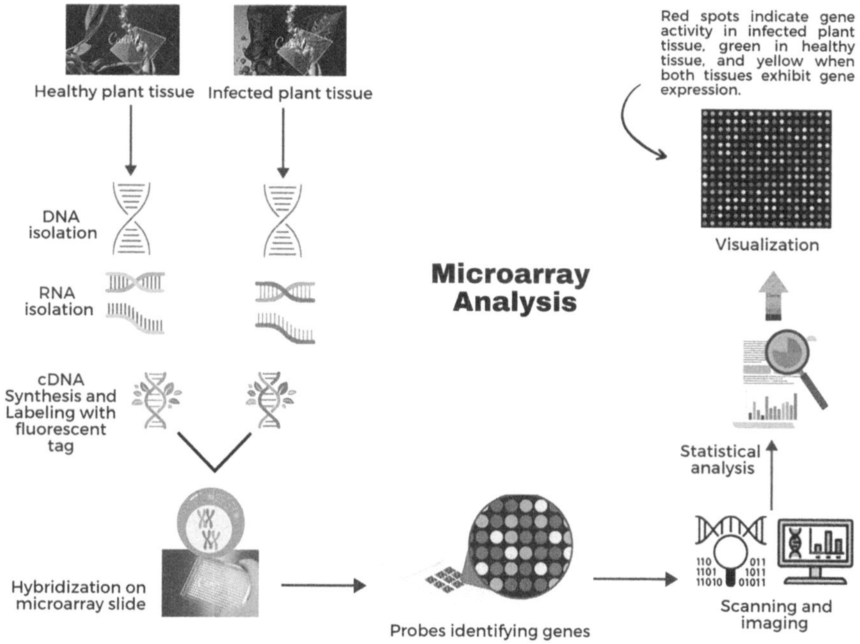

Fig. 10.14 Overview of the microarray data analysis workflow. This figure illustrates the key steps involved in microarray data analysis in plants. Experimental design is the initial stage, where researchers carefully plan the choice of plant species, tissues, treatments, and replicates to ensure statistical analysis. Subsequently, RNA isolation methods extract high-quality RNA from plant samples. Following RNA isolation, cDNA synthesis and labeling with fluorescent dyes were performed. The labeled cDNA samples are then subjected to hybridization on a microarray chip containing gene-specific probes. Scanning and imaging capture fluorescent signals corresponding to gene expression levels. Data preprocessing steps, including background correction and normalization, enhance data quality. Statistical analysis identifies differentially expressed genes (DEGs), while functional enrichment analysis provides insights into DEG biological significance. Finally, the results are visually represented through heatmaps and other graphical tools to facilitate interpretation. This workflow is integral to studying gene expression patterns and their biological implications in plant research

noninformative signals and transcripts that are not expressed is also included in the preprocessing of the microarray data.

After assessing our raw data, the data are subjected to a normalization approach. Normalization is a mathematical transformation of the data that guarantees that we can accurately determine that the channel on the same array could be compared either with each other or with the different arrays in the same experiment.

Unequal quantities of the input RNA, differences in fluorescent dye labeling and detection efficiencies, and emergence of systematic biases in the gene expression measurements are important reasons for which the normalization process is a prerequisite. Robust multiarray averaging (RMA) is the main normalization approach primarily used for background subtraction or correction, normalization, and summarization of the raw data. This approach does not have an advantage for the mismatch spot and does not take particular account of nonspecific probe hybridization in the probe set background calculation.

In RMA, background subtraction or correction is carried out through a multifarious algorithm that removes both the multiplicative and additive noise components. After applying background corrections to the microarray individual probes, RMA applies quantile normalization to the resultant data.

Quantile normalization is a widely used method for analyzing microarray data generated from Affymetrix GeneChips. This approach makes the distributions of two datasets statistically similar. In other words, "quantile normalization is a technique for making two distributions identical in statistical properties." This approach primarily uses the quantile-quantile plot that leads to the same distribution of gene expression pooled from each array. The main objective of normalization is to remove and minimize the technical variability among the samples under study. It is done by redistributing the signal intensities across all the samples so that they now have the same distribution, i.e., mean values or a minimum standard deviation of the mean.

If the background correction option is turned off in the microarray analysis, it will lead to high background noise that will result in unnecessary transcript abundance in addition to nonspecific binding (NSB) and autofluorescence from the surface of the gene chip. Ultimately, the background noise leads to unreliable results. Likewise, if the quantile normalization is turned off in the microarray analysis, this will lead to false positives, negatives, and outliers that may result in high standard deviation with less significant or unreliable gene expression results. If background correction and quantile normalization are turned off, the gene expression results will not be reliable whether we increase the number of biological and technical replicates of the samples or use highly efficient probes.

Mapping of Genes and Loci on Chromosomes

Mapping of genes or quantitative trait loci (QTL) serves as valuable data among different disciplines, such as agriculture, medicine, and evaluation, enabling researchers to establish connections between intricate phenotypic traits and specific chromosomal segments. It aims to pinpoint the precise locations, actions, interactions, and quantity of the regions involved (Myles and Wayne, 2008). Significantly employed statistical software programs for QTL mapping include QTL network (Yang et al., 2008), QGENE (Nelson, 1997), R (Broman et al., 2003), PLABQTL (Utz and Melchinger, 1996), and MapChart (Nadeem et al., 2018: Voorrips, 2002). Van Ooijen's 1994 program, DrawMap, is among the advanced software tools to create genetic linkage maps. Specific mapping software such as Mapmaker (Lander et al., 1987: Lincoln et al., 1990) displays linkage maps using locus distances, which need a conversion to determine map positions (Voorrips, 2002).

MapChart is a tool used to visualize linkage group data through charts. The workflow of the localization of genes/loci on chromosomes is given in Fig. 10.15. Such charts consist of vertical bars that display the locations and labels of loci. Next

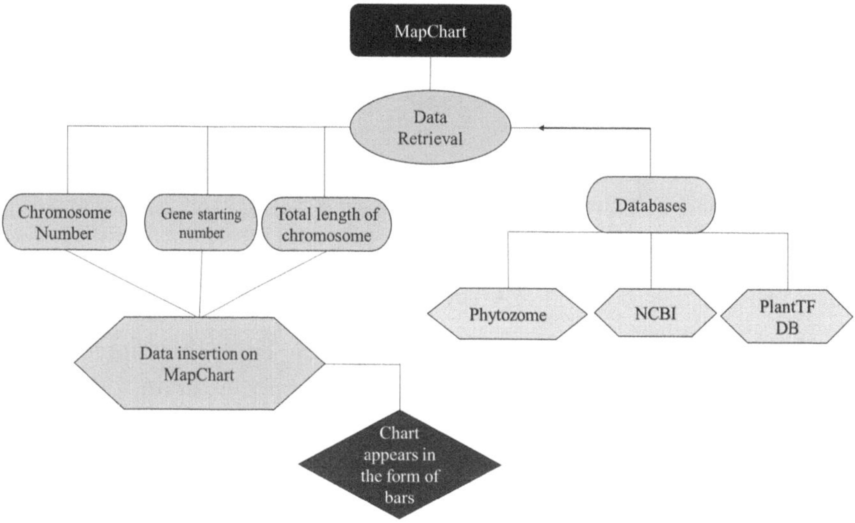

Fig. 10.15 Workflow of localization of gene/loci on chromosomes. It includes six steps: (1) retrieve data of the desired crop from the desired gene family; (2) obtain data using different databases, i.e., NCBI, Phytozome, and Plant TFDB; (3) prepare an Excel sheet by obtaining the number of chromosomes, total length of chromosome, and gene starting numbers and align all the data according to chromosome numbers; (4) download MapChart, a statistical software by using the link https://www.wur.nl/en/show/mapchart.htm; (5) open MapChart and paste all the data as shown in the picture; and (6) click on chart and bars. Chromosomes containing gene positions will appear, as shown in the picture

to the toolbar, you can incorporate QTL intervals and QTL graphs, and you can also emphasize particular bar segments. Additionally, if the charts are too large for a single page, they can be divided into segments. Each element within the charts can be individually tailored with unique colors, fill styles, text fonts, and additional customization options (Voorrips, 2002). The interface of the MapChart software for the development of chromosome maps with different gene/loci positions is given in Fig. 10.16.

Genome-Wide Association Studies

Genome-wide association studies (GWASs), or genome-wide association mapping (Baloch et al., 2015), are more modern approaches for revealing natural variations and connecting certain features to genotypes. Although biparental QTL mapping has been successful, developing a mapping population requires a considerable time investment. Additionally, its ability to uncover new genes is still limited by the genetic context of just two parent lines (Cockram and Mackay, 2018). The full characterization of many crops from various geographical areas has been made possible

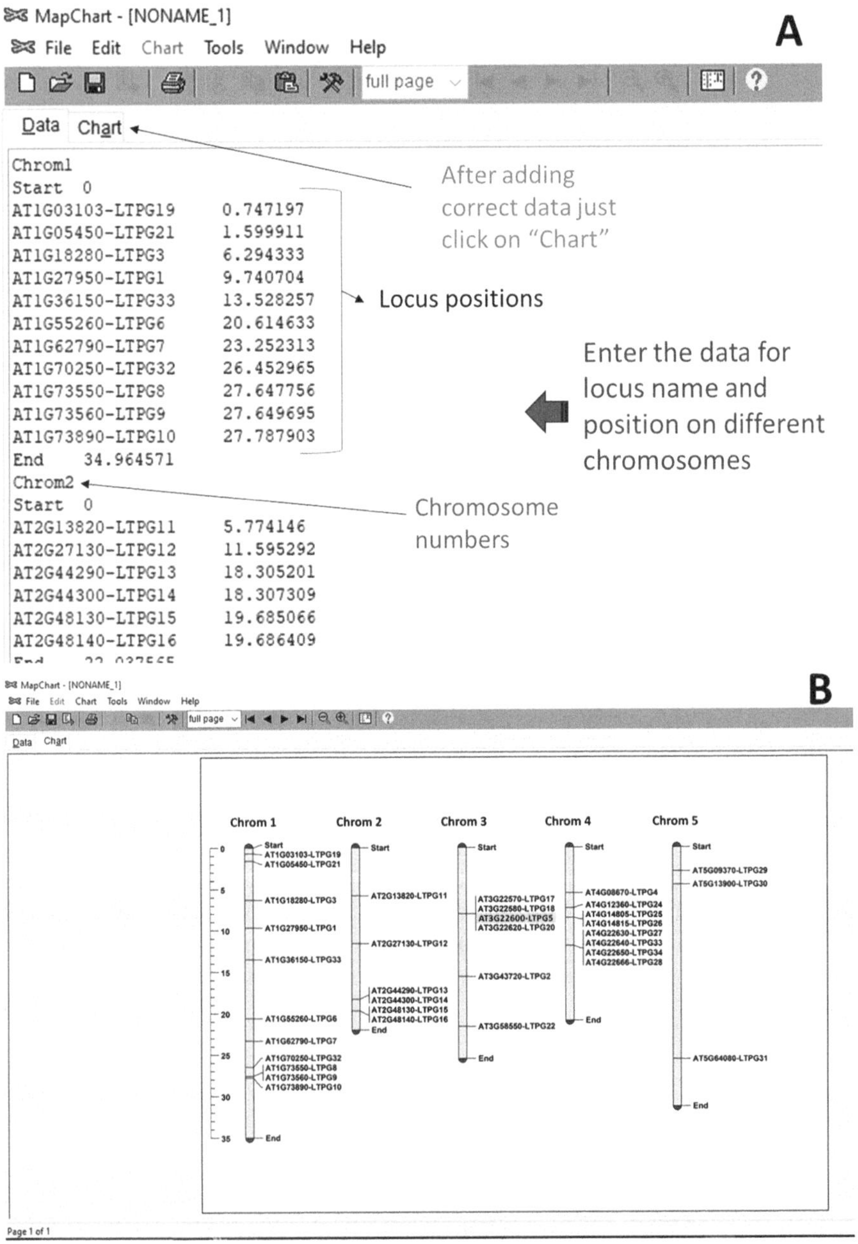

Fig. 10.16 The interface of MapChart software for developing chromosome maps with different gene/loci positions. Part A shows the input data, while Part B displays the output data as chromosome maps

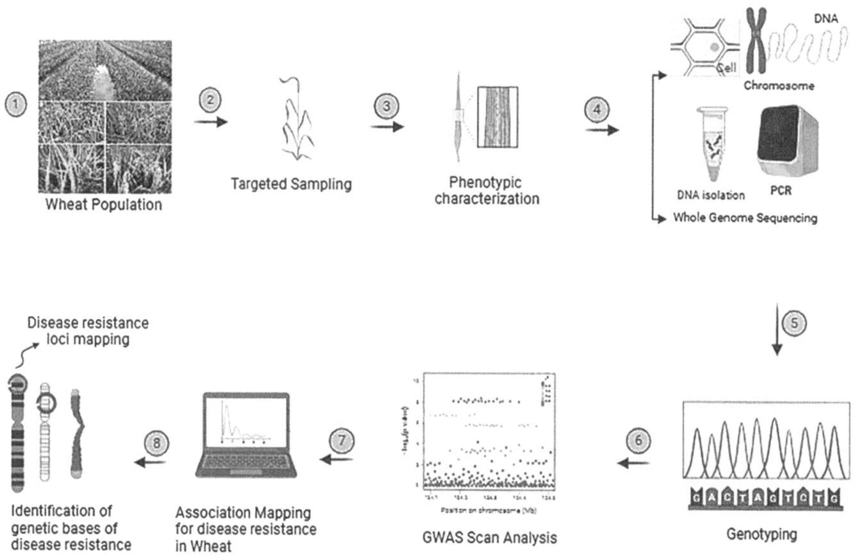

Fig. 10.17 Overview of genome-wide association study (GWAS) analysis steps in plants. This figure visually represents the essential steps involved in a genome-wide association study (GWAS) analysis applied to plants. It outlines the processes of germplasm selection, genotyping, phenotype measurement, statistical analysis, and identifying significant genetic loci associated with specific traits. This systematic approach helps solve the genetic basis of various traits in plant populations

by recent improvements in genotyping through sequencing (GBS) techniques. Due to this advancement, single nucleotide polymorphisms (SNPs) are used more frequently in GWAS (Muhammad et al., 2020). Following sequencing operations, SNP identification was performed using a variety of methods. For instance, in genome-wide association studies, SNP arrays, such as 90 K arrays, have been used to identify genomic regions and markers linked to a variety of traits in wheat, such as grain asparagine content, resistance to Hessian flies, disease resistance, grain yield, and frost tolerance (Muhu-Din Ahmed et al., 2022).

GWAS is a practical approach for pinpointing significant trait-associated markers (Verges et al., 2021). For instance, scientists have conducted GWAS investigations concerning the three major fungal diseases that threaten wheat production: rust, smut, and bunt (Goutam et al., 2015). Genome-wide association studies have been frequently employed to establish connections between specific genomic regions and the resistance or susceptibility of crop plants to different pathogens. These advancements in food plant genetics have facilitated such developments (Ogbonnaya et al., 2001). Many resistance genes (R-genes) are linked to genetic interactions influenced by specific relationships, i.e., gene-for-gene interactions (Flor, 1971). These interactions are easily observable in the field, as pathogens can modify their homologous avirulence genes to avoid detection by their host plants. For example, the effective utilization of high-density single nucleotide polymorphism (SNP) maps has resulted in the identification of numerous significant marker features associated with *Fusarium* head blight (FHB) characteristics by using GWAS (Khan et al., 2020). Figure 10.17 illustrates the key processes involved in GWAS for disease resistance in crop plants.

Bioinformatics Tools Used for GWAS

Data preparation: GWAS primarily requires two data types: genotypic and phenotypic. The former data can be obtained through various genotyping technologies, including genotyping arrays and whole-genome sequencing. Phenotypic data are collected for the traits of interest, i.e., disease resistance, using standardized and accurate phenotyping methods.

Data preprocessing: Data preprocessing includes two main steps: quality control (QC) and imputation. QC is performed on genotype data to filter out low-quality markers and samples using tools such as PLINK or VCFtools. However, imputation tools such as Beagle or IMPUTE2 impute missing genotype data. Population structure analysis: This step uses tools such as STRUCTURE, PCA, or ADMIXTURE to assess and correct population structure and relatedness among individuals (Pritchard et al., 2000).

Association analysis: This bioinformatics analysis associates the genotypic data with the phenotypic data. Association tests, such as single marker analysis (SMA), mixed linear model (MLM), and generalized linear model (GLM), are used for this purpose (Yu et al., 2006). Using tools such as PLINK or TASSEL, SMA tests each SNP individually for association with phenotypic traits. Similarly, MLM is utilized for population structure and relatedness using tools such as GAPIT or EMMAX, while GLM is generally used for binary or count traits.

The next step is multiple testing correction, which applies multiple testing corrections such as Bonferroni, false discovery rate (FDR), or permutation testing to control for false positives (Benjamini & Hochberg, 1995). Multiple testing correction is followed by candidate gene identification, which annotates significant SNPs and identifies candidate genes using resources such as Phytozome or Ensembl Plants (Goodstein et al., 2012). Functional enrichment analysis can help prioritize candidate genes. The final step is to validate the identified candidate genes and SNPs through additional experiments, such as functional genomic studies or transgenic plant experiments, to confirm their role in the trait of interest.

Molecular Docking, Classifications, Techniques, and Methods

Basics of Molecular Docking

Molecular docking is a computational technique that predicts the binding orientation of small molecule ligands to protein receptors to form stable complexes (Kitchen et al., 2004). It combines principles of molecular recognition and thermodynamics with search algorithms to evaluate the strength and geometry of molecular interactions (Cheng et al., 2012). Docking employs knowledge of molecular surfaces, electrostatic potential, and hydrophobic properties governing molecular association between ligands and receptors (Cheng et al., 2012).

The prediction of a ligand's binding mode and affinity is achieved through a sequential process. First, different ligand orientations are generated within the receptor's binding site via conformational searching. Each orientation is then scored and ranked based on the interaction energy between the molecules and geometric complementarity. Postdocking analysis refines complexes and identifies interactive residues (Cheng et al., 2012). Accuracy depends on forcefield parameterization, sampling algorithms, and conformational flexibility treatment (Meng et al., 2011).

Molecular docking is widely applied in computer-aided drug design to facilitate hit identification, lead optimization, and target profiling. It is also a valuable tool in agricultural biotechnology for elucidating pathogen virulence strategies and host resistance mechanisms at the molecular level. In plant pathology, molecular docking provides insights into host-pathogen interactions crucial for designing more targeted disease control strategies (Berthiller et al., 2013).

To better understand docking studies, one must familiarize oneself with the fundamental terminology associated with this field. Table 10.2 outlines the frequently employed terms in the context of docking research.

Ligand-Based Classification of Molecular Docking

Based on the ligand, molecular docking can be classified into three main types (Roy et al., 2015):

- Protein-small molecule docking: This is the most common type of docking and is used to predict the binding of small molecules to proteins.
- Protein-nucleic acid docking: This type predicts the binding of small molecules to nucleic acids, such as DNA and RNA.
- Protein-protein docking: This type of docking is used to predict the binding of two proteins.

The docking performance depends on several factors, including the accuracy of the protein and ligand structures, the choice of search algorithm, and the scoring function (Kirkpatrick, 2004). The first step in docking is to generate a set of possible

Table 10.2 Summary of the basic terminology employed in molecular docking

Term	Definition
Receptor	A protein molecule or polymeric structure that recognizes and binds a ligand, leading to a cellular response
Ligand	A molecule that binds to a receptor facilitates a bimolecular response. Common ligands include small drug molecules, effectors, hormones, and more
Docking	A molecular modeling technique designed to find the best fit between a ligand and its binding site (receptor)
Dock pose	Different positions, conformations, and orientations in which a ligand can bind to a receptor
Binding mode	The orientation of the ligand relative to the receptor, along with their conformations when bound to each other
Dock score	An evaluation process that quantifies favorable intermolecular interactions (e.g., hydrogen bonds, hydrophobic contacts) to determine the most energetically favorable pose
Ranking	Classifying ligands based on their predicted free energy of binding resulted in a rank-ordered list for further study
Pose prediction	The search for a ligand's accurate binding mode often involves multiple trials to identify the energetically best poses
Scoring	The application of scoring functions to evaluate each ligand's energetically best pose(s) allows for rank ordering based on affinity scores
Scoring functions	Methods used to calculate the energy associated with ligand-receptor binding are typically divided into knowledge-based and physical interaction-based scoring schemes
ΔGbind	The change in free energy upon binding of a ligand to its target decomposed into specific contributions: ΔGint, ΔGsolv, ΔGconf, and ΔGmotion
ΔGint	The energy associated with specific ligand-receptor interactions
ΔGsolv	The energy is related to ligand, receptor, and solvent interactions
ΔGconf	The energy is associated with conformational changes in the ligand and receptor during binding
ΔGmotion	The energy is related to movements in the protein and ligand during complex formation

binding poses for the ligand in the protein's binding pocket. It can be done using a variety of approaches, such as rigid docking, flexible docking, and fragment-based docking. The second step is to score the binding poses. The scoring function considers the ligand and protein interactions, such as van der Waals, electrostatic, and hydrogen bonds. The pose with the highest score is the most likely binding pose.

The first step is to find the X-ray crystallized structure of the protein-ligand complex in the Protein Data Bank (PDB) to identify the active site of a protein. If the bound ligand is absent in the crystal structure, then the active site can be identified using the following methods (Roy et al., 2015):

- Literature review: A comprehensive literature review of the source papers can be used to identify the active site of residues.
- Experimental methods: Experimental methods, such as mutagenesis and site-directed mutagenesis, can be used to identify the active site of residues.

- Computational methods: Computational methods, such as molecular docking, can be used to identify the active site of residues.

Once the active site has been identified, the docking process can be used to predict the binding of potential drug candidates to the protein.

Methods for Molecular Docking

In general, docking can be categorized into three main classes: protein-ligand docking, protein-nucleic acid docking, and protein-protein docking (Kolb et al., 2009). Protein-ligand docking is particularly important in structure-based drug design. Systematic classification based on techniques, algorithms, and resources to perform

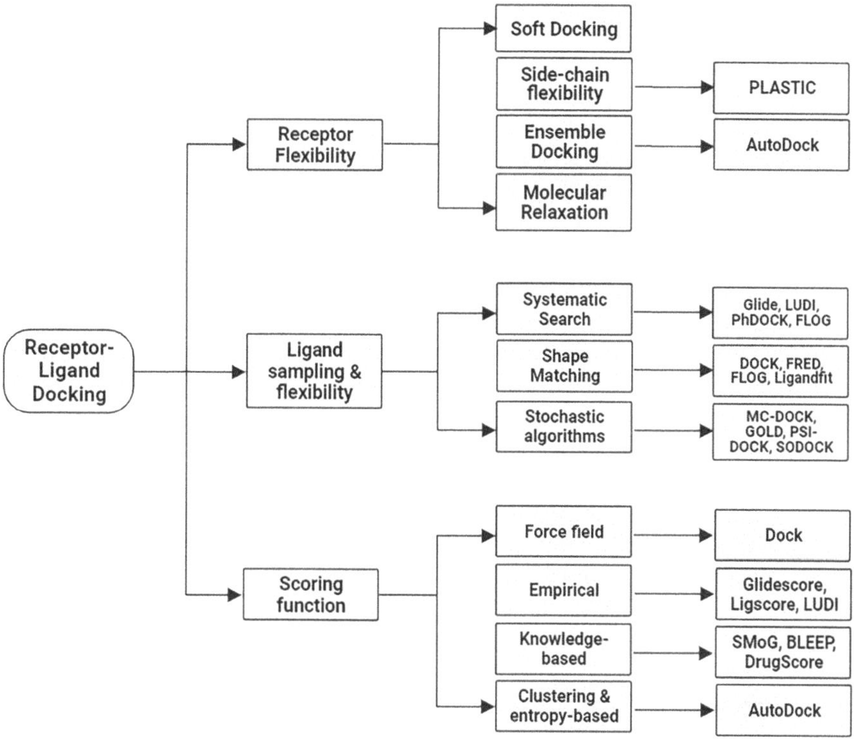

Fig. 10.18 Illustration of the systematic classification of molecular docking methods into three essential categories: protein flexibility, ligand sampling and flexibility, and docking scoring functions. Each category encompasses a variety of approaches, techniques, software, and online resources crucial for predicting molecular interactions and optimizing ligand binding configurations. These classifications are the foundation for advancing various applications, including drug discovery and structural biology

protein-ligand docking is provided in Fig. 10.18. It can be further classified into three types:

1. Rigid-body docking: Both the receptor and ligand are treated as rigid structures.
2. Flexible ligand docking: The receptor remains rigid, but the ligand is allowed to be flexible.
3. Flexible docking: Both the receptor and ligand can undergo flexibility changes.

These categories are essential for understanding protein flexibility, ligand sampling, and scoring functions in docking studies.

Receptor/Protein Flexibility

When ligands bind to proteins, they often induce conformational changes in the protein structure. These changes optimize interactions with the ligand. Receptor flexibility poses a significant challenge due to the large size and many degrees of freedom of proteins (Leulliot & Varani, 2001). Receptor flexibility can be categorized into four main types:

1. Soft docking: This basic method allows minimal ligand-protein overlap and employs additional energy terms, such as van der Waals forces, to accommodate slight conformational changes. It is computationally efficient but limited in capturing substantial structural shifts (Ferrari et al., 2004).
2. Side-chain flexibility: This approach enhances flexibility modeling by keeping protein backbones rigid and sampling side-chain conformations. Optimizing ligand-side-chain orientations through algorithms such as the dead-end elimination method improves accuracy (Leach, 1994).
3. Molecular relaxation: Combining rigid-body docking with subsequent relaxation of the protein structure, including backbone and nearby side-chain atoms, adds more flexibility. Although computationally demanding, such as molecular dynamics simulations (MDSs), they offer insights into dynamic protein behavior (Davis & Baker, 2009).
4. Ensemble docking: This method employs a collection of protein structures to represent potential conformational changes. Ensembles can be generated experimentally or theoretically and are used to select the most fitting protein conformation (Cozzini et al., 2008).

Ligand Sampling and Flexibility

Ligand sampling and flexibility play a pivotal role in predicting molecular interactions. This critical aspect of the docking process involves generating and exploring various conformations and orientations of the ligand molecule within the binding site of the target protein. The goal is to identify the most energetically favorable binding pose, representing the optimal ligand arrangement within the protein's active site. Ligand sampling is crucial in protein-ligand docking. Three types of ligand sampling algorithms are employed:

1. Shape matching: Places the ligand in orientations that match the molecular surface of the protein's binding site (Lorber & Shoichet, 2005).
2. Systematic search: Explores all degrees of freedom of the ligand to generate possible binding conformations.
3. Stochastic algorithms: Randomly sample ligand orientations and conformations based on probabilistic criteria (Huang & Zou, 2010).

Docking Scoring Functions

Docking scoring functions are a fundamental component of molecular docking studies, serving as the metric by which the quality of potential ligand-protein interactions is assessed. These scoring functions are pivotal in determining which binding poses are energetically favorable and, therefore, more likely to represent biologically relevant interactions (Jain, 2006). Their primary purpose is assigning a numerical score or energy value to each docking pose, enabling ranking and selecting the most promising ligand binding configurations. Scoring functions evaluate the accuracy of a docking algorithm. They can be divided into five categories based on their derivation methods:

1. FF scoring functions: Partition the ligand-binding energy into interaction terms using force field parameters.
2. Empirical scoring functions: Calculate binding energy scores based on weighted empirical energy terms.
3. Knowledge-based scoring functions: Use structural information from experimentally determined protein-ligand complexes (Verkhivker et al., 1995).
4. Consensus scoring: Combines scoring information from multiple functions to improve accuracy (Charifson et al., 1999).
5. Clustering- and entropy-based scoring methods: These methods account for entropic effects by dividing ligand-binding modes into clusters (Lee & Seok, 2008).

Basic Steps of Molecular Docking

Molecular docking is a systematic three-step process, regardless of the software or algorithms employed. These steps are crucial for predicting molecular interactions accurately and reliably (Venkatesan et al., 2010). The process unfolds as follows:

Ligand preparation: The initial step involves preparing the ligands for docking. Duplicate structures are eliminated, and adjustments are made to ensure optimal ligands. It addresses ionization changes, tautomer and isomer generation, and generating three-dimensional structures. These modifications are essential for the

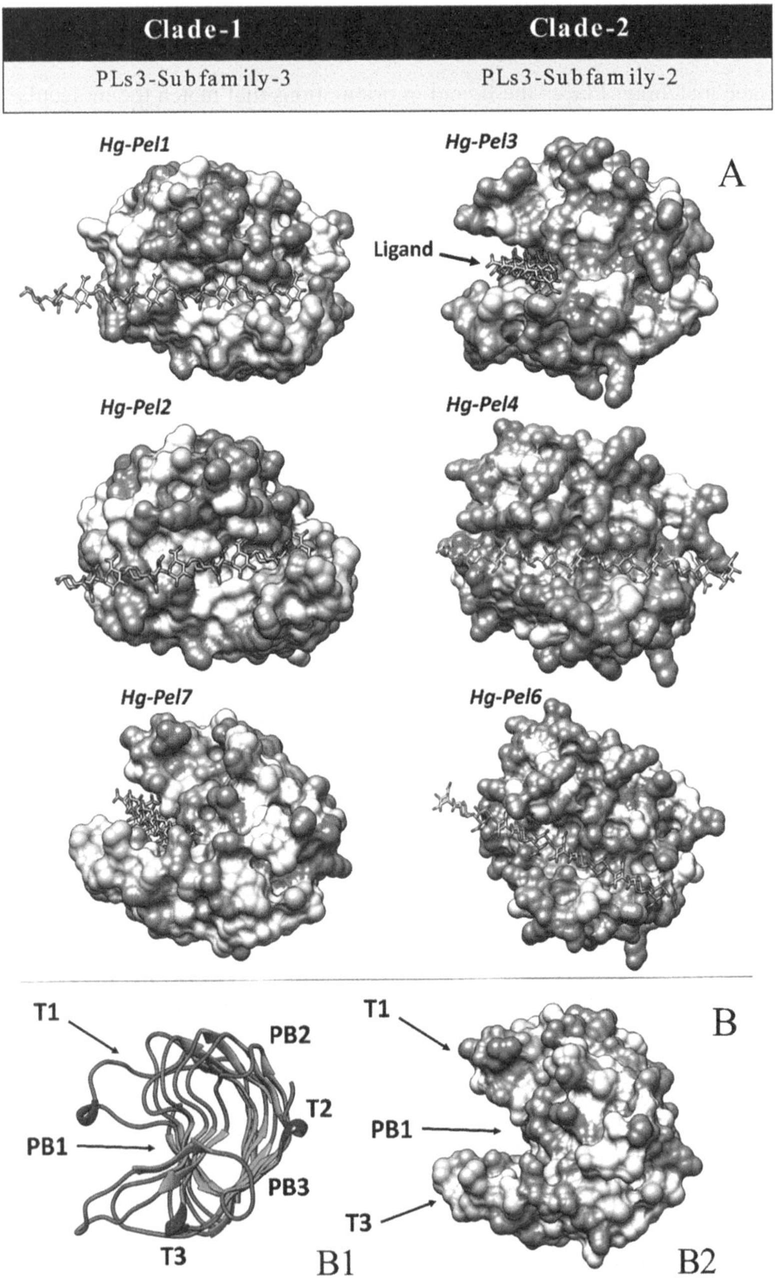

Fig. 10.19 Representation of the molecular docking of six *H. glycines* pectate lyases (Pels) with a pectic acid ligand molecule. (A) It depicts the interaction of each Pel with the ligand molecule in the 3D plane. The clades of Clade-1 and Clade-2 are labeled accordingly. At the bottom of Fig. (B), B1 shows the different folds of turns and strands in the secondary structure of a Pel. B2 displays the molecular surface view to show the T1-PB1-T3 groove as a binding and catalytic site. The figure is adapted from our open-access article (Zahoor et al., 2023)

ligands to be compatible with the chosen docking software. Figure 10.19 shows the interaction of pectic acid molecules (ligand) with pectate lyase (receptor) cell wall-degrading enzymes (CWDEs) released by plant-parasitic nematodes.

Protein preparation: The protein structure is meticulously prepared to create a suitable environment for docking. Hydrogen atoms are added, and the protein is minimized using a software-specific force field. Water molecules are typically removed from the protein except those in the active site. Any errors or incomplete residues near the active site are corrected. Charges and atom types are properly set, especially for metal atoms. In cases where bonds to metal ions exist, these bonds are deleted, and formal charges of the involved atoms and the metal ion are adjusted. The prepared protein structure is now ready to serve as the receptor for docking, i.e., pectate lyase (receptor) in Fig. 10.19.

Ligand-protein docking: The docking process begins with ligands and the receptor in the appropriate format. Receptor grid files are generated for grid-based docking, often centered at the ligand's active site. Alternatively, active pockets within the protein can be identified for docking. Ligands from the dataset are docked into the receptor's active site(s), and interaction energies between each ligand and the receptor are computed. The results are compared to those of the bound ligand in the crystallized protein structure to assess whether the ligands fit into the specified active site. Multiple docking poses are saved for each ligand conformation, ranked based on their dock score functions. Subsequently, the interactions between the ligands and the receptor are analyzed. Docking studies provide insights into the receptor-ligand interactions, which can be correlated with the biological activity of the compounds in the dataset. For validating the docking procedure, the cocrystallized ligand is extracted from the receptor's active site and reduced to ensure that it binds to the same active site and interacts with the same amino acid residues as previously observed.

Protein Interaction Networks

Protein interaction data from both individual, small-scale experiments and automated, high-throughput screenings could be summarized as protein interaction networks (Bork et al., 2004); in other words, a protein interaction network represents a collection of possible physical connections that proteins can establish with one another (Bossi & Lehner, 2009). The biological significance of protein interaction networks would also apply to protein sets that are believed to be crucial for regulating protein interaction networks. Regulatory, phosphorylation, and genetic interactions, as well as bottleneck interactions, all involve protein complexes in a substantial way (Wuchty, 2014).

Applications in Plant Pathology

Proteins in a cell physically interact to produce multisubunit protein complexes that function as complex multimolecular machines, as opposed to acting independently or in isolation. Physical interactions between proteins and this multimolecular machinery are essential for cell activity. It takes much work to identify protein complexes and disclose the arrangement of protein interactions in a cell. The highly connected static PPI networks that identified functional protein complexes and modules do not consider spatial or temporal features and, as a result, do not accurately reflect the actual situation in a cell. As a result, the accuracy of protein complex prediction is still far from ideal. The study of protein interaction networks may offer a helpful resolution to the issue (Wang et al., 2014).

Protein interaction networks integrated with disease datasets enable:

1. Discovery of genes and proteins linked to various diseases.
2. Examination of network characteristics.
3. Detection of subnetworks and their classification (Sevimoglu & Arga, 2014).

Resources Used for Studying Interaction Networks

Several Web-based resources are used for studying protein interaction networks, as given in Table 10.3. However, Fig. 10.20 shows the output of the STRING database for *Arabidopsis* auxin response factor 2 (ARF2) protein interacting with other proteins in the network.

How to Do It?

Get hit protein ID from databases like Plant TFDB, NCBI, and Pytozome, for which you want to study the interaction networks. Access any Web tool from the above-given links, such as the STRING database, and search that protein ID in the

Table 10.3 Resources used for studying interaction networks

Web tools	Links	References
STRING	https://string-db.org/	Szklarczyk et al. (2015)
PathBLAST	https://www.hsls.pitt.edu/obrc/index.php?page=URL1097769497	Kelley et al. (2004)
BioGRID	https://thebiogrid.org/	Oughtred et al. (2021)
Cytoscape	https://cytoscape.org/	Shannon et al. (2003)

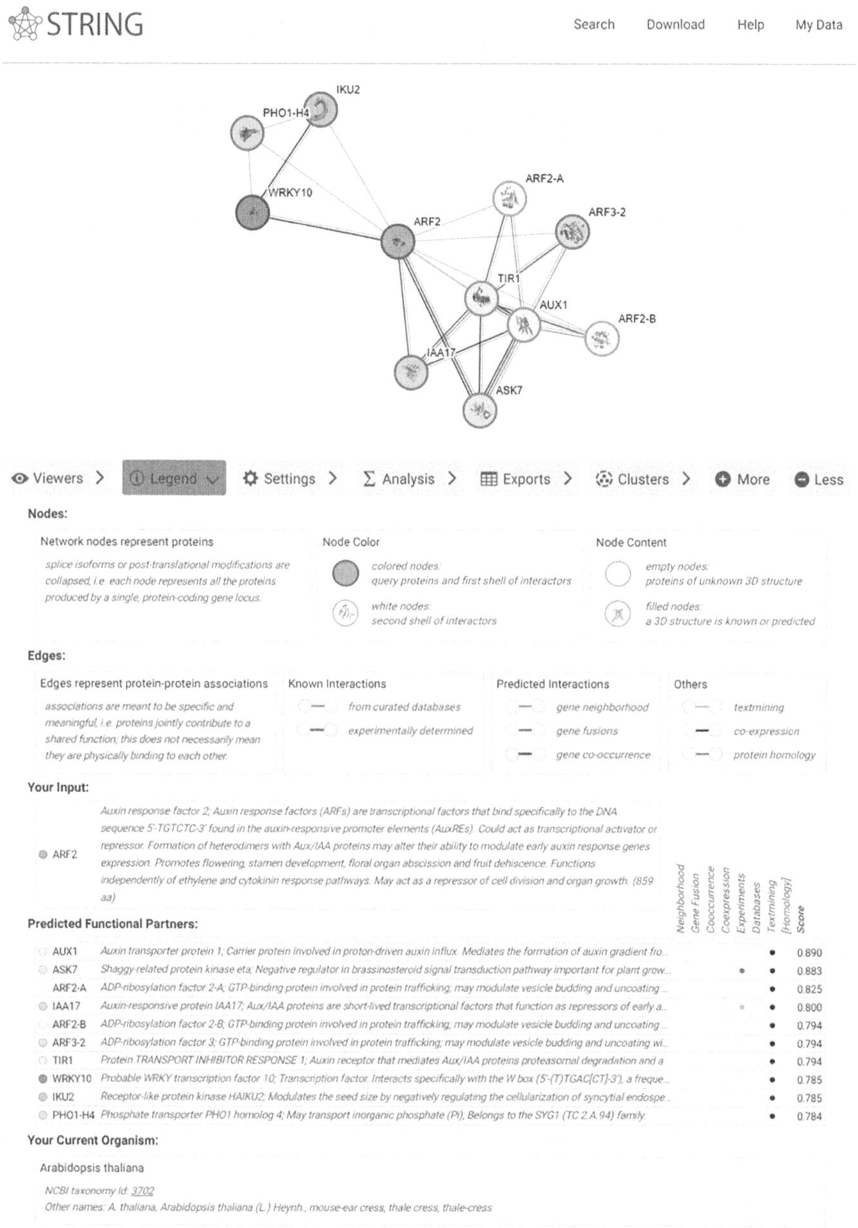

Fig. 10.20 Output of the STRING database for *Arabidopsis* auxin response factor 2 (ARF2) protein interacting with other proteins in the network

database by pasting it on the search bar. Then, the database will provide a visual of the protein interaction network, as shown in Fig. 10.20.

References

Abad, P., Gouzy, J., Aury, J. M., Castagnone-Sereno, P., Danchin, E. G. J., Deleury, E., Perfus-Barbeoch, L., Anthouard, V., Artiguenave, F., Blok, V. C., Caillaud, M. C., Coutinho, P. M., Dasilva, C., De Luca, F., Deau, F., Esquibet, M., Flutre, T., Goldstone, J. V., Hamamouch, N., Hewezi, T., Jaillon, O., Jubin, C., Leonetti, P., Magliano, M., Maier, T. R., Markov, G. V., McVeigh, P., Pesole, G., Poulain, J., Robinson-Rechavi, M., Sallet, E., Segurens, B., Steinbach, D., Tytgat, T., Ugarte, E., van Ghelder, C., Veronico, P., Baum, T. J., Blaxter, M., Bleve-Zacheo, T., Davis, E. L., Ewbank, J. J., Favery, B., Grenier, E., Henrissat, B., Jones, J. T., Laudet, V., Maule, A. G., Quesneville, H., Rosso, M. N., Schiex, T., Smant, G., Weissenbach, J., & Wincker, P. (2008). Genome sequence of the metazoan plant-parasitic nematode Meloidogyne incognita. *Nature Biotechnology, 26*, 909–915.

Ahmed, H. G. M. D., Naeem, M., Zeng, Y., Rashid, M. A. R., Ullah, A., Saeed, A., & Qadeer, A. (2022). Genome-wide association mapping for high temperature tolerance in wheat through 90k SNP array using physiological and yield traits. *Plos One, 17*(1), e0262569.

Alemu, K. (2015). The role and application of bioinformatics in plant disease management. *Advances in Life Science and Technology, 28*, 28–33.

Ali, M. A., Azeem, F., Abbas, A., Joiya, F. A., Li, H., & Dababat, A. A. (2017a). Transgenic strategies for enhancement of nematode resistance in plants. *Frontiers in Plant Sciences., 8*, 750. https://doi.org/10.3389/fpls.2017.00750

Ali, M. A., Azeem, F., Li, H., & Bohlmann, H. (2017b). Smart parasitic worms use multifaceted strategies to parasitize plants. *Frontiers in Plant Sciences, 8*, 1699. (ISI Impact Factor 4.459). https://doi.org/10.3389/fpls.2017.01699

Ali, M. A., Anjam, M. S., Nawaz, M. A., Lam, H.-M., & Chung, G. (2018). Signal transduction in plant–nematode interactions. *International Journal Molecular Sciences, 19*, 1648. https://doi.org/10.3390/ijms19061648

Almaghrabi, B., Ali, M. A., Zahoor, A., Shah, K. H., & Bohlmann, H. (2019). Arabidopsis thionin-like genes are involved in resistance against beet-cyst nematode (*Heterodera schachtii*). *Plant Physiology and Biochemistry, 140*, 55–67. https://doi.org/10.1016/j.plaphy.2019.05.005

Altschul, S. F., Gish, W., Miller, W., Myers, E. W., & Lipman, D. J. (1990). Basic local alignment search tool. *Journal of Molecular Biology, 215*, 403–410.

Amil-Ruiz, F., Blanco-Portales, R., Munoz-Blanco, J., & Caballero, J. L. (2011). The strawberry plant defense mechanism: A molecular review. *Plant and Cell Physiology, 52*(11), 1873–1903.

Anonymous. (1999). It's sink or swim as a tidal wave of data approaches. *Nature, 399*(6736), 517–518.

Antonovics, J., Boots, M., Ebert, D., Koskella, B., Poss, M., & Sadd, B. M. (2013). The origin of specificity by means of natural selection: Evolved and nonhost resistance in host–pathogen interactions. *Evolution, 67*(1), 1–9.

Aono, A. H., Costa, E. A., Rody, H. V. S., Nagai, J. S., Pimenta, R. J. G., Mancini, M. C., dos Santos, F. R. C., Pinto, L. R., de Andrade Landell, M. G., & de Souza, A. P. (2020). Reginaldo Massanobu Kuroshu. *Scientific Reports*. https://doi.org/10.1038/s41598-020-77063-5

Azeem, F., Hussain, M., Hussain, S., Zubair, M., Nadeem, H., Ali, M. A., Afzal, M., & Siddique, M. H. (2021). Genome-wide analysis and expression profiling of potassium transport related genes in *Solanum tuberosum. Pakistan Journal of Agricultural Sciences, 58*(1), 81–94. https://doi.org/10.21162/PAKJAS/21.9703

Bailey, T. L., Boden, M., Buske, F. A., Frith, M., Grant, C. E., Clementi, L., Ren, J., Li, W. W., & Noble, W. S. (2009). MEME SUITE: Tools for motif discovery and searching. *Nucleic Acids Res, 37*(Suppl. 2), W202–W208.

Bailey, T., et al. (2013). Practical guidelines for the comprehensive analysis of ChIP-seq data. *PLoS Computational Biology, 9*(11), e1003326.

Baloch, F. S., Cömertbay, G. Ö. N. Ü. L., & Özkan, H. A. K. A. N. (2015). DNA molecular markers for disease resistance in plant breeding with example in wheat. *Nematodes of Small Grain Cereals, 159.*

Baltrus, D. A., et al. (2011). Dynamic evolution of pathogenicity revealed by sequencing and comparative genomics of 19 Pseudomonas syringae isolates. *PLoS Pathogens, 7*(7), e1002132.

Bawono, P., & Heringa, J. (2014). Phylogenetic analyses. *Comprehensive Biomedical Physics, 6*, 93–110. https://doi.org/10.1016/b978-0-444-53632-7.01108-4

Bayat, A. (2002). Science, medicine, and the future: Bioinformatics. *BMJ: British Medical Journal, 324*(7344), 1018–1022. https://doi.org/10.1136/bmj.324.7344.1018

Benjamini, Y., & Hochberg, Y. (1995). Controlling the false discovery rate: A practical and powerful approach to multiple testing. *Journal of the Royal Statistical Society: Series B (Methodological), 57*(1), 289–300.

Berthiller, F., Crews, C., Dall'Asta, C., Saeger, S. D., Haesaert, G., Karlovsky, P., et al. (2013). Masked mycotoxins: A review. *Molecular Nutrition & Food Research, 57*(1), 165–186.

Biłas, R., Szafran, K., Hnatuszko-Konka, K., & Kononowicz, A. K. (2016). Cis-regulatory elements used to control gene expression in plants. *Plant Cell, Tissue and Organ Culture (PCTOC), 127*, 269–287.

Bocos-Asenjo, I. T., Niño-Sánchez, J., Ginésy, M., & Diez, J. J. (2022). New insights on the integrated management of plant diseases by RNA strategies: Mycoviruses and RNA interference. *International Journal of Molecular Sciences, 23*(16), 9236.

Bogusz, M., & Whelan, S. (2017). Phylogenetic tree estimation with and without alignment: New distance methods and benchmarking. *Systematic Biology, 66*(2), 218–231.

Bork, P., Jensen, L. J., Von Mering, C., Ramani, A. K., Lee, I., & Marcotte, E. M. (2004). Protein interaction networks from yeast to human. *Current Opinion in Structural Biology, 14*(3), 292–299.

Bosserhoff, A., & Kappelmann-Fenzl, M. (2021). Next generation sequencing (NGS): What can be sequenced? In *Next generation sequencing and data analysis* (pp. 1–15). Springer International Publishing.

Bossi, A., & Lehner, B. (2009). Tissue specificity and the human protein interaction network. *Molecular Systems Biology, 5*(1), 260.

Brazma, A., & Vilo, J. (2000). Gene expression data analysis. *FEBS Letters, 480*(1), 17–24.

Broman, K. W., Wu, H., Sen, Ś., & Churchill, G. A. (2003). R/qtl: QTL mapping in experimental crosses. *Bioinformatics, 19*(7), 889–890.

Brown, T. A. (2023). *Genomes 5*. CRC Press.

Carlton, J. M. (2003). Genome sequencing and comparative genomics of tropical disease pathogens. *Cellular Microbiology, 5*(12), 861–873.

Carroll, S. B., Grenier, J. K., & Weatherbee, S. D. (2013). *From DNA to diversity: Molecular genetics and the evolution of animal design*. John Wiley & Sons.

Charifson, P. S., Corkery, J. J., Murcko, M. A., & Walters, W. P. (1999). Consensus scoring: A method for obtaining improved hit rates from docking databases of three-dimensional structures into proteins. *Journal of Medicinal Chemistry, 42*(25), 5100–5109.

Chatzou, M., Magis, C., Chang, J. M., Kemena, C., Bussotti, G., Erb, I., & Notredame, C. (2016). Multiple sequence alignment modeling: Methods and applications. *Briefings in Bioinformatics, 17*(6), 1009–1023.

Chen, C., Chen, H., Zhang, Y., Thomas, H. R., Frank, M. H., Heand, Y., & Xia, R. (2020). TBtools: An integrative toolkit developed for interactive analyses of big biological data. *Molecular Plant, 13*, 1194–1202.

Cheng, T., Li, Q., Zhou, Z., Wang, Y., & Bryant, S. H. (2012). Structure-based virtual screening for drug discovery: A problem-centric review. *The AAPS Journal, 14*, 133–141.

Choudhuri, S. (2014). Phylogenetic analysis. In *Bioinformatics for beginners* (pp. 209–218). Elsevier. https://doi.org/10.1016/b978-0-12-410471-6.00009-8

Cockram, J., & Mackay, I. (2018). Genetic mapping populations for conducting high-resolution trait mapping in plants. *Plant Genetics and Molecular Biology*, 109–138.

Conesa, A., Madrigal, P., Tarazona, S., et al. (2016). A survey of best practices for RNA-seq data analysis. *Genome Biology, 17*, 13. https://doi.org/10.1186/s13059-016-0881-8

Cozzini, P., Kellogg, G. E., Spyrakis, F., Abraham, D. J., Costantino, G., Emerson, A., Fanelli, F., Gohlke, H., Kuhn, L. A., Morris, G. M., & Orozco, M. (2008). Target flexibility: An emerging consideration in drug discovery and design. *Journal of Medicinal Chemistry, 51*(20), 6237–6255.

Davis, I. W., & Baker, D. (2009). Rosetta Ligand docking with full ligand and receptor flexibility. *Journal of Molecular Biology, 385*(2), 381–392.

Diederichs, S., Bartsch, L., Berkmann, J. C., Fröse, K., Heitmann, J., Hoppe, C., et al. (2016). The dark matter of the cancer genome: Aberrations in regulatory elements, untranslated regions, splice sites, non-coding RNA and synonymous mutations. *EMBO Molecular Medicine, 8*(5), 442–457.

Doolittle, W. F. (2013). Is junk DNA bunk? A critique of ENCODE. *Proceedings of the National Academy of Sciences, 110*(14), 5294–5300.

Edgar, R. C. (2004). MUSCLE: Multiple sequence alignment with high accuracy and high throughput. *Nucleic Acids Research, 32*(5), 1792–1797. https://doi.org/10.1093/nar/gkh340

Eun, C. H., Kim, S. U., & Kim, I. J. (2017). Regulatory cis-elements on citrus peel-specific expressed gene, CuCRTISO-like, promoter respond to hormones and abiotic stresses in transgenic Arabidopsis. *Plant Biotechnology Reports, 11*, 63–69.

Feitelson, D. G., & Treinin, M. (2002). The blueprint for life? *Computer, 35*(7), 34–40.

Fei, Z., Joung, J. G., Tang, X., Zheng, Y., Huang, M., Lee, J. M., et al. (2010). Tomato functional genomics database: A comprehensive resource and analysis package for tomato functional genomics. *Nucleic Acids Research, 39*(suppl_1), D1156–D1163.

Flor, H. H. (1971). Current status of the gene-for-gene concept. *Annual Review of Phytopathology, 9*(1), 275–296.

Ferrari, A. M., Wei, B. Q., Costantino, L., & Shoichet, B. K. (2004). Soft docking and multiple receptor conformations in virtual screening. *Journal of Medicinal Chemistry, 47*(21), 5076–5084.

Frazer, K. A., et al. (2003). Cross-species sequence comparisons: A review of methods and available resources. *Genome Research, 13*(1), 1–12.

Gabaldón, T., et al. (2013). Comparative genomics of emerging pathogens in the Candida glabrata clade. *BMC Genomics, 14*(1), 623.

Gama-Castro, S., Salgado, H., Santos-Zavaleta, A., Ledezma-Tejeida, D., Muñiz-Rascado, L., García-Sotelo, J. S., et al. (2016). RegulonDB version 9.0: High-level integration of gene regulation, coexpression, motif clustering and beyond. *Nucleic Acids Research, 44*(D1), D133–D143.

Gilbert, G. S., & Parker, I. M. (2010). Rapid evolution in a plant-pathogen interaction and the consequences for introduced host species. *Evolutionary Applications, 3*(2), 144–156. https://doi.org/10.1111/j.1752-4571.2009.00107.x

Glover-Cutter, K., Kim, S., Espinosa, J., & Bentley, D. L. (2008). RNA polymerase II pauses and associates with pre-mRNA processing factors at both ends of genes. *Nature Structural & Molecular Biology, 15*(1), 71–78.

Goodstein, D. M., Shu, S., Howson, R., Neupane, R., Hayes, R. D., Fazo, J., et al. (2012). Phytozome: A comparative platform for green plant genomics. *Nucleic Acids Research, 40*(D1), D1178–D1186.

Goutam, U., Kukreja, S., Yadav, R., Salaria, N., Thakur, K., & Goyal, A. K. (2015). Recent trends and perspectives of molecular markers against fungal diseases in wheat. *Frontiers in Microbiology, 6*, 154636.

Grienberg, I., & Benayahu, D. (2005). Osteo-promoter database (OPD)–promoter analysis in skeletal cells. *BMC Genomics, 6*(1), 1–7.

Guo, A. Y., Chen, X., Gao, G., Zhang, H., Zhu, Q. H., Liu, X. C., et al. (2007). PlantTFDB: A comprehensive plant transcription factor database. *Nucleic Acids Research, 36*(suppl_1), D966–D969.

Guttman, D. S., McHardy, A. C., & Schulze-Lefert, P. (2014). Microbial genome-enabled insights into plant–microorganism interactions. *Nature Reviews Genetics, 15*(12), 797–813.

Hartwell, L. H., Hopfield, J. J., Leibler, S., & Murray, A. W. (1999). From molecular to modular cell biology. *Nature, 402*(Suppl. 6761), C47–C52.

Hawamda, A. I. M., Reichert, S., Ali, M. A., Nawaz, M. A., Austerlitz, T., Schekahn, P., Abbas, A., Tenhaken, R., & Bohlmann, H. (2022). Characterization of an Arabidopsis defensin-like gene conferring resistance against nematodes. *Plants, 11*, 280. https://doi.org/10.3390/plants11030280

Hernandez-Garcia, C. M., & Finer, J. J. (2014). Identification and validation of promoters and cis-acting regulatory elements. *Plant Science, 217*, 109–119.

Higo, K., Ugawa, Y., Iwamoto, M., & Korenaga, T. (1999). Plant cis-acting regulatory DNA elements (PLACE) database: 1999. *Nucleic Acids Research, 27*(1), 297–300.

Hong, M., Tao, S., Zhang, L., et al. (2020). RNA sequencing: New technologies and applications in cancer research. *Journal of Hematology & Oncology, 13*, 166. https://doi.org/10.1186/s13045-020-01005-x

Hu, B., Jin, J., Guo, A., Zhang, H., Luo, J., & Gao, G. (2015). GSDS 2.0: An upgraded gene feature visualization server. *Bioinformatics, 31*(8), 1296–1297.

Huang, S. Y., & Zou, X. (2010). Advances and challenges in protein-ligand docking. *International Journal of Molecular Sciences, 11*(8), 3016–3034.

Hunjan, M. S., & Lore, J. S. (2020). Climate change: Impact on plant pathogens, diseases, and their management. *Crop Protection Under Changing Climate*, 85–100.

Jabbar, A., Javed, N., Munir, A., Khan, S. A., Moosa, A., Jabran, M., Adams, B. J., & Ali, M. A. (2021). Occurrence and molecular characterization of *Meloidogyne graminicola* on rice in Central Punjab, Pakistan. *Journal of Nematology, 52*, e2020–e2123. https://doi.org/10.21307/jofnem-2020-123

Jabran, M., Ali, M. A., Zahoor, A., Muhae-Ud-Din, G., Liu, T., Chen, W., & Gao, L. (2023). Intelligent reprogramming of wheat for enhancement of fungal and nematode disease resistance using advanced molecular techniques. *Frontiers in Plant Sciences, 14*, 1132699. https://doi.org/10.3389/fpls.2023.1132699

Jackson, A. O., & Taylor, C. B. (1996). Plant-microbe interactions: Life and death at the interface. *The Plant Cell, 8*(10), 1651.

Jain, A. N. (2006). Scoring functions for protein-ligand docking. *Current Protein and Peptide Science, 7*(5), 407–420.

Katoh, K., & Standley, D. M. (2013). MAFFT multiple sequence alignment software version 7: Improvements in performance and usability. *Molecular Biology and Evolution, 30*(4), 772–780. https://doi.org/10.1093/molbev/mst010

Katoh, K., Misawa, K., Kuma, K., & Miyata, T. (2002). MAFFT: A novel method for rapid multiple sequence alignment based on fast Fourier transform. *Nucleic Acids Research, 30*(14), 3059–3066. https://doi.org/10.1093/nar/gkf436

Kaur, A., Pati, P. K., Pati, A. M., & Nagpal, A. K. (2017). In-silico analysis of cis-acting regulatory elements of pathogenesis-related proteins of Arabidopsis thaliana and Oryza sativa. *PLoS One, 12*(9), e0184523.

Kelley, B. P., Yuan, B., Lewitter, F., Sharan, R., Stockwell, B. R., & Ideker, T. (2004). PathBLAST: A tool for alignment of protein interaction networks. *Nucleic Acids Research, 32*(Suppl. 2), W83–W88.

Kessler, A. C., & Maraia, R. J. (2021). The nuclear and cytoplasmic activities of RNA polymerase III, and an evolving transcriptome for surveillance. *Nucleic Acids Research, 49*(21), 12017–12034.
Khan, M. K., Pandey, A., Athar, T., Choudhary, S., Deval, R., Gezgin, S., et al. (2020). *Fusarium head blight in wheat: Contemporary status and molecular approaches., 3 Biotech, 10*, 1–17.
Khan, A., Singh, S., & Singh, V. K. (2021). Bioinformatics in plant pathology. *Emerging Trends in Plant Pathology*, 725–844.
Kirkpatrick, P. (2004). Gliding to success. *Nature Reviews Drug Discovery, 3*(4), 299–299.
Kitchen, D. B., Decornez, H., Furr, J. R., & Bajorath, J. (2004). Docking and scoring in virtual screening for drug discovery: Methods and applications. *Nature Reviews. Drug Discovery, 3*, 935–949.
Kolb, P., Ferreira, R. S., Irwin, J. J., & Shoichet, B. K. (2009). Docking and chemoinformatic screens for new ligands and targets. *Current Opinion in Biotechnology, 20*(4), 429–436.
Kukurba, K. R., & Montgomery, S. B. (2015). RNA sequencing and analysis. *Cold Spring Harbor Protocols*, (11), pdb-top084970. https://doi.org/10.1101/pdb.top084970
Kushwaha, U. K. S., et al. (2017). Role of bioinformatics in crop improvement. *Global Journal of Science Frontier Research D Agriculture & Veterinary, 17*, 13–24.
Lander, E. S., Green, P., Abrahamson, J., Barlow, A., Daly, M. J., Lincoln, S. E., & Newburg, L. (1987). Mapmaker: An interactive computer package for constructing primary genetic linkage maps of experimental and natural populations. *Genomics, 1*(2), 174–181.
Langmead, B., & Salzberg, S. L. (2012). Fast gapped-read alignment with Bowtie 2. *Nature Methods, 9*, 357–359.
Leach, A. R. (1994). Ligand docking to proteins with discrete side-chain flexibility. *Journal of Molecular Biology, 235*(1), 345–356.
Lee, J., & Seok, C. (2008). A statistical rescoring scheme for protein–ligand docking: Consideration of entropic effect. *Proteins: Structure, Function, and Bioinformatics, 70*(3), 1074–1083.
Lee, S. W., Han, S. W., Bart, R., Wang, Y., Gim, B. S., Jeong, B. R., et al. (2011). The rice Pika gene enhances antibacterial defenses via a jasmonate/lycopene-dependent pathway. *The Plant Cell, 23*(2), 925–941.
Lescot, M., Déhais, P., Thijs, G., Marchal, K., Moreau, Y., Van de Peer, Y., et al. (2002). PlantCARE, a database of plant cis-acting regulatory elements and a portal to tools for in silico analysis of promoter sequences. *Nucleic Acids Research, 30*(1), 325–327.
Lesk, A. (2019). *Introduction to bioinformatics*. Oxford University Press.
Leulliot, N., & Varani, G. (2001). Current topics in RNA–protein recognition: Control of specificity and biological function through induced fit and conformational capture. *Biochemistry, 40*(27), 7947–7956.
Li, R., Zhu, F., & Duan, D. (2020). Function analysis and stress-mediated cis-element identification in the promoter region of VqMYB15. *Plant Signaling & Behavior, 15*(7), 1773664.
Li, B., Liu, Y., Cui, X. Y., Fu, J. D., Zhou, Y. B., Zheng, W. J., et al. (2019). Genome-wide characterization and expression analysis of soybean TGA transcription factors identified a novel TGA gene involved in drought and salt tolerance. *Frontiers in Plant Science, 10*, 549.
Lincoln, S. E., Daly, M. J., & Lander, E. S. (1990). Constructing genetic linkage maps with mapmaker: A tutorial and reference manual. A Whitehead Institute for Biomedical Research
Longabaugh, W. J., Davidson, E. H., & Bolouri, H. (2005). Computational representation of developmental genetic regulatory networks. *Developmental Biology, 283*(1), 1–16.
Lorber, D. M., & Shoichet, B. K. (2005). Hierarchical docking of databases of multiple ligand conformations. *Current Topics in Medicinal Chemistry, 5*(8), 739–749.
Luo, X., & Wei, Y. (2019). Batch effects correction with unknown subtypes. *Journal of the American Statistical Association, 114*(526), 581–594. https://doi.org/10.1080/01621459.2018.1497494
Macke, T. J., et al. (2001). RNAMotif, an RNA secondary structure definition and search algorithm. *Nucleic Acids Research, 29*(22), 4724–4735.
Martinelli, F., Scalenghe, R., Davino, S., Panno, S., Scuderi, G., Ruisi, P., et al. (2015). Advanced methods of plant disease detection. A review. *Agronomy for Sustainable Development, 35*, 1–25.

Mathelier, A., Fornes, O., Arenillas, D. J., Chen, C. Y., Denay, G., Lee, J., et al. (2016). JASPAR 2016: A major expansion and update of the open-access database of transcription factor binding profiles. *Nucleic Acids Research, 44*(D1), D110–D115.

Matoulkova, E., Michalova, E., Vojtesek, B., & Hrstka, R. (2012). The role of the 3'untranslated region in post-transcriptional regulation of protein expression in mammalian cells. *RNA Biology, 9*(5), 563–576.

Matys, V., Fricke, E., Geffers, R., Gößling, E., Haubrock, M., Hehl, R., et al. (2003). TRANSFAC®: Transcriptional regulation, from patterns to profiles. *Nucleic Acids Research, 31*(1), 374–378.

Meng, X. Y., Zhang, H. X., Mezei, M., & Cui, M. (2011). Molecular docking: A powerful approach for structure-based drug discovery. *Current Computer-Aided Drug Design, 7*(2), 146–157.

Meng, J., et al. (2014). A protocol for RNA methylation differential analysis with MeRIP-Seq data and exomePeak R/Bioconductor package. *Methods, 69*(3), 274–281.

Morgulis, A., Coulouris, G., Raytselis, Y., Madden, T. L., Agarwala, R., & Schaffer, A. A. (2008). Database indexing for production MegaBLAST searches. *Bioinformatics, 15*, 1757–1764.

Muhammad, A., Hu, W., Li, Z., Li, J., Xie, G., Wang, J., & Wang, L. (2020). Appraising the genetic architecture of kernel traits in hexaploid wheat using GWAS. *International Journal of Molecular Sciences, 21*(16), 5649.

Munjal, G., Hanmandlu, M., & Srivastava, S. (2019). Phylogenetics algorithms and applications. In *Ambient communications and computer systems: RACCCS-2018* (pp. 187–194). Springer Singapore.

Myles, C., & Wayne, M. (2008). Quantitative trait locus (QTL) analysis. *Nature Education, 1*(1), 208.

Nadeem, M. A., Nawaz, M. A., Shahid, M. Q., Doğan, Y., Comertpay, G., Yıldız, M., et al. (2018). DNA molecular markers in plant breeding: Current status and recent advancements in genomic selection and genome editing. *Biotechnology & Biotechnological Equipment, 32*(2), 261–285.

Needleman, S. B., & Wunsch, C. D. (1970). A general method applicable to the search for similarities in the amino acid sequence of two proteins. *Journal of Molecular Biology, 48*(3), 443–453. https://doi.org/10.1016/0022-2836(70)90057-4

Neik, T. X., Amas, J., Barbetti, M., Edwards, D., & Batley, J. (2020). Understanding host–pathogen interactions in Brassica napus in the omics era. *Plants, 9*(10), 1336.

Nelson, J. C. (1997). QGENE: Software for marker-based genomic analysis and breeding. *Molecular Breeding, 3*, 239–245.

Notredame, C., Higgins, D. G., & Heringa, J. (2000). T-Coffee: A novel method for fast and accurate multiple sequence alignment. *Journal of Molecular Biology, 302*(1), 205–217. https://doi.org/10.1006/jmbi.2000.4042

Ogbonnaya, F. C., Seah, S., Delibes, A., Jahier, J., Lopez-Brana, I., Eastwood, R. F., & Lagudah, E. S. (2001). Molecular-genetic characterisation of a new nematode resistance gene in wheat. *Theoretical and Applied Genetics, 102*, 623–629.

Osuna-Cruz, C. M., Paytuvi-Gallart, A., Di Donato, A., Sundesha, V., Andolfo, G., Aiese Cigliano, R., et al. (2018). PRGdb 3.0: A comprehensive platform for prediction and analysis of plant disease resistance genes. *Nucleic Acids Research, 46*(D1), D1197–D1201.

Oughtred, R., Rust, J., Chang, C., Breitkreutz, B. J., Stark, C., Willems, A., et al. (2021). The BioGRID database: A comprehensive biomedical resource of curated protein, genetic, and chemical interactions. *Protein Science, 30*(1), 187–200.

Ozsolak, F., & Milos, P. (2011). RNA sequencing: Advances, challenges and opportunities. *Nature Reviews. Genetics, 12*, 87–98. https://doi.org/10.1038/nrg2934

Palaniswamy, S. K., James, S., Sun, H., Lamb, R. S., Davuluri, R. V., & Grotewold, E. (2006). AGRIS and AtRegNet. a platform to link cis-regulatory elements and transcription factors into regulatory networks. *Plant Physiology, 140*(3), 818–829.

Priest, H. D., Filichkin, S. A., & Mockler, T. C. (2009). Cis-regulatory elements in plant cell signaling. *Current Opinion in Plant Biology, 12*(5), 643–649.

Pritchard, J. K., Stephens, M., & Donnelly, P. (2000). Inference of population structure using multilocus genotype data. *Genetics, 155*(2), 945–959.

Rani, V. (2007). Computational methods to dissect cis-regulatory transcriptional networks. *Journal of Biosciences, 32*, 1325–1330.
Rao, V. S., et al. (2008). Recent developments in life sciences research: Role of bioinformatics. *African Journal of Biotechnology, 7*(5).
Rehm, B. (2001). Bioinformatic tools for DNA/protein sequence analysis, functional assignment of genes and protein classification. *Applied Microbiology and Biotechnology, 57*, 579–592.
Roccaro, M., Ahmadinejad, N., Colby, T., & Somssich, I. E. (2013). Identification of functional cis-regulatory elements by sequential enrichment from a randomized synthetic DNA library. *BMC Plant Biology, 13*(1), 1–11.
Rombauts, S., Déhais, P., Van Montagu, M., & Rouzé, P. (1999). PlantCARE, a plant cis-acting regulatory element database. *Nucleic Acids Research, 27*(1), 295–296.
Röttjers, L., & Faust, K. (2018). From hairballs to hypotheses–biological insights from microbial networks. *FEMS Microbiology Reviews, 42*(6), 761–780.
Roy, K., Kar, S., & Das, R. N. (2015). Other related techniques. In *Understanding the basics of QSAR for applications in pharmaceutical sciences and risk assessment* (p. 357). Academic Press.
Saeed, F., Perez-Rathke, A., Gwarnicki, J., Berger-Wolf, T., & Khokhar, A. (2012). A high performance multiple sequence alignment system for pyrosequencing reads from multiple reference genomes. *Journal of Parallel and Distributed Computing, 72*(1), 83–93.
Sangwan, N. S., Jadaun, J. S., Tripathi, S., Mishra, B., Narnoliya, L. K., & Sangwan, R. S. (2018). Plant metabolic engineering. In *Omics technologies and bioengineering* (pp. 143–175). Academic Press.
Satpathy, R. (2022). Application of bioinformatics in the plant pathology research. In *Advances in agricultural and industrial microbiology: Volume-2: Applications of microbes for sustainable agriculture and in-silico strategies* (pp. 191–205). Singapore.
Sayers, E. W., Bolton, E. E., Brister, J. R., Canese, K., Chan, J., Comeau, D. C., et al. (2022). Database resources of the national center for biotechnology information. *Nucleic Acids Research, 50*(D1), D20.
Sazegari, S., Niazi, A., & Ahmadi, F. S. (2015). A study on the regulatory network with promoter analysis for Arabidopsis DREB-genes. *Bioinformation, 11*(2), 101.
Schaffer, A. A., Aravind, L., Madden, T. L., Shavirin, S., Spouge, J. L., Wolf, Y. I., Koonin, E. V., & Altschul, S. F. (2001). Improving the accuracy of PSI-BLAST protein database searches with composition-based statistics and other refinements. *Nucleic Acids Research, 29*(14), 2994–3005.
Sevimoglu, T., & Arga, K. Y. (2014). The role of protein interaction networks in systems biomedicine. *Computational and Structural Biotechnology Journal, 11*(18), 22–27.
Shannon, P., Markiel, A., Ozier, O., Baliga, N. S., Wang, J. T., Ramage, D., et al. (2003). Cytoscape: A software environment for integrated models of biomolecular interaction networks. *Genome Research, 13*(11), 2498–2504.
Sievers, F., Wilm, A., Dineen, D., Gibson, T. J., Karplus, K., Li, W., Lopez, R., McWilliam, H., Remmert, M., Söding, J., Thompson, J. D., & Higgins, D. G. (2011). Fast, scalable generation of high-quality protein multiple sequence alignments using Clustal Omega. *Molecular Systems Biology, 7*(1), 539. https://doi.org/10.1038/msb.2011.75
Sivashankari, S., & Shanmughavel, P. (2007). Comparative genomics—A perspective. *Bioinformation, 1*(9), 376–378. https://doi.org/10.6026/97320630001376
Smith, T. F., & Waterman, M. S. (1981). Identification of common molecular subsequences. *Journal of Molecular Biology, 147*(1), 195–197. https://doi.org/10.1016/0022-2836(81)90087-5
Soltis, D. E., & Soltis, P. S. (2003). The role of phylogenetics in comparative genetics. *Plant Physiology, 132*(4), 1790–1800.
Steffens, N. O., Galuschka, C., Schindler, M., Bülow, L., & Hehl, R. (2004). AthaMap: An online resource for in silico transcription factor binding sites in the Arabidopsis thaliana genome. *Nucleic Acids Research, 32*(suppl_1), D368–D372.
Sturgeon, X., & Gardiner, K. J. (2011). Transcript catalogs of human chromosome 21 and orthologous chimpanzee and mouse regions. *Mammalian Genome, 22*, 261–271.

Swarbreck, D., Wilks, C., Lamesch, P., Berardini, T. Z., Garcia-Hernandez, M., Foerster, H., et al. (2007). The Arabidopsis information resource (TAIR): Gene structure and function annotation. *Nucleic Acids Research, 36*(suppl_1), D1009–D1014.

Szklarczyk, D., Franceschini, A., Wyder, S., Forslund, K., Heller, D., Huerta-Cepas, J., et al. (2015). STRING v10: Protein–protein interaction networks, integrated over the tree of life. *Nucleic Acids Research, 43*(D1), D447–D452.

Tan, Y. C., Kumar, A. U., Wong, Y. P., & Ling, A. P. K. (2022). Bioinformatics approaches and applications in plant biotechnology. *Journal, Genetic Engineering & Biotechnology, 20*(1), 106. https://doi.org/10.1186/s43141-022-00394-5

Tang, C. N., Wan Abdullah, W. M. A. N., Wee, C. Y., Balia Yusof, Z. N., Yap, W. S., Cheng, W. H., et al. (2023). Promoter Cis-element analyses reveal the function of αvpe in drought stress response of Arabidopsis. *Biology, 12*(3), 430.

Thompson, J. D., Higgins, D. G., & Gibson, T. J. (1994). CLUSTAL W: Improving the sensitivity of progressive multiple sequence alignment through sequence weighting, position-specific gap penalties and weight matrix choice. *Nucleic Acids Research, 22*(22), 4673–4680. https://doi.org/10.1093/nar/22.22.4673

Tian, F., Yang, D. C., Meng, Y. Q., Jin, J., & Gao, G. (2020). PlantRegMap: Charting functional regulatory maps in plants. *Nucleic Acids Research, 48*(D1), D1104–D1113.

Toktay, H., Evlice, E., Imren, M., Özer, G., Ali, M. A., & Dababat, A. A. (2020). Characterization of potato golden cyst nematode populations (Globodera rostochiensis) in Turkey. *International Journal of Agriculture and Biology., 23*(6), 1095–1100. https://doi.org/10.17957/IJAB/15.1390

Tompa, M., et al. (2005). Assessing computational tools for the discovery of transcription factor binding sites. *Nature Biotechnology, 23*(1), 137–144.

Toufighi, K., Brady, S. M., Austin, R., Ly, E., & Provart, N. J. (2005). The botany array resource: E-Northerns, expression angling, and promoter analyses. *The Plant Journal, 43*(1), 153–163.

Trapnell, C., Williams, B. A., Pertea, G., Mortazavi, A., Kwan, G., Van Baren, M. J., Salzberg, S. L., Wold, B. J., & Pachter, L. (2010). Transcript assembly and quantification by RNA-Seq reveals unannotated transcripts and isoform switching during cell differentiation. *Nature Biotechnology, 28*, 511–515.

Utz, H. F., & Melchinger, A. E. (1996). PLABQTL: A program for composite interval mapping of QTL. *Journal Quantitative Trait Loci, 2*(1), 1–5.

Van Ooijen, J. W. (1994). DrawMap: a computer program for drawing genetic linkage maps. *The Journal of Heredity, 85*(1), 66–66.

Vandepoele, K., Quimbaya, M., Casneuf, T., De Veylder, L., & Van de Peer, Y. (2009). Unraveling transcriptional control in Arabidopsis using cis-regulatory elements and coexpression networks. *Plant Physiology, 150*(2), 535–546.

Venkatesan, S. K., Shukla, A. K., & Dubey, V. K. (2010). Molecular docking studies of selected tricyclic and quinone derivatives on trypanothione reductase of Leishmania infantum. *Journal of Computational Chemistry, 31*(13), 2463–2475.

Verges, V. L., Brown-Guedira, G. L., & Van Sanford, D. A. (2021). Genome-wide association studies combined with genomic selection as a tool to increase Fusarium head blight resistance in wheat. *Crop Breeding, Genetics and Genomics, 3*(4).

Verkhivker, G., Appelt, K., Freer, S. T., & Villafranca, J. E. (1995). Empirical free energy calculations of ligand-protein crystallographic complexes. I. Knowledge-based ligand-protein interaction potentials applied to the prediction of human immunodeficiency virus 1 protease binding affinity. *Protein Engineering, Design and Selection, 8*(7), 677–691.

Vitulo, N., et al. (2014). A deep survey of alternative splicing in grape reveals changes in the splicing machinery related to tissue, stress condition and genotype. *BMC Plant Biology, 14*, 1–16.

Vogel, C., et al. (2004). Supra-domains: Evolutionary units larger than single protein domains. *Journal of Molecular Biology, 336*(3), 809–823.

Voorrips, R. (2002). MapChart: Software for the graphical presentation of linkage maps and QTLs. *Journal of Heredity, 93*(1), 77–78.

Wang, Z., Zhang, J., & Zhang, X. (2009). Distribution of gene expression in the genome. *Gene, 446*(1), 8–12.

Wang, J., Peng, X., Peng, W., & Wu, F. X. (2014). Dynamic protein interaction network construction and applications. *Proteomics, 14*(4-5), 338–352.

Weirauch, M. T., Yang, A., Albu, M., Cote, A. G., Montenegro-Montero, A., Drewe, P., et al. (2014). Determination and inference of eukaryotic transcription factor sequence specificity. *Cell, 158*(6), 1431–1443.

Wuchty, S. (2014). Controllability in protein interaction networks. *Proceedings of the National Academy of Sciences, 111*(19), 7156–7160.

Yang, J., Hu, C., Hu, H., Yu, R., Xia, Z., Ye, X., & Zhu, J. (2008). QTLNetwork: Mapping and visualizing genetic architecture of complex traits in experimental populations. *Bioinformatics, 24*(5), 721–723.

Yao, X., Wang, Y., Yao, Y., An, L., Bai, Y., Li, X., et al. (2021). Use of gene family analysis to discover argonaut (AGO) genes for increasing the resistance of Tibetan hull-less barley to leaf stripe disease. *Plant Protection Science, 57*(3), 226–239.

Ye, J., Coulouris, G., Zaretskaya, I., Cutcutache, I., Rozen, S., & Madden, T. L. (2012). Primer-BLAST: A tool to design target-specific primers for polymerase chain reaction. *BMC Bioinformatics, 13*, 134.

Yilmaz, A., Nishiyama, M. Y., Jr., Fuentes, B. G., Souza, G. M., Janies, D., Gray, J., & Grotewold, E. (2009). GRASSIUS: A platform for comparative regulatory genomics across the grasses. *Plant Physiology, 149*(1), 171–180.

Yu, J., Pressoir, G., Briggs, W. H., Vroh Bi, I., Yamasaki, M., Doebley, J. F., & Buckler, E. S. (2006). A unified mixed-model method for association mapping that accounts for multiple levels of relatedness. *Nature Genetics, 38*(2), 203–208.

Yuan, T., & Wang, S. (2011). Pathogen-responsive cis-elements. In *Plant defence: Biological control* (pp. 363–378). Springer Netherlands.

Zahoor, A., Niaz, Z., Seo, H., Jabran, M., Azeem, F., Joyia, F. A., Nawaz, M. A., Yang, S. H., & Ali, M. A. (2023). Computational insights into evolutionary and functional biology of plant cell wall degrading pectate lyases from plant parasitic nematodes of the order Tylenchida. *Pakistan Journal of Agricultural Sciences., 60*, 95–106. https://doi.org/10.21162/PAKJAS/23.210

Chapter 11
Chemoinformatics

Mahrukh Babar, Faiza Hassan, Misbah Ijaz, and Muhammad Tahir Mohyuddin

Abstract Chemoinformatics combines chemistry and computer tools to collect, analyze, and interpret data. It allows extracting significant information from chemical datasets by combining molecular modeling, data mining, and machine learning. Chemoinformatics is critical for comprehending structure-activity relationships (SARs) and provides insights into the properties required for a compound's usefulness by connecting chemical structures with biological activities. This understanding informs the rational design of innovative compounds with increased efficacy and decreased toxicity. The importance of the topic extends to predictive modeling, in which machine learning algorithms understand complicated patterns in chemical data. These models may predict various outcomes, including a molecule's bioactivity and toxicity, as well as its solubility and stability. Such predictive capabilities simplify compound selection and optimization decision-making, possibly saving time and money in experimental operations. The combination of computational approaches and chemical insights enables to make better judgments, propelling advances in drug discovery, compound optimization, and predictive modeling. This chapter explains its uses, notably in drug discovery and predictive modeling, which is positioned to catalyze innovation across sectors depending on molecular design and chemical analysis as technology advances.

Keywords Algorithms · Machine learning · Chemoinformatics · Predictive modeling

M. Babar (✉) · F. Hassan
Institute of Physiology and Pharmacology, University of Agriculture, Faisalabad, Pakistan

M. Ijaz · M. T. Mohyuddin
Department of Clinical Medicine and Surgery, University of Agriculture, Faisalabad, Pakistan

S. Ijaz et al. (eds.), *Trends in Plant Biotechnology*,
https://doi.org/10.1007/978-981-97-0814-7_11

Introduction

Chemoinformatics, an interdisciplinary area of chemistry, computer science, and informatics, has evolved as a potent and necessary toolkit in the current era of drug discovery, molecular design, and chemical data processing. This complicated confluence of scientific disciplines seeks to extract useful information from enormous stores of chemical data, allowing researchers to understand the complex links between molecular structures and their behaviors. Chemoinformatics has become a cornerstone in the exploration of chemical space, guiding researchers toward innovative avenues for developing new drugs, materials, and molecular entities, thanks to the exponential growth of chemical information and the ever-expanding landscape of computational techniques. Chemoinformatics, which we define as "the use of informatics methods to solve chemical problems," is just starting to be recognized as one of these aspects (Gasteiger, 2006).

The phrase seems to have been used for the first time in 1998 (Brown, 1998). Chemoinformatics is the systematic organization, analysis, and interpretation of chemical data using computer methods. Chemoinformatics bridges the gap between experimental chemistry and computer analysis by utilizing information science, statistical analysis, and computational modeling concepts, providing unique insights into the complexities of molecular behavior and accelerating the discovery and creation of new compounds with desirable features, therefore catalyzing developments in fields ranging from medicine to materials research. Regardless of the name, the main objective of chemoinformatics is to manipulate data about chemical structures, whether they take the form of planar two-dimensional (2D) structure diagrams or (increasingly) three-dimensional (3D) atomic coordinates. Large or exceptionally large collections of molecules are subjected to modifications, necessitating the application of techniques that can handle even the largest chemical datasets.

Additionally, many molecules under consideration display biological properties that interest the industrial sectors of agrochemicals, foods, and pharmaceuticals. Researchers in computer science and information science have made significant contributions to chemoinformatics research because of the need for computational efficiency. In contrast, molecular biologists and other life science researchers have become increasingly involved because of the search for biological activity (Agrafiotis et al., 2007; Willett, 2008).

Although our understanding of biological systems has significantly improved, drug discovery research is only now beginning to apply this knowledge. Consequently, the single-target paradigm, an example of the reductionist methodology, is still a hallmark of drug development today (Maggiora, 2011).

A Milestone in Drug Discovery

Chemoinformatics, which combines computer approaches with chemical data analysis, is a driving factor in current drug development. It accelerates the discovery of prospective lead compounds by rapidly mining chemical databases using descriptors and similarity metrics. Quantitative structure-activity relationship (QSAR) models enable the prediction of biological activity in this sector, while virtual screening and structure-based design help in rational compound selection and optimization. Furthermore, chemoinformatics aids in the evaluation of a compound's absorption, distribution, metabolism, excretion, and toxicity (ADME/Tox) qualities, which helps to streamline the drug discovery process and revolutionize medicinal development (Xu & Hagler, 2002).

(a) **Optimizing chemical potency, selectivity, and ADMET characteristics simultaneously**

By understanding the relationship and connections among key concepts in this field, a nuanced understanding of the synergies and interactions can be gained that shape the landscape of chemoinformatics. The intricate interplay between various terms within the realm of chemoinformatics is given in Table 11.1.

(b) **The conflict between predictability and structural variability**

A powerful contradiction arises in chemoinformatics between predictability and structural variability. This duality arises from the difficulty of reconciling the goal for precise predictions with the intrinsic variety of chemical structures.

On the one hand, predictability is a pillar of chemoinformatics. Developing links between chemical structures and their qualities, activities, or behaviors is critical to the field's success. To drive drug development and design efforts, techniques such as quantitative structure-activity relationship (QSAR) modeling and molecular docking rely significantly on precise predictions. Predictive models are based on the assumption that molecules with similar structures would have similar characteristics, allowing researchers to effectively filter through enormous chemical spaces to uncover prospective lead compounds. On the other hand, structural variability is a basic property of chemical substances. Molecules can have a wide range of structural characteristics, resulting in a wide range of physicochemical qualities and biological activities. The problem is to accommodate this diversity inside prediction models while maintaining accuracy. Overly rigid models may fail to represent structural diversity subtleties, resulting in limited application and generalization. Overly flexible models, on the other hand, may create noise and misleading correlations, reducing their predictive value. Finding the correct balance between predictability and structural variability is a difficult task. Chemoinformatics researchers must contend with issues such as how much molecular information to put into models and how to account for small structural differences *(vide infra)*.

(c) **The route from knowledge discovery to data mining**

Table 11.1 The interplay of chemoinformatics with related terms

Term	Interplay with chemoinformatics
Chemistry	Chemoinformatics is a branch of chemistry that analyzes and interprets chemical structures, characteristics, and interactions using computer methods. It assists drug discovery, chemical synthesis, molecular modeling, and chemical behavior prediction
Computational chemistry	Chemoinformatics is a branch of computational chemistry that provides tools for processing and understanding enormous chemical datasets such as molecular databases, chemical processes, and compound libraries
Bioinformatics	Chemoinformatics is comparable to bioinformatics, which utilizes computational tools to analyze biological data. Both areas entail working with enormous datasets, predicting attributes, and creating models. Chemoinformatics can work with bioinformatics to analyze the interactions between medications and biological molecules during drug discovery
Molecular modeling	Chemoinformatics contributes to molecular modeling by giving data and methods for creating realistic models of molecular structures, studying their interactions, and predicting features such as binding affinities and biological activities
Drug discovery	Chemoinformatics, including virtual screening, molecular docking, and quantitative structure-activity relationship (QSAR) research, plays an important part in drug development. It aids in the identification of prospective drug candidates by predicting how they will interact with target proteins and analyzing their chemical characteristics
Chemical databases	Chemoinformatics is establishing and administrating chemical databases containing information about molecule structures, characteristics, and activities. These databases make data retrieval, analysis, and comparison more efficient
Machine learning	Chemoinformatics mainly depends on machine learning methods to construct prediction models for chemical characteristics and activities. Machine learning may be used to detect patterns in chemical data, anticipate outcomes, and optimize molecular designs
Data mining	Chemoinformatics employs data mining techniques to extract important insights and patterns from vast chemical datasets, which can assist in studying structure-activity correlations and identifying novel chemical entities
Structural informatics	This chemoinformatics specialty analyzes and compares molecular structures to discover correlations between structure and function. Techniques such as molecular alignment, similarity searching, and clustering are used
Chemical informatics software	For chemoinformatics applications, many software tools and platforms have been created. These technologies help with molecular visualization, prediction of properties, molecular docking, and other computational activities
Quantitative structure-activity relationship (QSAR)	QSAR is a chemoinformatics approach that links a compound's chemical structure with its biological activity. It aids in the prediction of novel chemical activity based on structural properties

(continued)

Table 11.1 (continued)

Term	Interplay with chemoinformatics
Chemical visualization	Chemoinformatics is the study and use of tools and techniques for visualizing chemical structures, reactions, and characteristics. Visualization assists in comprehending complicated chemical interactions and efficiently expressing findings

It has been explained how the knowledge discovery on databases (KDD) process works. The development of massive volumes of data highlights the need for new methods to extract knowledge from various databases. The reality that most database data cannot be considered of uniform quality and are extremely sparse and dispersed widely (Maggiora & Gokhale, 2016; Mestres et al., 2008) is the main cause of difficulties with this technique. Even if data quality was not a problem, sparsity and dispersion are problematic since creating efficient QSAR models with such data is challenging. Compared to those found in the original database, they are far denser. However, since greater data density results in information loss, problems still exist. Whitehead et al. (2019) have devised an intriguing deep learning approach for directly utilizing the data in sparse databases, but its effectiveness may be jeopardized if the data are widely spread. A potential technique has also been identified: support vector machines (SVMs). There is much work to be done in this crucial area.

Different Roles of Chemoinformatics

The following subsections offer a comprehensive overview of the fundamental techniques and some of the more specialized areas of chemoinformatics, including encoding chemical compounds, database storage and search, data generation, and data analysis.

Chemical Compound and Reaction Representation

Chemical structures must be represented in a computer-readable way for both the input and output of databases and computer programs. Structure representations are the foundation for database (sub)structure searching techniques. Scientists might have various ideas and viewpoints when discussing the structural details of chemical compounds. A compound's name, two-dimensional picture, or three-dimensional molecular model can all be used to describe it. Additionally, this hierarchy illustrates the many degrees of sophistication for describing chemical structures in their electronic forms, starting with linear representations and progressing through two-dimensional chemical graphs, three-dimensional structural representations, and surfaces of molecules. In linear notations, chemical compound structures are represented by linear sequences of letters and digits.

The most widely used linear notations are:

- Wiswesser Line Notation (WLN)
- Representation of Organic Structures Description Arranged Linearly (ROSDAL)
- Simplified Molecular Input Line Entry Specification (SMILES)
- SYBYL Line Notation (SLN), the most widely used linear notation

Today, SMILES is still widely utilized. In contrast to WLN and ROSDAL, which are more distantly related to organic chemists' natural comprehension, SMILES is a commonly used linear notation that uses ASCII letters to express chemical structures. It encodes atom types, connectivity, and bond types using letters, allowing for simple exchange and manipulation of molecular information (Bone et al., 1999, Weininger, 1988). The SMILES notation has been expanded and improved multiple times since 1988 (e.g., SMARTS, SMIRKS).

The naturally occurring graphical language structure diagram is the world's most popular option for depicting chemical structures. Its appeal stems from its natural ability to graphically portray the complex features of molecules with amazing clarity and universal comprehension. A structural diagram may be seen mathematically as a topological graph of nodes or vertices (atoms) linked with edges and arcs (bonds). As a result, many issues encountered while processing structural data may be connected to the mathematical discipline of graph theory (Balaban, 1995; Tarjan, 1977). In graph theory, distance and adjacency matrices are frequently employed. They may also be used to depict chemical structures. However, they often do not provide details regarding atom kinds and bond ordering and instead merely show the connections within a structure. A BEM, instead of adjacency and distance matrices, lists the number of free valence electrons and bonds present (Dugundji & Ugi, 2006; Spialter, 1964).

Connection Table

A molecular connectivity table or graph, also known as a connection table, is a structured representation that illustrates the interactions between atoms and bonds inside a molecule. The connection table (CT) is an adaptation of the matrix format. It contains critical information on atomic composition and connectivity and serves as the foundation for producing numerous molecular representations and analyses. Each row in a connection table corresponds to each atom in the molecule, while the columns contain information about the atom's characteristics and connections to other atoms via bonds. Atom types (e.g., carbon, oxygen), bond types (single, double, triple), bond connections (identifying atom pairs linked by bonds), and perhaps additional features such as atom charges and stereochemistry are commonly given. The connection table has been the most popular technique for representing chemical structures in computer systems since the early 1980s.

For molecular file, Molfile is a popular file format for describing chemical structures and related data. MDL was created to store and communicate molecular data in a standardized and machine-readable format and is a common sharing format based on a connection table (Dalby et al., 1992; Helma & Kramer, 2003). Representations of a molecule with a connection table can be done just as effectively using graph theoretical techniques. Canonical numbering is a significant difficulty when processing structural data in database systems or chemoinformatics applications. Morgan algorithms (Morgan, 1965) presented categorize congeneric atoms in a compound and choose atoms with invariant labels. This serves as the foundation for a distinctive and clear structural representation, also a feature of constitutional symmetry perception (Jochum & Gasteiger, 1977).

Additional chemical structure representations were created for the following purposes:

(i) Markush structures are widely used as generic structure diagrams in patents to cover groupings of chemical compounds, broadening their protective reach beyond individual molecules (Barnard, 1991).
(ii) Alphanumeric strings, represented by fragment codes such as SMARTS, serve an important role in chemical information systems by precisely designating substructure fragments (Bayada et al., 1999; Wild & Blankley, 2000).
(iii) Fingerprints or bitwise representations encode substructure components in compounds using binary bits (1 or 0), allowing for distinguishing characteristics in similarity searches (Lewis et al., 2000; Willett et al., 1998).
(iv) Hashcoding provides a quick and effective approach for creating compact digital "fingerprints" that help in registration and tracking while providing quick access to relevant chemical information (Ihlenfeldt a& Gasteiger, 1994; Wipke et al., 1978).

A connection table, which captures the 2D molecular makeup of a chemical, may be used to generate an automated 3D structural representation. The atom and bond connectivity information from the 2D connection table is converted into a three-dimensional arrangement that precisely matches the molecule's spatial shape (Zheng et al., 2017). Among the prominent 3D structure generators, CORINA, CONCORD, and other well-known programs are the most extensively used. Because of the intrinsic rotating movement of molecules around single bonds, the difficulty of conformational flexibility emerges during the construction of 3D molecular models. This process has far-reaching ramifications, including the ability of pharmaceutical compounds to adhere to the structural needs of receptor proteins. CORINA, CONCORD, and similar techniques are critical in tackling this difficulty since they forecast the many conformations that a molecule can take. These generators use molecular mechanic concepts and computer techniques to investigate the potential energy surfaces of molecules and determine the most stable and representative three-dimensional forms. This skill is critical in revealing molecules' complex dynamics and providing insight into their preferred conformations. Pharmacological compounds benefit from exceptional conformational flexibility. This innate capacity to switch between conformations allows for successful interactions with complex receptor protein complexes. Pharmacological compounds can optimize their binding affinity and biological activity by accepting differences in the receptor's binding pocket and spatial layout. This dynamic interaction between chemicals and proteins highlights the significance of conformational studies in drug discovery and rational design (Zheng et al., 2017).

Thus, it is possible to describe various chemical and physical characteristics of compounds using two-dimensional structure diagrams and three-dimensional molecular structures. The models presented thus far only depict the molecular 3D skeleton and not the real appearance in space, particularly the molecular surface.

The technique of automated knowledge acquisition from reaction databases has received increased interest, owing principally to the exponential growth of reaction databases. These collections include millions of reactions, making manual information extraction unfeasible and demanding automated data extraction approaches.

Because of these response databases' immense quantity and complexity, academics and data scientists have developed automated approaches for sifting through and extracting meaningful insights from the plethora of accessible data. Researchers hope to extract significant patterns, trends, and links from these massive collections of responses by using computational approaches and machine learning algorithms. Sorting response occurrences into reaction categories is crucial in learning from individual reactions and acquiring new information. Addition, subtraction, elimination, and rearrangement comprise the standard and most basic categorization scheme for chemical processes (Engel, 2006).

The emphasis on categorizing chemical reactions has shifted from identifying broad reaction classes within a complete system to creating hierarchical algorithms with the arrival of computers. All reactions are created by deriving them from their fundamental forms, aided by computer-generated representations.

Chemists sought to construct generic reaction classes encompassing various chemical changes in the early phases of reaction classification. These categories gave a basic grasp of chemical reactivity but were frequently hampered by their inability to handle the complexities of individual reaction cases. The paradigm change toward hierarchical algorithms represents a trend toward more detailed and adaptive response classification systems. Computer-generated representations of basic reaction templates are used in these methods. By methodically altering these templates, the algorithms can effectively build a range of reaction instances that appropriately depict the heterogeneous chemical landscape.

This method has several advantages, including capturing broad trends and subtle changes within reaction types. It enables researchers to investigate various response possibilities, from traditional textbook examples to innovative and complicated transformations. Furthermore, the hierarchical nature of these algorithms allows the organization and navigation of response data [82]. Reactions are categorized using a predetermined model or classification scheme in a model-driven approach (Dugundji & Ugi, 2006; Hendrickson & Sander, 1995). Rather than depending on established models, data-driven approaches such as the Hierarchical Organization of Reactions via Attribute and Condition Education (HORACE) system use computers to analyze response datasets autonomously. These techniques produce categorization outputs dynamically by analyzing input response data, eliminating the requirement for predetermined structures. The benefit of data-driven techniques is their capacity to catch subtleties and complexity that predetermined models may miss. Because these approaches autonomously understand the response data's fundamental properties, they can detect fresh trends and classifications that were not predicted in advance.

HORACE, for example, uses this strategy to construct hierarchical organization within reaction datasets, allowing for the automated discovery of reaction classes and associations based on the data's features and circumstances. This data-driven approach corresponds to the dynamic nature of chemical reactions and enables researchers to uncover hidden insights within reaction datasets, thereby driving innovation and enhancing our understanding of chemical reactivity (Gelernter et al., 1990; Rose & Gasteiger, 1994).

The Chemistry Data Types

Analysis of experimental data contributes significantly to our understanding of chemistry. As a result, gathering and analyzing data are crucial activities that demand careful thought.

The complexity of the data types that chemistry works with makes it one of the most complicated scientific fields. Chemical structure data (such as 2D/3D structures, bit vectors, hash codes, structural keys, fingerprints) and different molecular spectra data (such as NMR, UV/VIS, IR, MS) are more challenging to describe and process than processing numerical and textual data (Engel, 2006). However, the architecture of chemical databases and data analysis both require these data formats. Only a few common data formats are necessary to interchange data between the numerous proprietary formats used by different software programs, such as spectral data.

Database and Data Sources for Chemicals

Only electronic techniques can manage the complex information on chemical compounds (more than 30 million compounds are now known in the CAS registry) and reactions, such as literature data, physicochemical parameters, and spectra. More than 1 million new compounds are discovered each year, and there are over 700,000 publications that add to the body of knowledge on chemicals (Engel, 2006). As a result, chemistry was among the first scientific fields to use databases to preserve its wealth of knowledge. Currently, the only ways to thoroughly understand a subject are through database searches and data mining. It is only feasible to provide a summary of this complex subject here and to make references to in-depth analyses (Engel, 2006).

Chemistry and its associated fields have a staggering array of databases containing helpful or even essential information. By formally classifying databases into three main categories according to the type of material they include, it is possible to evaluate and choose the most relevant information sources:

- Literary (text)
- Factual (alphanumeric)
- Structural (topological)

Literature Database

Individual works from the original literature are in literary databases separated into bibliographic and full-text databases. These databases use character strings, such as alphanumeric letters, numerals, and special characters, to represent these works as objects. These databases may be used in particular ways to obtain bibliographic and content descriptions (indexing) of publications, including author names, article titles, publication years, keywords, and abstracts.

Literature databases play a crucial role in chemoinformatics research by providing access to scientific publications, patents, and other scholarly sources. These databases contain vast information, including chemical structures, properties, synthesis methods, and biological activities. Here are some widely used literature databases for chemoinformatics:

- **PubMed (https://pubmed.ncbi.nlm.nih.gov/):** PubMed is a free database maintained by the National Library of Medicine (NLM). It predominantly emphasizes on biomedical literature, including research articles in medicine, pharmacology, and biochemistry. PubMed provides access to abstracts and full-text articles from various journals and publications.
- **Scopus (https://www.scopus.com/)**: Scopus is an Elsevier comprehensive abstract and citation database. It covers various scientific disciplines, including chemistry, and offers access to research articles, conference papers, and patents. Scopus provides tools for literature searching, citation analysis, and tracking research impact.
- **Web of Science (https://www.webofscience.com/)**: Web of Science, also known as WoS, is a multidisciplinary database offered by Clarivate Analytics. It indexes a vast collection of scholarly literature, including articles from scientific journals, conference proceedings, and patents. WoS provides citation metrics, author profiles, and powerful search capabilities.
- **SciFinder (https://www.cas.org/products/scifinder)**: SciFinder is a subscription-based database provided by the Chemical Abstracts Service (CAS). It offers comprehensive coverage of scientific literature in chemistry and related fields. SciFinder enables searching by chemical structure, reaction, or topic and provides access to millions of references, patents, and substance information.

- **Google Scholar (https://scholar.google.com/)**: Google Scholar is a freely accessible search engine for scholarly literature. It indexes various sources, such as journal articles, conference papers, theses, and patents. While it may not have the same level of precision as specialized databases, Google Scholar can be a useful tool for finding relevant literature in chemoinformatics.

These literature databases allow researchers to explore the existing knowledge in chemoinformatics, discover new research findings, and obtain valuable data for their studies. Researchers can search for specific chemical compounds, investigate their properties and activities, analyze trends in the literature, and gather insights for various chemoinformatics applications, such as drug discovery, molecular modeling, and quantitative structure-activity relationship (QSAR) analysis.

Bibliographic patent databases are specialized resources that provide information and data related to patents. These databases are valuable for chemoinformatics research, as they offer access to patent documents containing chemical structures, synthetic methods, formulations, and other relevant information. Here are some specific bibliographic patent databases commonly used in chemoinformatics:

Patent Database

Chemoinformatics research heavily relies on accessing and analyzing patent information containing valuable chemical and structural data. These databases specialize in extracting chemical information from patent documents and provide powerful search interfaces for efficient retrieval and analysis:

- **SureChem** is a notable chemical patent database known for its expertise in extracting chemical structures and related information from patents. It offers advanced search capabilities, including chemical structure, substructure, and similarity searching, enabling researchers to retrieve chemical information for analysis and research purposes.
- **Derwent Innovation**, formerly the Derwent World Patents Index (DWPI), is another comprehensive patent database widely used in chemoinformatics. It covers global patents and provides detailed chemical indexing and searching features. Researchers benefit from chemoinformatics' chemical structure, fragment-searching capabilities, and advanced search options (Simmons, 2006).
- **INPADOC**, a patent database provided by the European Patent Office (EPO), is particularly relevant for chemoinformatics researchers. It offers access to a wide range of patent information, including chemical structures, classifications, and legal status. Researchers can utilize its chemical structure and keyword-searching functionalities to retrieve patents relevant to their research interests (Lingua, 2005).
- **PatentInspiration** is a patent search and analysis platform that specifically focuses on providing comprehensive chemical information from patent documents. Chemoinformatics researchers find value in its chemical structure, substructure, and similarity search features. It allows access to a vast collection of

patent documents and facilitates the extraction of chemical data for further analysis.

- While not solely a patent database, **SciFinder**, provided by the Chemical Abstracts Service (CAS), includes patent information alongside scientific literature. It offers researchers comprehensive coverage of patents from various jurisdictions and provides powerful search tools for chemical structures, reactions, and other chemical information within patents. SciFinder supports patent landscape analysis and facilitates extracting relevant chemical data for chemoinformatics research.

Factual Database

A factual database is a structured repository of information that organizes real-world facts, statistics, and objective details. As opposed to subjective content or speculative information, factual databases include verifiable and objective facts that may be quantified, measured, or proven by observation or empirical proof.

Scientific research, historical records, economic data, geographical information, and other fields are all covered by factual databases. They are critical in delivering accurate and trustworthy information for decision-making, analysis, research, and reference. These databases are created to ensure data integrity, consistency, and correctness. Factual data are often organized using a preset schema or data model, which allows for effective data storage, retrieval, and modification. Depending on the nature of the information being saved, data in factual databases might be organized, semistructured, or unstructured.

Scientific databases carrying empirical research data, census databases documenting demographic information, financial databases tracking economic indicators, and geographical databases offering spatial information about locations and landmarks are examples of factual databases. Factual databases are useful tools for gaining access to credible and trustworthy information, fostering informed decision-making, and increasing knowledge in various disciplines of study and industry. Numerical databases, meta-databases, research project databases, and chemical component catalogs are the different factual databases (Engel, 2006).

Numerical Database

Physicochemical values and the outcomes of numerous measurements are two examples of the numerical data of chemical substances that are primarily found in **numerical databases**. Beilstein (Lawson et al., 2014), Gmelin (Nebel et al., 1992), SpecInfo (Barth, 1992), KnowItAll, DETHERM, Cambridge Structural Database (CSD) (Cox, 2015), and other common numerical databases are examples. There are numerous specialized numerical databases available in addition to the well-known databases Beilstein and Gmelin, such as those in the fields of thermodynamics, safety, and toxicity (RTECS, EINECS, HSDB, etc.), or biochemical sources such as the enzyme database BRENDA (Wexler, 2004).

The database content is described in detail by **meta-databases**. Examples include the Gale Directory of Databases, STNGuide, Dialog Dialindex, and others (Schomburg et al., 2004).

Research project databases include data on reports and abstracts organized by research initiatives. Federal Research in Progress (FEDRIP) and UFORDAT (Environment Research in Progress) are two common databases for research projects (Engel, 2006).

Although **catalog databases** of chemical compounds, including catalogs of several chemical suppliers (containing package sizes, purity information, pricing, and ordering information), do not offer very exciting material, they provide helpful examples of factual databases. The CHEMCATS, CHEmSources, and Available Chemicals Directory from MDL are some of the most significant catalog databases.

Structure Database

Structure and reaction databases are important in chemistry because they contain information on chemical structures as individual molecules and reaction participants. The structural diagrams are represented, for instance, as connection tables rather than being kept as visuals (i.e., images that cannot be searched by structure). The topological arrangement of atoms, their connections, and their stereochemistry are included in this depiction of two- and three-dimensional systems.

Computer-Aided Techniques Used in Drug Design

Computer-aided techniques have revolutionized the drug design and discovery field, providing researchers with powerful tools to expedite the identification and optimization of potential drug candidates (Liao et al., 2011). These techniques leverage computational algorithms, modeling, and simulation methods to analyze and predict the properties and behavior of molecules. Here, we explore some of the key computer-aided techniques used in drug design and discovery:

1. **Molecular Modeling and Virtual Screening**: Molecular modeling techniques, such as molecular docking and molecular dynamics simulations, play a crucial role in drug discovery. Molecular docking predicts the binding modes and affinities of small molecules to a target protein, enabling the identification of potential drug leads. Virtual screening involves the computational screening of large compound libraries to identify molecules with desired properties or binding interac-

tions, saving time and resources compared to experimental screening (Xue & Bajorath, 2000).

2. **Quantitative Structure-Activity Relationship (QSAR) Analysis:** QSAR models establish correlations between the chemical structures of molecules and their biological activities or properties. These models utilize statistical and machine learning approaches to predict and prioritize the biological activities of new compounds based on their structural features. QSAR analysis aids in optimizing lead compounds and identifying key structural elements responsible for desired activities (Campbell et al., 1999).
3. **Pharmacophore Modeling**: Pharmacophore modeling involves identifying key features or spatial arrangements of atoms important for a molecule to interact with a biological target. Researchers can design or screen compound libraries by mapping the pharmacophore features to identify molecules with the desired pharmacophoric properties. Pharmacophore modeling helps understand ligand-receptor interactions and rational drug design (Seidel et al., 2017).
4. **De Novo Design**: Using computational algorithms, novel molecules with desired properties or activities can be generated through de novo design methods. These techniques explore chemical space by assembling molecular fragments or building blocks to create new compounds. De novo design facilitates the exploration of chemical diversity and the generation of drug-like molecules with specific target interactions and optimized physicochemical properties (Kutchukian & Shakhnovich, 2010).
5. **Machine Learning and Artificial Intelligence**: Machine learning and artificial intelligence (AI) algorithms are increasingly used in drug discovery. These techniques can analyze large datasets, identify patterns, and predict compound properties, activities, and toxicities. Machine learning models aid in virtual screening, compound optimization, toxicity prediction, and target identification, among other applications (Mitchell, 2014).
6. **Structure-Based Drug Design**: Structure-based drug design utilizes three-dimensional structural information of the target protein to guide the design and optimization of small molecules. This approach involves molecular docking, virtual screening, and structure-activity relationship analysis to identify compounds that bind to the target with high affinity and selectivity. Structure-based drug design is particularly effective when the three-dimensional structure of the target is known or can be reliably predicted.
7. **Fragment-Based Drug Design**: Fragment-based drug design involves the screening and optimization of small molecular fragments that bind to a target protein. These fragments are starting points for developing larger compounds with improved affinity and potency. Fragment-based approaches are valuable in cases where traditional high-throughput screening methods may be challenging due to compound size or complexity (Sheng & Zhang, 2013).

Different Methods of Representing Chemical Structures

There are several methods for representing chemical structures, each with advantages and disadvantages (Warr, 2011). Here are some commonly used methods:

1. **Chemical Line Notation (SMILES)**
   Advantages

- Compact representation using ASCII letters.
- Human-readable and modifiable.
- Enables the creation of unique chemical compound identification.
- SMILES strings can be used to feed into various chemoinformatics methods.
- **Disadvantages**
- SMILES strings lack specific stereochemistry information, necessitating extra annotations or conventions.
- Ambiguity can develop in situations of complicated structures or many resonance forms.
- Some uncommon or exotic structures may lack a conventional SMILES representation.

2. **Chemical Structure Diagrams**
   Advantages

- Simple and intuitive graphic representation that may convey structural information, stereochemistry, and connectivity.
- Commonly used in articles, textbooks, and chemical databases.
- Enables visualization of complicated molecular structures.
- **Disadvantages**
- Diagrams can be subjective and are dependent on the artist's abilities.
- Restrictions on the capacity to encode precise electronic and steric information.
- Complicated to manage computationally, necessitating conversion to other formats for some purposes.

3. **Molecular Graphs**
   Advantages

- Graph-based models capture the connections and bonds between atoms.
- Enable the use of graph theory methods and mathematical analysis.
- It is simple to convert to various formats for computer analysis.
- Make it possible to extract molecular descriptors and fingerprints for similarity-based searches.
- **Disadvantages**
- Do not express spatial layout or three-dimensional structures directly.
- Inadequate knowledge of atom locations and bond angles.
- In the absence of further annotations, the capacity to encode stereochemistry is limited.

4. **InChI (International Chemical Identifier)**
   Advantages

   - Standardized and unique chemical compound representation; contains structural information, including connectivity, stereochemistry, and tautomeric forms.
   - Machine-readable and suitable for use as a search key in chemical databases.
   - Resistant to atom order and tautomeric form alterations.
   - **Disadvantages**
   - InChI strings might be long and difficult to read.
   - No visual depiction of the molecule is provided.
   - More computationally demanding to create than other approaches.
   - It may be challenging to establish a correct InChI representation for some complicated structures.

5. **Molecular Fingerprints**
   Advantages
   - Compact binary or hashed representations for similarity searching and clustering.
   - Enable rapid comparison of large chemical databases.
   - It can capture structural features, substructures, and molecular properties.
   - Useful for virtual screening and compound prioritization.
   - **Disadvantages**
   - Loss of specific structural information.
   - Fingerprint collisions, in which various compounds have the same fingerprint.
   - Sensitivity to minor structural changes, resulting in false negatives or positives.
   - The capacity to encode three-dimensional information is limited.

It is worth mentioning that various representation approaches are frequently employed in tandem or converted into one another to capitalize on their strengths while correcting their weaknesses. The specific application and the information necessary for analysis or modeling determine the representation used.

Applications of Chemoinformatics

1. **Drug Discovery and Design**: Chemoinformatics is important in drug development since it helps identify, optimize, and design possible therapeutic candidates. It enables virtual screening of vast compound databases to discover compounds with desired features, biological activity prediction using quantitative structure-activity relationship (QSAR) models, and chemical space exploration utilizing de novo design approaches. Chemoinformatics technologies help in lead optimization, drug-likeness prediction, structure-activity connection analysis, and identifying possible off-target effects (Gillet, 2019).

2. **Virtual Screening**: The process of computationally screening compound libraries to find compounds with the potential to interact with a target protein or biological target is known as virtual screening. Chemoinformatics technologies, such as molecular docking and pharmacophore modeling, enable the effective screening of large chemical databases, saving time and money compared to experimental screening. Virtual screening is frequently utilized in hit detection, lead optimization, and repurposing current medicines for new indications (Varnek & Tropsha, 2008).
3. **Quantitative Structure-Property Relationship (QSPR)**: QSPR models identify relationships between chemical structures and physicochemical parameters such as solubility, partition coefficient, toxicity, and bioavailability. These models help in compound property prediction and optimization, compound selection, and prioritization of synthetic efforts. In the early stages of drug development, materials science, and environmental research, QSPR models are quite useful (Ivanciuc, 2013).
4. **ADME/Tox Prediction**: The features of ADME (absorption, distribution, metabolism, and excretion) and toxicity are critical concerns in drug development. ADME/Tox characteristics, including oral bioavailability, blood-brain barrier penetration, metabolic stability, and toxicity endpoints, may be predicted and prioritized using chemoinformatics. These predictions help to optimize compound attributes early in the drug development process, minimizing the number of compounds that must be evaluated experimentally (Avram et al., 2020).
5. **Chemical Property Prediction**: Physicochemical characteristics (e.g., molecular weight, logP), spectroscopic properties (e.g., UV–vis absorption, NMR shifts), and reactivity parameters are all predicted and modeled using chemoinformatics approaches. These predictions aid in compound selection and optimization for specific applications such as materials design, catalysis, and environmental investigations (Varnek & Baskin, 2011).
6. **Chemogenomics and Target Identification**: Chemical and biological data are used in chemoinformatics methodologies to find new therapeutic targets and study the interactions between chemicals and target proteins. One of these strategies is mining chemical and biological databases to find patterns, similarities, and linkages that can assist target discovery, illuminate biological mechanisms, and ease the development of multitargeted medicines (Bredel & Jacoby, 2004).
7. **Materials Design and Discovery**: Materials science increasingly uses chemoinformatics approaches to create and find new materials. Researchers can speed the identification of materials with desired features, such as energy storage materials, catalysts, polymers, and electronic materials, by researching and analyzing the links between chemical structures and material properties (Xu et al., 2022).
8. **Chemoinformatics Databases and Software:** Chemoinformatics has created extensive chemical databases and software tools that provide access to massive volumes of chemical data, fast data mining and analysis, and decision-making in various study domains. These tools are critical for organizing, searching for, and retrieving chemical information and encouraging cooperation and knowledge exchange among researchers (Engel, 2006).

Applications Beyond Drug Discovery

Chemoinformatics has uses that extend beyond drug discovery. Here are a few examples of where chemoinformatics has had a substantial impact (Engel & Gasteiger, 2018):

Environmental Chemistry

Chemoinformatics is the study of chemical environmental fate, toxicity, and bioavailability. It assists in synthesizing ecologically friendly substances, detecting contaminants, and developing models for assessing environmental risk (Gasteiger, 2016).

Materials Science

Techniques in chemoinformatics aid in creating and optimizing materials with desired qualities. It assists in developing new catalysts, polymers, and energy storage materials. Chemoinformatics approaches such as property prediction models and data mining aid in creating efficient materials. Chemoinformatics is used to find novel luminescent materials for organic light-emitting diodes (OLEDs), which create efficient and high-performance OLEDs (Le & Winkler, 2018).

Food Science

Chemoinformatics contributes to food safety and quality management. It aids in predicting and evaluating food components' biological activity, flavor, and scent. Food-related data are analyzed, food products are classified, and food attributes are predicted using chemoinformatics. Chemoinformatics was utilized to create models for forecasting wine sensory qualities, assisting in the optimization of wine production and quality (Martinez-Mayorga & Medina-Franco, 2009).

Agrochemicals and Crop Science

Chemoinformatics aids in the research and development of effective agrochemicals such as insecticides and herbicides. It aids in the prediction and modeling of their interactions with target pests or plants, as well as the optimization of their characteristics and the assessment of their environmental effects. Chemoinformatics was used to create and optimize new pesticides that had better action against a specific pest while having low toxicity to nontarget species (Karade, 2021).

Cosmetics and Personal Care Products

Chemoinformatics is used to formulate and develop cosmetics and personal care products. It assists in predicting cosmetic ingredient physicochemical qualities, biological activity, and safety profiles. Chemoinformatics models help with chemical screening and selection for desired aesthetic effects. Chemoinformatics approaches were utilized to estimate cosmetic components' skin sensitization potential, which helped produce safer and hypoallergenic products (Anzali et al., 2018).

Patent Analysis and Intellectual Property

Chemoinformatics is the analysis of chemical patent databases to identify new substances, trends, and technical developments. It assists in protecting intellectual property, patent landscaping, and competitive analysis. Chemoinformatics technologies help with patent database exploration for drug development, materials science, and other applications (Aristodemou & Tietze, 2018).

These are only a few examples of the many uses of chemoinformatics. Chemoinformatics has resulted in remarkable success stories in various sectors, allowing academics and industry to make more informed decisions, optimize operations, and expedite innovation.

Future Perspectives of Chemoinformatics

Integration of Artificial Intelligence (AI) and Machine Learning (ML): Combining AI and machine learning approaches with chemoinformatics has enormous promise. Advanced algorithms can aid in the prediction of chemical characteristics, identification of trends in large datasets, optimization of molecular structures, and development of new drugs. Integrating AI/ML with chemoinformatics will almost certainly result in faster and more accurate predictions, speeding up the discovery and creation of novel chemicals.

Big Data Analytics: The capacity to successfully analyze and derive meaningful insights from large datasets becomes increasingly important as the number and diversity of chemical data grow. Chemoinformatics will increasingly focus on creating efficient algorithms, data mining approaches, and visualization tools to handle and comprehend large-scale chemical datasets.

Integration of Experimental and Computational Approaches: The future of chemoinformatics lies in its seamless integration with experimental methodologies. Researchers can increase the accuracy and dependability of results by integrating computational predictions with experimental validation, decreasing the time and expense required in drug discovery and materials development.

Multiobjective Optimization: Chemoinformatics will shift to multiobjective optimization, which will take into account several parameters at the same time, such as biological activity, toxicity, pharmacokinetics, and synthetic feasibility. Multiobjective optimization methods will aid in discovering compounds with the best balance of several desirable features.

Personalized Medicine and Precision Drug Design: Chemoinformatics can be crucial in personalized medicine by designing tailored medicines using patient-specific data such as genetics and proteomics. Chemoinformatics techniques can help adapt medicinal therapies to individual patients by merging molecular data with patient information, optimizing effectiveness, and minimizing unwanted effects.

Challenges of Chemoinformatics

1. **Data Quality and Availability:** It is still difficult to ensure high-quality, dependable, standardized chemical data availability. Errors, inconsistencies, or biased data in databases can influence the accuracy and resilience of chemoinformatics models and predictions. Efforts must be taken to enhance data curation, set data-sharing guidelines, and address data bias concerns.
2. **Including Complexity and Flexibility**: Many chemical systems, such as protein-ligand interactions, conformational changes, and dynamic behavior, have intrinsic complexity and flexibility. Chemoinformatics has difficulty in effectively modeling and forecasting these complex systems. A current topic of study is the development of techniques that effectively capture and depict the dynamic character of chemical systems.
3. **Computationally Demanding Activities**: Computationally demanding chemoinformatics activities include molecular dynamics simulations, large-scale virtual screening, and de novo design. The requirement for powerful computational resources and efficient methods to tackle these tasks continues to be a problem, particularly for researchers with limited access to high-performance computer resources.
4. **Interpreting and Understanding Black Box Models**: AI and ML algorithms frequently operate as black-box models, making it difficult to explain their judgments and comprehend the underlying logic. In chemoinformatics, interpretable and explainable AI/ML models are required to build trust and confidence in predictions and to promote the discovery of new insights.
5. **Legal and Ethical Issues**: The application of chemoinformatics, particularly in drug development, poses legal and ethical concerns about intellectual property,

privacy, and responsible data use. Addressing these challenges and maintaining proper rules and ethical frameworks are critical for responsible chemoinformatics methodology development and use.

Despite these obstacles, ongoing advances in computer power, data availability, and algorithmic techniques hold enormous promise for the future of chemoinformatics. Chemoinformatics has the potential to revolutionize drug development, materials science, and related areas by solving these obstacles and pushing the frontiers of innovation, resulting in improved healthcare outcomes and scientific advances.

References

Agrafiotis, D. K., Bandyopadhyay, D., Wegner, J. K., & van Vlijmen, H. (2007). Recent advances in chemoinformatics. *Journal of Chemical Information and Modeling, 47*(4), 1279–1293.

Anzali, S., Pflücker, F., Heider, L., & Jonczyk, A. (2018). Computational approaches to cosmetics products discovery. In *Applied chemoinformatics* (pp. 527–546).

Aristodemou, L., & Tietze, F. (2018). The state-of-the-art on Intellectual Property Analytics (IPA): A literature review on artificial intelligence, machine learning and deep learning methods for analysing intellectual property (IP) data. *World Patent Information, 55*, 37–51.

Avram, S., Mernea, M., Limban, C., Borcan, F., & Chifiriuc, C. (2020). Potential therapeutic approaches to Alzheimer's disease by bioinformatics, cheminformatics and predicted Adme-Tox tool. *Current Neuropharmacology., 18*(8), 696–719.

Balaban, A. T. (1995). Chemical graphs: Looking back and glimpsing ahead. *Journal of Chemical Information and Computer sciences., 35*(3), 339–350.

Barnard, J. M. (1991). A comparison of different approaches to Markush structure handling. *Journal of Chemical Information and Computer Sciences, 31*(1), 64–68.

Barth, A. (1992). Specinfo—An integrated spectroscopic information system. *Journal of Chemical Information and Computer Sciences, 32*, 291.

Bayada, D. M., Hamersma, H., & van Geerestein, V. J. (1999). Molecular diversity and representativity in chemical databases. *Journal of Chemical Information and Computer Sciences, 39*(1), 1–10.

Bone, R. G., Firth, M. A., & Sykes, R. A. (1999). SMILES extensions for pattern matching and molecular transformations: Applications in chemoinformatics. *Journal of Chemical Information and Computer Sciences, 39*(5), 846–860.

Bredel, M., & Jacoby, E. (2004). Chemogenomics: an emerging strategy for rapid target and drug discovery. *Nature Reviews Genetics, 5*(4), 262–275.

Brown, F. K. (1998). Chemoinformatics: What is it and how does it impact drug discovery. *Annual Reports in Medicinal Chemistry, 33*, 375–384.

Campbell, P., Srinivasan, R., Knoell, T., Phipps, D., Ishida, K., Safarik, J., Cormack, T., & Ridgway, H. (1999). Quantitative structure–activity relationship (QSAR) analysis of surfactants influencing attachment of a Mycobacterium sp. to cellulose acetate and aromatic polyamide reverse osmosis membranes. *Biotechnology and Bioengineering., 64*(5), 527–544.

Cox, A. N. (2015). *Allen's astrophysical quantities*. Springer.

Dalby, A., Nourse, J. G., Hounshell, W. D., Gushurst, A. K., Grier, D. L., Leland, B. A., & Laufer, J. (1992). Description of several chemical structure file formats used by computer programs developed at Molecular Design Limited. *Journal of Chemical Information and Computer Sciences, 32*(3), 244–255.

Dugundji, J., & Ugi, I. (2006). An algebraic model of constitutional chemistry as a basis for chemical computer programs. In *Computers in chemistry*. Springer.

Engel, T. (2006). Basic overview of chemoinformatics. *Journal of Chemical Information and Modeling., 46*(6), 2267–2277.

Engel, T., & Gasteiger, J. (2018). *Applied chemoinformatics: achievements and future opportunities*. John Wiley & Sons.

Gasteiger, J. (2006). The central role of chemoinformatics. *Chemometrics and Intelligent Laboratory Systems, 82*(1–2), 200–209.

Gasteiger, J. (2016). Chemoinformatics: Achievements and challenges, a personal view. *Molecules, 21*(2), 151.

Gelernter, H., Rose, J. R., & Chen, C. (1990). Building and refining a knowledge base for synthetic organic chemistry via the methodology of inductive and deductive machine learning. *Journal of Chemical Information and Computer Sciences, 30*(4), 492–504.

Gillet, V. J. (2019). Applications of chemoinformatics in drug discovery. In *Biomolecular and bioanalytical techniques: Theory, methodology and applications* (pp. 17–36).

Helma, C., & Kramer, S. (2003). A survey of the predictive toxicology challenge 2000–2001. *Bioinformatics, 19*(10), 1179–1182.

Hendrickson, J. B., & Sander, T. (1995). COGNOS: A Beilstein-type system for organizing organic reactions. *Journal of Chemical Information and Computer Sciences, 35*(2), 251–260.

Ihlenfeldt, W. D., & Gasteiger, J. (1994). Hash codes for the identification and classification of molecular structure elements. *Journal of Computational Chemistry., 15*(8), 793–813.

Ivanciuc, O. (2013). Chemical graphs, molecular matrices and topological indices in chemoinformatics and quantitative structure-activity relationships §. *Current Computer-Aided Drug Design, 9*(2), 153–163.

Jochum, C., & Gasteiger, J. (1977). Canonical numbering and constitutional symmetry. *Journal of Chemical Information and Computer Sciences, 17*(2), 113–117.

Karade, D. (2021). Chemoinformatics based investigation of plant metabolites for their medicinal and crop protection values.

Kutchukian, P. S., & Shakhnovich, E. I. (2010). De novo design: balancing novelty and confined chemical space. *Expert Opinion on Drug Discovery, 5*(8), 789–812.

Lawson, A. J., Swienty-Busch, J., Géoui, T., & Evans, D. (2014). The making of reaxys—towards unobstructed access to relevant chemistry information. In *The future of the history of chemical information* (pp. 127–148). ACS Publications.

Le, T. C., & Winkler, D. A. (2018). Applications in materials science. In *Applied chemoinformatics* (pp. 547–569).

Lewis, R. A., Pickett, S. D., & Clark, D. E. (2000). Computer-aided molecular diversity analysis and combinatorial library design. *Reviews in Computational Chemistry.*, 1–51.

Liao, C., Sitzmann, M., Pugliese, A., & Nicklaus, M. C. (2011). Software and resources for computational medicinal chemistry. *Future Medicinal Chemistry, 3*(8), 1057–1085.

Lingua, D. G. (2005). INPADOC: 30 years of endeavours yet unmapped territories remain! *World Patent Information., 27*(2), 105–111.

Maggiora, G. M. (2011). The reductionist paradox: are the laws of chemistry and physics sufficient for the discovery of new drugs? *Journal of Computer-Aided Molecular Design, 25*, 699–708.

Maggiora, G., & Gokhale, V. (2016). Non-specificity of drug-target interactions–consequences for drug discovery. In *Frontiers in molecular design and chemical information science-Herman Skolnik Award Symposium 2015: Jürgen Bajorath* (pp. 91–142).

Martinez-Mayorga, K., & Medina-Franco, J. L. (2009). Chapter 2 Chemoinformatics—Applications in food chemistry. In *Advances in food and nutrition research* (pp. 33–56). Academic Press.

Mestres, J., Gregori-Puigjane, E., Valverde, S., & Sole, R. V. (2008). Data completeness—The Achilles heel of drug-target networks. *Nature Biotechnology, 26*(9), 983–984.

Mitchell, J. B. (2014). Machine learning methods in chemoinformatics. *Wiley Interdisciplinary Reviews: Computational Molecular Science, 4*(5), 468–481.

Morgan, H. L. (1965). The generation of a unique machine description for chemical structures—A technique developed at chemical abstracts service. *Journal of Chemical Documentation., 5*(2), 107–113.

Nebel, A., Toelle, U., Maass, R., Olbrich, G., Deplanque, R., & Lister, P. (1992). The integrated Gmelin information system: new developments in information processing. *Analytica Chimica Acta, 265*(2), 305–312.
Rose, J. R., & Gasteiger, J. (1994). HORACE: an automatic system for the hierarchical classification of chemical reactions. *Journal of Chemical Information and Computer Sciences, 34*(1), 74–90.
Schomburg, I., Chang, A., Ebeling, C., Gremse, M., Heldt, C., Huhn, G., & Schomburg, D. (2004). BRENDA, the enzyme database: updates and major new developments. *Nucleic Acids Research, 32*(Suppl. 1), D431–D433.
Seidel, T., Bryant, S. D., Ibis, G., Poli, G., & Langer, T. (2017). 3D Pharmacophore modeling techniques in computer-aided molecular design using LigandScout. *Tutorials in Chemoinformatics*, 279–309.
Sheng, C., & Zhang, W. (2013). Fragment informatics and computational fragment-based drug design: an overview and update. *Medicinal Research Reviews, 33*(3), 554–598.
Simmons, E. S. (2006). Intellectual property. In *Using the pharmaceutical literature* (pp. 213–248). CRC Press.
Spialter, L. (1964). The Atom Connectivity Matrix (ACM) and Its Characteristic Polynomial (ACMCP). *Journal of Chemical Documentation, 4*(4), 261–269.
Tarjan, R. E. (1977). *Graph algorithms in chemical computation*. ACS Publications.
Varnek, A., & Baskin, I. I. (2011). Chemoinformatics as a theoretical chemistry discipline. *Molecular Informatics., 30*(1), 20–32.
Varnek, A., & Tropsha, A. (2008). *Chemoinformatics approaches to virtual screening*. Royal Society of Chemistry.
Warr, W. A. (2011). Representation of chemical structures. *Wiley Interdisciplinary Reviews: Computational Molecular Science, 1*(4), 557–579.
Weininger, D. (1988). SMILES, a chemical language and information system. 1. Introduction to methodology and encoding rules. *Journal of Chemical Information and Computer Sciences, 28*(1), 31–36.
Wexler, P. (2004). The US National Library of Medicine's Toxicology and Environmental Health Information Program. *Toxicology, 198*(1-3), 161–168.
Whitehead, T. M., Irwin, B. W., Hunt, P., Segall, M. D., & Conduit, G. J. (2019). Imputation of assay bioactivity data using deep learning. *Journal of Chemical Information and Modeling, 59*(3), 1197–1204.
Wild, D. J., & Blankley, C. J. (2000). Comparison of 2D fingerprint types and hierarchy level selection methods for structural grouping using Ward's clustering. *Journal of Chemical Information and Computer Sciences, 40*(1), 155–162.
Willett, P. (2008). A bibliometric analysis of the literature of chemoinformatics. In *Aslib Proceedings* (pp. 4–17).
Willett, P., Barnard, J. M., & Downs, G. M. (1998). Chemical similarity searching. *Journal of Chemical Information and Computer Sciences, 38*(6), 983–996.
Wipke, W. T., Krishnan, S., & Ouchi, G. I. (1978). Hash functions for rapid storage and retrieval of chemical structures. *Journal of Chemical Information and Computer Sciences, 18*(1), 32–37.
Xu, J., & Hagler, A. (2002). Chemoinformatics and drug discovery. *Molecules, 7*(8), 566–600.
Xu, P., Chen, H., Li, M., & Lu, W. (2022). New opportunity: Machine learning for polymer materials design and discovery. *Advanced Theory and Simulations, 5*(5), 2100565.
Xue, L., & Bajorath, J. (2000). Molecular descriptors in chemoinformatics, computational combinatorial chemistry, and virtual screening. *Combinatorial Chemistry & High Throughput Screening, 3*(5), 363–372.
Zheng, Y., Tice, C. M., & Singh, S. B. (2017). Conformational control in structure-based drug design. *Bioorganic & Medicinal Chemistry Letters, 27*(13), 2825–2837.

Index

S. Ijaz et al. (eds.), *Trends in Plant Biotechnology*,
https://doi.org/10.1007/978-981-97-0814-7

SPRINGER NATURE

GPSR Compliance

The European Union's (EU) General Product Safety Regulation (GPSR) is a set of rules that requires consumer products to be safe and our obligations to ensure this.

If you have any concerns about our products, you can contact us on ProductSafety@springernature.com

In case Publisher is established outside the EU, the EU authorized representative is:

Springer Nature Customer Service Center GmbH
Europaplatz 3
69115 Heidelberg, Germany

The manufacturer's authorised representative in the EU is Springer Nature Customer Service Centre GmbH, Europaplatz 3, 69115 Heidelberg, Germany. If you have any concerns regarding our products, please contact ProductSafety@springernature.com

Printed and bound by CPI Group (UK) Ltd, Croydon, CR0 4YY
01/10/2026
02225938-0004